Springer Proceedings in Mathematics & Statistics

Volume 540

This book series features volumes composed of selected contributions from workshops and conferences in all areas of current research in mathematics and statistics, including data science, operations research and optimization. In addition to an overall evaluation of the interest, scientific quality, and timeliness of each proposal at the hands of the publisher, individual contributions are all refereed to the high quality standards of leading journals in the field. Thus, this series provides the research community with well-edited, authoritative reports on developments in the most exciting areas of mathematical and statistical research today.

George Halkos · Teresa A. Oliveira ·
Christos Kitsos · Amilcar Oliveira
Editors

Environmental Risk Analysis

ENVECON, Volos, Greece, December 6–7, 2024

Springer

Editors
George Halkos
University of Thessaly
Thessaly, Greece

Christos Kitsos
University of West Attica
Athens, Greece

Teresa A. Oliveira
Universidade Aberta
Lisbon, Portugal

Amilcar Oliveira
Universidade Aberta
Lisbon, Portugal

ISSN 2194-1009 ISSN 2194-1017 (electronic)
Springer Proceedings in Mathematics & Statistics
ISBN 978-3-032-24154-2 ISBN 978-3-032-24155-9 (eBook)
https://doi.org/10.1007/978-3-032-24155-9

Mathematics Subject Classification: 62P20, 62P12, 91, 91B84, 91B76, 91A23

This Springer imprint is published by the registered company Springer Nature Switzerland AG
The registered company address is: Gewerbestrasse 11, 6330 Cham, Switzerland

Preface

In December 2024 and for the 10th Anniversary Conference on the *Economics of Natural Resources and the Environment* (***ENVECON***), it was a joy and an honour to mark a decade of steady growth, hard-won insights, and shared purpose. The 10th Anniversary recognized the people and partnerships that made those years matter, and, especially, all of the colleagues, partners, students, and the wider academic community.

Ten years ago, in 2014, a simple and ambitious idea came into reality, to bring together the many strands of environmental science under one holistic framework that combines issues from environmental and ecological economics, marine and water studies, psychology and education, circular economy, transportation, tourism, and other relevant scientific strands. The initial idea began with a small team, bold questions, and a belief that evidence and dialogue could move practice as well as theory. Moreover, the ENVECON conference faced several challenges and opportunities, but the success measures outcomes, resilience, and innovation.

The ENVECON has four milestones. First, the *national phase*: four Greek conferences hosted by the Department of Economics at the University of Thessaly, organized by the Laboratory of Operations Research within the projects *COOPERATION 2011* and "Greenhouse Gas Emission Scenarios and Policies to Combat them by 2030 in Energy, Transport, and Industry". Two meetings on Climate Change (March and October 2014) were followed by two on Environmental Economics (2015 and 2016) at our former premises in the old French Institute. Second, the *international turn*: the 5th ENVECON in 2018 opened our doors wider, welcoming global perspectives to Volos. Third, *strategic cooperation for excellence and openness*: the 6th ENVECON (June 2021, online) with Harokopio University's Department of Economy and Sustainable Development; the 7th (November 2021, online) with our Interdepartmental Postgraduate Program Education for Sustainability and the Environment, highlighting environmental awareness and education; the 8th (December 2022, Athens and online) with the Research Unit of Environmental Communication and Education, University of West Attica, on Environmental Research Activities: Progress and Trends; and the 9th (December 2023, Thessaloniki and online) with the Laboratory of Forest Economics, Aristotle University of Thessaloniki.

Today, we celebrate the fourth milestone: the *10th Anniversary ENVECON*, hosted by our Laboratory of Operations Research in the Department of Economics at the University of Thessaly. This year's scope is true to our roots: to confront the central questions in the economics of natural resources and the environment, spotlight pressing problems, and debate management and policy solutions. The heart of ENVECON has always been interdisciplinarity, by exchanging experiences and consultations across fields; we have discovered common ground—and new pathways—to address environmental risk and opportunity.

Thessaly, Greece

Prof. George Halkos
Editor-in-Chief

Introduction

Environmental risk analysis has become a central theme in recent research as it sits at the crossroads of climate science, economics, policy, and human behaviour. Essentially, it does not ask only what the challenges in natural and socio-technical systems are, but also who bears the costs, when losses unfold, and which interventions shift outcomes towards resilience. In this book, the environmental risk is treated as systemic and dynamic, shaped by hazards, exposures, and vulnerabilities.

Moreover, the present book covers a wide range of theoretical and practical applications that span causal econometrics, game theory, time series, and panel models, as well as, information theory, probability modelling, and policy evaluation, reflecting the field's pluralism and its need for both generalizable tools and context-aware evidence. Overall, the aim of the book is to offer relevant policy implications for dealing with environmental risks under the scope of sustainable development principles.

The book consists of 16 chapters and because the chapters are independent, readers can enter through multiple choices. First, the readers can inspect how hazard type (e.g., extreme weather) can affect productivity and efficiency performance or other economic mechanisms (e.g., taxation, subsidies, markets, and corporate reporting). Second, it is also possible to observe novel methodological applications such as data envelopment analysis (DEA), γ-order distributions, and Quantile-VAR. Third, it is also possible for readers to gain insights based on sectoral (e.g., fisheries in Sri Lanka, smallholder farmers in Vietnam) or regional (e.g., air quality in Greek cities, SMEs in Europe, inclusive development in South Asia) levels.

Interestingly, across cases, several cross-cutting themes recur: (i) nonlinearity and feedback, from cascading hazards to policy-behaviour loops; (ii) distribution and fairness, regarding who pays, who benefits, and how equity is measured; and (iii) information and incentives, how data, reporting, and institutional design change risk perception and action. To summarize, these studies build beyond diagnosing risks to designing implementable policies and actions that are efficient, equitable, and credible under uncertainty and adequate for strengthening the achievement of sustainable development.

The first chapter, authored by *S. Weerasekara, C. Wilson, B. Lee, V. Hoang*, and *K. Mohottala Gedara*, is entitled "Extreme Weather Events and Their Impact on Total Factor Productivity and Efficiency in Sri Lanka's Fisheries Sector". This chapter gives a first empirical look at how disasters shift technical efficiency and productivity across inland, coastal, and deep-sea fisheries using SFA and DEA with double bootstrap. The results show that high rainfall and strong winds depress efficiency (with heterogeneous drought effects), implying policy must be tailored to agro-ecological contexts to protect food security and livelihoods.

The second chapter, authored by *G. Halkos*, and *A. Zisiadou*, is entitled "The Cascading Effects among Natural Hazards: Interaction Maps as Tool for Risk Analysis and Management". The chapter maps how primary hazards (e.g., cyclones, floods, earthquakes) propagate secondary crises (e.g., landslides, infrastructure, and technological accidents), and it argues for integrated, cross-sectoral risk governance aligned with the Sendai Framework.

The third chapter, authored by *A. Seijas-Macias, A. Oliveira*, and *T. A. Oliveira*, is entitled "Analysis of the Environmental Tax in EU. A Study of its Equity through Gini Index and Lorenz Curve". This chapter employs a Gini–Lorenz assessment that compares environmental tax revenues with GHG emissions, finding alignment for CO_2 but weaker equity for CH_4 and N_2O. The policy implication is to sharpen tax design in order to match climate targets and accelerate clean technology diffusion.

The fourth chapter, authored by *G. E. Halkos, G. J. Papageorgiou, E. G. Halkos*, and *J. G. Papageorgiou*, is entitled "Dynamic Environmental Subsidy Policies". This chapter showcases that for a monopolistic extractor (renewable or non-renewable resource), a performance-indexed dynamic subsidy outperforms a static per-unit scheme in present-value terms. Overall, the feedback design aligns long-run incentives using only local information, an implementable risk-management lever.

The fifth chapter, authored by *C. P. Kitsos, C.-S. Nisiotis, and I. S. Stamatiou*, is entitled "The γ-Order Generalized Distributions for Environmental Economics Applications". This chapter provides a concise overview of γ-order families suited to fat-tailed phenomena and K-S testing, with air-pollution examples from Athens and Thessaloniki, Greece. Additionally, it equips analysts to pick probability models that better capture risk extremes.

The sixth chapter, authored by *G. Halkos, and C. Kitsos*, is entitled "Measuring Uncertainty, Transfer Entropy, and G-Causality in Environmental Economics". The chapter motivates entropy-based measures, especially transfer entropy, for policy contexts with model uncertainty and heavy tails, linking to γ-order generalized normal distribution and Logarithm Sobolev Inequality (LSI) foundations. Overall, it frames information flows as key to diagnosing environmental risk and causality.

The seventh chapter, authored by *N. A. Kyriazis*, is entitled "Environmental Risk Footprint on Cryptocurrencies and Money Evolution in Inflationary Crises". This chapter applies Quantile-VAR analysis to assess systemic risk spillovers among major cryptocurrencies (e.g., BTC, LTC, XRP, and ZEC) during the pandemic and war periods. The analysis reveals that different cryptocurrencies become dominant sources of systemic risk transmission, as physical and transition risks evolve. The empirical findings guide regulators and investors on how environmental shocks reshape digital-asset dynamics.

The eighth chapter, authored by *C. Kitsos*, and *C.-S. Nisiotis*, is entitled "A Time Series Analysis Comparing the Main Air Pollutants of Athens and Salonica in Greece". The chapter applies EDA and ARMA-type models, which reveal broadly declining pollutant trends (CO, NO/NO_2, SO_2, O_3) across central and peripheral zones, with city- and site-specific turning points. The work underscores the value of localized modelling for urban air-quality risk.

The ninth chapter, authored by *G. Ekonomou*, and *George Halkos*, is entitled "Identifying Impacts and Causalities Regarding Air Quality, Environment, and Tourism in the Eurozone Economic Space: A Panel Data Analysis". This chapter tests the EKC with energy-sector CH_4 and CO_2 intensity, plus tourism and GDP/population. The analysis supports the EKC and identifies unidirectional causal links from growth to pollution proxies and policy takeaways, stress energy efficiency, and eco-friendly practices for sustainable tourism-led development.

The tenth chapter, authored by *P. Koronaios, P. Kalimeris*, and *G. Maroulis*, is entitled "Global Bioeconomy: History, Visions, and Limitations". This chapter presents a bibliometric review that traces the evolution of bioeconomy thinking, biotechnology, bioresource, and bioecology visions, and surfaces implementation constraints. Moreover, the overview clarifies conceptual pathways and realistic guardrails for bioeconomy transition strategies.

The eleventh chapter, authored by *K. Kucheruk*, and *K. Evangelinos*, is entitled "Exploring the Impact of Corporate Social Responsibility on Employee Mental Health, Resilience, and Job Satisfaction: Evidence from a Survey in the NGO". The present chapter employs a survey with evidence linking CSR practices (environmental and social actions, leadership, communication) with resilience, satisfaction, and turnover intentions among NGO employees. Ultimately, the chapter offers actionable guidance to integrate CSR for workforce well-being, an often-overlooked dimension of organizational climate risk.

The twelfth chapter, authored by *P.-S. C. Aslanidis, J. Moll de Alba*, and *George Halkos*, is entitled "Overcoming Environmental Risks in European SMEs: A Policy Review of EMS Integration Within ESG". The chapter reviews 47 studies, mapping ISO 14001/EMAS to ESG and the emerging ESRS standards, arguing that synergistic adoption strengthens SME resilience to environmental risks while aligning with SDGs and market expectations. Overall, the paper provides a roadmap for compliance-ready, risk-aware management.

The thirteenth chapter, authored by *P.-P. Nana*, and *A. Skouloudis*, is entitled "A Note on Climate Change Accounting and Reporting in Greece: Insights from the Adoption of the TCFD Recommendations Among Large-Caps". The chapter presents a content analysis of ATHEX large-caps that shows stronger disclosure on governance/risk assessment than on strategy, metrics, and targets. The policy implications call for more effective regulation to avoid box-ticking and prepare firms for ESRS/CSRD requirements.

The fourteenth chapter, authored by *G. E. Halkos, G. J. Papageorgiou, E. G. Halkos*, and *J. G. Papageorgiou*, is entitled "Ecologically Optimal Quantity Extraction: Conflicting Cases". This chapter provides within Nash and Stackelberg differential games, in which exploiters internalize only part of future environmental loss and policymakers may act more impatiently, shaping steady states and payoffs. The chapter identifies when regulator–exploiter conflicts intensify, informing durable extraction policies.

The fifteenth chapter, authored by *S. Chen*, and *S. Managi*, is entitled "Challenges in Achieving Inclusive Development in South Asia". This chapter monitors inclusive wealth evolution in India, Bangladesh, and Nepal and shows that per-capita wealth lags behind global trends. Consequently, restoring natural capital and strengthening social capital are pivotal for sustainable wealth accumulation. The study links social-capital conditions to component-wise wealth accumulation and introduces a CO_2-damage risk adjustment.

The sixteenth chapter, authored by *M. Donaldson, T.-A. Duc Nguyen*, and *T. T. Nguyen*, is entitled "Information and Communication Technology Adoption and Climate Risk Perception: Evidence from Smallholder Farmers in Rural Vietnam". The chapter employs a panel data analysis and a double-hurdle model that show ICT access does not raise the incidence of positive risk perception, but significantly increases intensity among those already perceiving risk, especially for income and asset threats. To conclude, the complementary awareness programs and gender-sensitive interventions are recommended for ICT adoption and climate risk perception.

The editors wish to express their sincere appreciation to all anonymous reviewers of this volume. Each contribution was evaluated by two independent experts, whose careful assessments and constructive recommendations have been invaluable in enhancing the scholarly rigor and overall quality of the chapters. Their commitment of time and expertise is gratefully acknowledged.

Prof. George Halkos
Editor-in-Chief

Assoc. Prof. Teresa A. Oliveira
Emer. Prof. Christos Kitsos
Assoc. Prof. Amilcar Oliveira

Contents

Extreme Weather Events and Their Impact on Total Factor Productivity and Efficiency in Sri Lanka's Fisheries Sector

Sajeevani Weerasekara, Clevo Wilson, Boon Lee, Vincent Hoang, and Kularatne Mohottala Gedara

Abstract This study extends the existing literature on the impact of natural disasters on agricultural sectors by providing the first empirical analysis of how such events affect technical efficiency (TE) and productivity in the Sri Lankan fisheries sector. The fisheries industry is one of the most vulnerable to the impact of natural disasters and extreme weather conditions which are posing formidable challenges to the work of agro-meteorologists globally. Agricultural productivity and efficiency are key drivers for ensuring the wellbeing of farmers, fishers, the agro-based industry as a whole and therefore for dependent societies. These drivers are strongly linked to food security, affordable prices and poverty alleviation in developing countries. It is therefore imperative that effective strategies are deployed and integrated in order to cope with the impact of natural disasters. This study presents a comprehensive multi-sectoral analysis of the fisheries industry, encompassing inland, coastal and deep-sea fisheries. We employ a stochastic production function and data envelopment analysis (DEA) double bootstrap approach to examine the effect on the TE of different decision-making units (DMUs), sectors, agro-ecological regions and seasons. The results reveal that disasters are a major factor in the decline in the efficiency and productivity of fisheries. In particular, high rainfall and strong winds have had a significant negative impact on efficiency although droughts have positively impacted some fisheries sectors. Overall, the impact of a combination of all disasters indicates a negative relationship with efficiency. This study highlights the need to change policies and resource allocation by taking into account heterogeneous environments and their specific impacts on fisheries. Such changes need to reflect the global trend of greater climatic variations resulting from extreme weather conditions.

S. Weerasekara (✉)
Central Bank of Sri Lanka, Colombo, Sri Lanka
e-mail: sajeevani@cbsl.lk

S. Weerasekara · C. Wilson · B. Lee · V. Hoang
School of Economics and Finance, Faculty of Business and Law, Queensland University of Technology, Brisbane, Australia

K. M. Gedara
Department of Economics, Faculty of Social Sciences, University of Kelaniya, Colombo, Sri Lanka

G. Halkos et al. (eds.), *Environmental Risk Analysis*, Springer Proceedings in Mathematics & Statistics 540, https://doi.org/10.1007/978-3-032-24155-9_1

Keywords Fisheries · Inland · Coastal · Deep sea · Efficiency · Productivity · SFA · DEA · Bootstrap

1 Introduction

Global fish production has grown phenomenally in the last four decades. According to the latest report of the FAO, total fisheries and aquaculture production reached its all-time peak in 2016 of 223.2 million tonnes [22]. Even though production is on an upward trend, the world's fisheries are not performing at an optimum level due to a number of causes, such as over-fishing, habitat loss and pollution [8, 11 ,32, 36]. Moreover, there is compelling evidence that climate change is causing significant changes to the global oceans [21, 30, 53, 61, 15, 50]. However, the impact of climate change on fisheries extends beyond its direct effects on the fishing industry, given its global cross-sectoral influence and interaction with other socio-economic and environmental drivers. For instance, environmental fluctuations influence fish population dynamics by modifying ecological processes, which in turn produce cascading effects on fishing activity and the economic productivity of commercial fisheries that operate in estuarine and coastal systems [44, 55]. For agriculture- and fisheries-dependent countries, livelihoods and infrastructure are impacted by different climate-related weather abnormalities, i.e. sea-level rise, sea surges, typhoons and rainfall intensity, and prolonged droughts. In recent years, these countries have faced increased frequency and intensity of these extreme weather conditions causing serious negative impacts on the lives and livelihoods of millions of people. These impacts are increasingly undermining the socio-economic progress achieved over the past several decades.

While there is a growing body of studies on the direct impact of prolonged climate change and resulting weather abnormalities on the fisheries sector see, for example, [7, 12, 42, 40], little or no attention has been given to the indirect impacts such as efficiency and productivity losses due to climate-related weather abnormalities on the fisheries sector. Rather, studies on efficiency and productivity of the fisheries sector mostly consider inefficiency arising from a number of different characteristics and in particular fishing vessels, skippers and equipment [17, 51, 52]. However, they overlook the influence of varying climatic conditions. Hence, understanding the impact of weather variations on the fisheries sector's efficiency and productivity is urgently needed for fisheries-dependent countries. Moreover, account needs to be taken of the fact that fisheries-dependent countries may exploit a number of different fisheries sectors which characterize their marine resources. For instance, many countries are engaged in, variously, freshwater inland fisheries, coastal fisheries, offshore deep-sea fisheries and aquaculture. To have informed decisions in relation to policy planning and resource allocation, studies on these multiple sectors are essential. More specifically, each sector, each region or agro-climatic zones and seasons have their unique characteristics and conditions contributing to the technical efficiency (TE) and productivity of the fisheries sector.

Hence, the aim of this study is to assess the efficiency and productivity of the fisheries industry by evaluating productivity change, technical change (TC) and efficiency change (EC), with particular emphasis on the influence of environmental risks, including disaster-related factors. The study draws on data from various fisheries sectors, regions and seasons across Sri Lanka. A two-stage analytical framework is adopted, incorporating the bootstrap Data Envelopment Analysis (DEA) approach proposed by [59]. In the second stage, the study examines the impact of different types of natural disasters, climate variability and extreme weather events on the performance of the fisheries sector, taking into account regional disparities, types of fisheries and temporal variations.

2 Data and Methodology

Data on inputs and outputs used in fisheries production and other climate-related variables are drawn from Sri Lankan[1] government agencies. Selection of variables follows the existing literature [41, 51,62]. The output variable is fish production (catch) in metric tons. Input variables are labour [41], number of tanks (reservoirs) and volume of water [56], the surface area of water used for inland fisheries [41] and gross tonnage and length of vessels [62].

The study uses balanced panel data relating to inland fisheries from records of 21 of 25 districts for the period 2005–2017. For coastal fisheries, the study uses data from the 15 coastal districts for the period of 2000–2017. For offshore fisheries, (deep-sea) data was collected from the Indian Ocean Tuna Commission (IOTC) which includes the vessel type, gear type, engine type and species and catch. Due to unavailability of data, the study only selects three gear types: gill nets (GIL), offshore gill nets (GIOF) and ring nets (RIN). According to the vessel type and gear type, we generate gross tonnage (GT) and crew number using the information available in the vessel information registry and last vessel's survey (2007). A summary of key data including outputs, inputs and their descriptive statistics is given in Table 1.

Second stage analysis involves analysing different disaster and weather-related variables relevant to the different fisheries sectors. A summary of disaster-related data is given in Table 2. The data on natural disasters was collected from Sri Lanka's Disaster Management Centre, which maintains disaster-related data in collaboration with 'DesInventar', the Disaster Information Management System of UNISDR—the United Nations Office for Disaster Risk Reduction. The data on wind dipoles and ocean pressure were collected from the working group on Global Climate Observing System (GCOS).

For the purpose of this study, disaster indices are constructed for assessing the severity of a disaster. Based on the indices used by Fomby [25] and Keerthiratne

[1] Ministry of Fisheries, Aquatic and Ocean Resources; Ministry of Disaster Management; Indian Ocean Tuna Commission (IOTC).

Table 1 Descriptive statistics of the three fisheries sectors

Variable	Obs	Mean	Std. Dev	Min	Max
Inland fisheries					
Production (MT)	273	2687.32	3308.48	10	16,325
Crafts	273	370.15	404.72	7	1790
Labour	273	1461.67	1308.10	90	5054
Water sq meter	273	120.24	116.20	13	515
Tanks	273	60.67	36.90	4	139
CPUE	273	8.45	9.52	0.34	89.14
Coastal fisheries					
Production (MT)	270	21,888.16	14,459.56	260	68,050
Labor	270	12,018.15	8309.52	740	39,250
Boats	270	2727.91	1780.12	193	7331
Landings	270	60.41	40.37	10	188
CPUE	270	10.12	9.56	0.60	68.17
IOTC fisheries					
Production (MT)	108	1233.57	1221.70	101	6760
Trips	108	1385.81	1302.58	11	7165
Length (M)	108	17.50	3.44	13	25
Gross Tonnage (GT)	108	47.58	29.04	20	124
Crew	108	3.67	0.71	3	5
CPUE	108	1.26	1.15	0.07	9.18

[37], we develop an index based on the percentage of population affected in each district (i) in each season (t) taking into account the total population of each region.

In addition to above-mentioned variables, ocean pressure and wind dipole data are added to the production frontier analysis for offshore deep-sea fisheries. Sustained changes in the difference between sea-surface temperatures of the tropical western and eastern Indian Ocean are known as the Indian Ocean dipole (IOD) [6]. It is also known as the Indian Niño which is an irregular oscillation of sea-surface temperatures in which the western Indian Ocean becomes alternately warmer and then colder than the eastern part of the ocean. The IOD involves an aperiodic oscillation of sea-surface temperatures between positive, neutral and negative phases. A positive phase sees greater than average sea-surface temperatures and greater precipitation in the western Indian Ocean region. There is a corresponding cooling of waters in the eastern Indian Ocean and which tends to cause droughts in adjacent land areas. The negative phase of the IOD brings about opposite conditions, with warmer water and greater precipitation in the eastern Indian Ocean and cooler and drier conditions in the west [6].

Table 2 Descriptive statistics of natural disasters: 2000–2017

Variable	Obs	Mean	Std. Dev	Min	Max
Disaster data used for inland fisheries					
Floods_i	273	4.727	14.196	0.000	88.030
Droughts_i	273	2.783	8.904	0.000	76.109
Strong winds_i	273	0.316	2.882	0.000	36.930
All disasters_i	273	7.826	18.255	0.000	80.120
Disaster data used for coastal fisheries					
Strong winds_i	270	0.341	2.464	0.000	0.032
Tsunami_i	270	0.592	4.377	0.000	50.740
Floods_i	270	0.423	12.583	0.000	8.364
Droughts_i	270	0.392	14.296	0.000	7.597
All disasters_i	270	5.988	12.890	0.000	51.234
Disaster data used for IOT fisheries					
Ocean pressure	108	1011.957	5.029	999.070	1020.140
Wind depol	108	0.191	0.299	−0.435	0.858
Flood_i	108	0.424	1.504	0.000	8.364
Droughts_i	108	0.393	1.352	0.000	7.597
Strong winds_i	108	0.003	0.006	0.000	0.028

2.1 *Methodology*

Both DEA and stochastic frontier analyses (SFA) have been used in past studies. A wide range of studies have highlighted the appropriateness of DEA in the calculation of TE in the fishing industry, given its capacity to measure vessels both individually and in multiple vessel fishing fleets [46]. DEA also allows for a simultaneous assessment of the number of inputs and outputs [18, 38]. As [65] note, DEA has been applied to the analysis of a number of aspects of the fisheries industry on a global scale, including all the major types of vessels such as long liners [51], purse seiners [46, 64], trawlers [63], artisanal vessels [26, 48] and recreational fishing [29, 48]. Considering the stochastic nature of fisheries data, many studies on efficiency in the fisheries sector have used SFA in their methodology (see [34, 39, 51, 56]).

2.1.1 Stochastic Frontier Analysis

SFA is a method of economic modelling used in many of the studies on fisheries and other industries. The model was first proposed by Aigner, Lovell, and Schmidt [2] and Meeusen and Van den Broeck [47]. The TE of an individual unit is defined as the ratio of observed outputs to the corresponding frontier outputs, given the level of inputs used. Most of the earlier studies using SFA have not explicitly formulated a model for the technical inefficiency effect in terms of appropriate explanatory variables. The model introduced by [10] allows estimation of a firm's specific resources and the factors explaining efficiency differentials among farms in a single procedure.

Accordingly, the model we use in this study is represented as follows:

$$Y_{it} = X_{it}\beta + (V_{it} + U_{it})$$

$$U_{it} = z_{it}\delta + w_{it}$$

where

- Y_{it} denotes the production at time t (t = 1, 2 ...T) for the ith (i = 1, 2...N) DMU.
- X_{it} is a (1 × k) vector of the logarithm of inputs of production.
- β is a (k × 1) vector of unknown parameters to be estimated.
- V_{it} s are assumed to be random errors, independently distributed of the U_{it}.
- U_{it} s are non-negative random variables associated with the TE of production which are assumed to be independently distributed such that U_{it} is obtained by truncation (at zero) of the normal distribution with mean, $z_{it}\delta$, and variance, σ^2.
- z_{it} is a (1 × m) vector of explanatory variables associated with technical inefficiency of production of firms over time.
- δ is an (m × 1) vector of unknown coefficients.
- w_{it} is defined by the truncation of the normal distribution with zero mean and variance,σ^2.

2.1.2 Data Envelopment Analysis

The theoretical foundations of the use of DEA methodology to measure efficiency were first developed by [24] and [57, 58]. The work by [13] extended efficiency measurement to the use of both multiple inputs and outputs and, as such, was the starting point for the modern development of empirical measurement of efficiency through the DEA methodology. This analysis applies an output-based measure of efficiency, as inputs allocated to the agricultural sector are more or less fixed.

The output-oriented DEA efficiency estimator δ can be derived by solving the following linear programming:

$$\widehat{\delta_i} = \max\{\delta > 0 | \widehat{\delta_i} \leq \sum\nolimits_{i=1}^{n} y_i\lambda; x_i \geq \sum\nolimits_{i=1}^{n} x_i\lambda; \lambda \geq 0\} i = 1 \ldots n$$

where y_i is a vector of outputs, $x_i isa$ vector of inputs, and λ is an I × 1 vector of constants. The value of $\widehat{\delta_i}$ obtained is the TE score for the i-th DMU. A measure of $\widehat{\delta_i}$=1 indicates that the DMU is technically efficient, and inefficient if $\widehat{\delta_i}$<1. This linear programming problem must be solved n times, once for each DMU in the sample.

Since the DEA model described above is a constant return to scale (CRS) model, a variable return to scale (VRS) assumption is imposed on the above model by introducing the constraint $\sum_{i=1}^{n} \lambda = 1$. Efficiency scores ranging from 0 to 1 define a ranking for the DMU such that 1 corresponds to the most efficient and 0 indicates the least efficient. As the purpose of this research is to measure the impact of disasters on different DMUs engaged in fish production, we need to test the hypotheses that efficiency of particular DMUs is determined by different disaster-related contextual variables. Accordingly, this study uses the two-step approach as suggested by [16].

2.1.3 Bias-Corrected DEA—Bootstrap Approach by [59]

Regardless of their popularity and intuitive appeal, two-stage estimators have been criticized by different authors. To overcome these problems, [59] developed a two-stage procedure that takes into account the underlying data generating process that is consistent with a two-stage estimation procedure.

Accordingly, the following specification is estimated:

$\widehat{\delta_i} = z_i\beta + \varepsilon_i$ where z_i is a vector of environmental variables which is expected to explain the efficiency variations; β refers to a vector of parameters to be estimated, and εi is an error term. We use the double bootstrapping truncated regression algorithm #2. Following [59], the study employs 2000 replications in simulation.

2.1.4 Measuring TFP-DEA Malmquist Productivity Index (MPI)

This study also employs the MPI to observe productivity changes overtime. Following Fare et al. [23], the Malmquist (output-oriented) TFP changes the index between periods (the base period) with period t given by

$$m_o(y_s, x_s, y_t, x_t) = [\frac{d_o^s(y_t, x_t)}{d_o^s, (y_s, x_s)} = \frac{d_o^t(y_t, x_t)}{d_o^t, (y_s, x_s)}]^{1/2}$$

where the notation $d_o^s(y_t, x_t)$ represents the distance from the period t observation to the period s technology. A value of m_o greater than 1 will indicate positive TFP growth from periods s to t, while a value less than 1 indicates a TFP decline. For decomposing purposes, the following equation is convenient:

$$m_o(y_s, x_s, y_t, x_t) = \frac{d_o^t(y_t, x_t)}{d_o^s, (y_s, x_s)}[\frac{d_o^s(y_t, x_t)}{d_o^t, (y_t, x_t)} = \frac{d_o^t(y_s, x_s)}{d_o^t, (y_s, x_s)}]^{1/2}$$

where the ratio outside the brackets measures the changes in the output-oriented measure of Farell's [24] TE between period s and t. That EC is equivalent to the ratio of TE in period t to TE in period s. The remaining part of the index is TC. It is the geometric mean of the shift in technology between the two periods, evaluated as at x_t andx_s.

3 Efficiency Analysis in Inland, Coastal and Deep-Sea Fisheries

As discussed in the methodology section, this study employs different methods under different scenarios to investigate the productivity and efficiency of the fisheries sector. Accordingly, first the study investigates the production frontiers and efficiency level of different fisheries sectors. Second, this study examines the TFP change, EC and TC during the observed time period in individual fishery sectors. Third, it investigates the disaster-related determinants of inefficiency in different fisheries sectors.

3.1 Production Frontier and Efficiency Scores

In order to estimate the potential contribution of inputs to fish production, we estimated the stochastic frontier using the approach by [9]. The Cobb–Douglas specification is used given two of the three fishery sectors show better results in this form. We removed 'capacity per unit' (CPUE) to avoid a collinearity problem. Most of the inputs considered in the three sectors show positive and significant results, except 'landings' in the coastal fishery and gross tonnage. The number of crew in the Indian Ocean Tuna (IOT) fishery also shows a negative relationship with fish production. When we add east-coast and west-coast dummies to the coastal fisheries frontier it is the west coast which shows a positive and significant sign, while the east coast has a negative and significant sign. This indicates west coast is more productive for coastal fisheries. Inland fisheries indicated on average a 94% TE, while coastal fisheries and IOT fisheries show a range from 52 to 59% efficiency. These scores do not, however, represent the TE obtained from a model explicitly formulated to measure inefficiency effects in terms of appropriate explanatory variables. These scores were obtained only for the purpose of comparing efficiency scores with the scores discussed later in this section which are obtained using different inefficiency models that employ different disaster variables as efficiency determinants [10, 59]. Moreover, it is not recommended to compare these scores as these results show the average efficiency of a particular sector according to their own frontier (Table 3).

3.2 Impact of Disasters on Efficiency Using SFA and DEA

In this section, the study examines the bias-corrected efficiency scores of different sectors of fisheries. As explained in the methodology section, the study employs both the SFA inefficiency model estimation [10] and bias-corrected bootstrap DEA [59] to obtain efficiency scores using the models that are explicitly formulated to understand the impact of disaster-related inefficiency determinants on fisheries. Accordingly, this study indicates that the above inefficiency models produced on average

Table 3 Production frontier results

Inland fishery				Coastal fishery				IOT fishery			
Variable	coef.		SE	Variable	coef.		SE	Variable	coef.		SE
Constant	−0.603		0.884	_cons	2.786		1.173	_cons	−1.343		1.831
Boats/crafts	0.380	***	0.079	Boats	0.386	**	0.129	Trips	0.520	***	0..060
Labour	0.424	***	0.088	Labour	0.613	***	0.151	Length	2.629	**	1.225
Water (sqm)	0.299	***	0.085	Landings	−0.230	*	0.127	Gross tonnage (GT)	−0.221		0.177
Tanks	0.369	***	0.090	East coast D West coast D	−0.965	***		Crew	−1.157		0.981
Log likelihood			−325.010	Log likelihood			−101.690	Log likelihood			−337.750
Mean efficiency			0.940	Mean efficiency			0.510	Mean efficiency			0.580

***p<0.01, **p<0.05, *p<0.1

lower efficiency scores compared to the previously estimated [9] efficiency model. That is, inland fisheries indicated on average an 85% efficiency score compared to the previous estimation of 94%. Coastal fisheries indicated a 50% efficiency level compared to the 52% of the previous model. The efficiency of the IOT fishery is shown to be on average at 56% compared to the previous model's 59%. See Figs. 1 and 2 for a comparison of DEA and SFA efficiency scores in inland and coastal fisheries sectors.

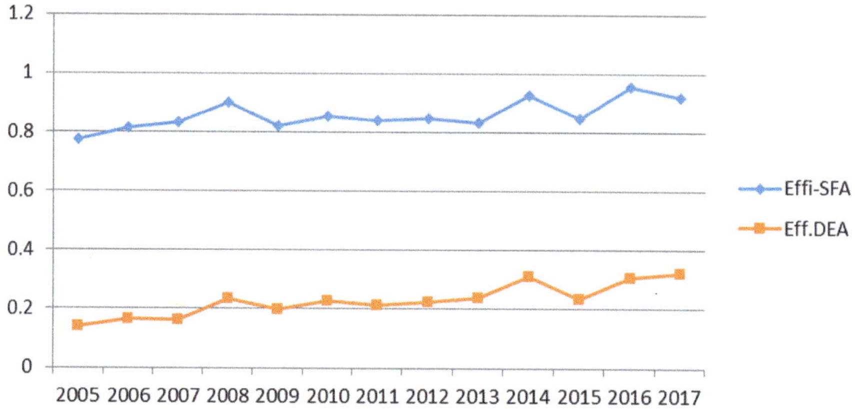

Fig. 1 Temporal trends in efficiency scores of inland fisheries: 2005–2017

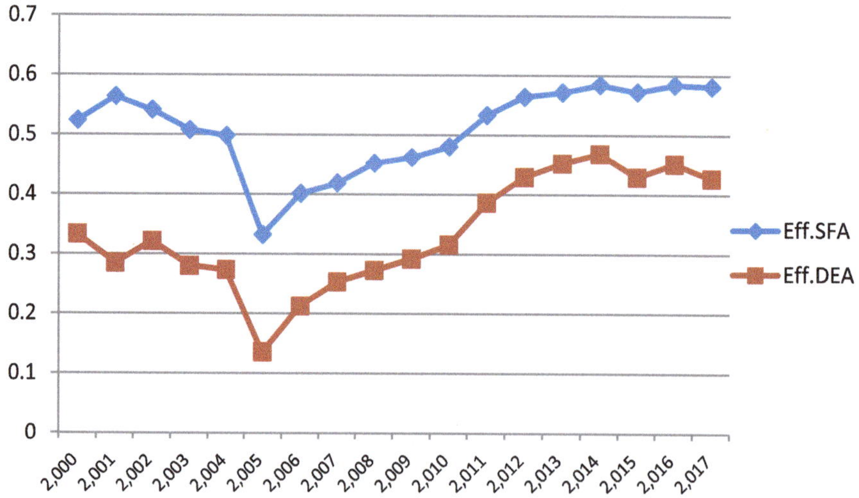

Fig. 2 Temporal trends in efficiency scores of coastal fisheries: 2000–2017

3.3 TFP Change, EC and TC

3.3.1 TFP Change, EC and TC in Inland Fisheries

This section examines the productivity change of the Sri Lankan inland fisheries sector for the period 2005–2017. The estimates of TFP growth, TC and EC based on the MPI are reported in Table 4. It shows that the mean TFP is 7.7% made up of gains in EC of 5.2% and gains in TC of 2.3%. This suggests productivity growth has been mainly in the form of EC. In decomposing efficiency, we note that PTE contributes 2.7% and EC 2.4% to the growth in efficiency. Moreover, it is shown that in the period 2005 to 2007, there were productivity losses mainly due to regressing TC. However, productivity loss appearing in the periods 2009–2010 and 2011–14 was mainly due to regressing EC. Technical regress and positive EC were revealed in the period of 2014–17. This indicates that the inland fisheries sector has been improving in terms of efficient use of resources and best practice management.

Due to the heterogeneous nature of different DMUs and to a better understanding of TFP and its decomposition, we present the growth rate of each component by different districts involved in inland fisheries in Table 5.

According to Table 5 in almost all districts, there is an improvement of TFP except for Badulla, Gampaha and Nuwara Eliya. Of the twenty-one districts, seventeen posted positive TFP growth being mainly attributed to improved EC in turn indicating movements towards the frontier. From the twenty-one districts, only four districts show efficiency regress, while only seven districts regressed technically.

Table 4 TFP change, efficiency change and technology change in the inland fisheries, 2005–2017

Year	TFP Change	Efficiency Change (EC)	Technical Change (TC)	Pure technical Efficiency (PTE)	Scale Efficiency (SE)
2005–06	0.943	1.054	0.894	0.97	1.087
2006–07	0.991	1.297	0.764	1.125	1.153
2007–08	1.298	1.095	1.186	1.306	0.838
2008–09	1.147	0.797	1.439	0.91	0.875
2009–10	0.942	0.878	1.073	0.926	0.949
2010–11	1.106	1.027	1.077	0.801	1.283
2011–12	1.625	1.712	0.949	1.318	1.299
2012–13	0.983	0.898	1.095	0.961	0.935
2013–14	0.97	0.661	1.469	0.806	0.819
2014–15	1.015	1.165	0.871	1.074	1.084
2015–16	0.957	1.138	0.841	1.078	1.055
2016–17	1.115	1.274	0.875	1.214	1.05
Mean	1.077	1.052	1.023	1.027	1.024

Table 5 TFP change, efficiency change and technology change in the inland fisheries districts

District	TFP change	Efficiency	Technical	Pure Eff	Scale Eff
		Change (EC)	Change (TC)	Change (PTE)	Change (SE)
Ampara	1.102	1.028	1.072	1.047	0.982
Anuradhapura	1.079	1.04	1.037	1.021	1.019
Badulla	0.999	0.94	1.062	0.924	1.018
Batticaloa	1.025	1.014	1.011	0.997	1.017
Colombo	1.015	0.992	1.023	1	0.992
Galle	1.299	1.321	0.983	1.29	1.024
Gampaha	0.911	0.958	0.952	1	0.958
Hambantota	1.195	1.094	1.093	1.089	1.004
Kalutara	1.1	1.104	0.997	1.054	1.047
Kandy	1.108	1.027	1.079	1.013	1.014
Kurunegala	1.089	1.059	1.028	1.059	1
Mannar	1.055	1.126	0.938	1.096	1.027
Matale	1.121	1.043	1.075	1.038	1.005
Matara	1.273	1.214	1.049	1	1.214
Monaragala	1.035	1.005	1.03	0.998	1.007
Nuwara Eliya	0.989	1.013	0.976	0.871	1.163
Polonnaruwa	1.047	0.991	1.056	1.008	0.984
Puttalam	0.976	1	0.976	1	1
Ratnapura	1.102	1.037	1.063	1.024	1.013
Trincomalee	1.101	1.071	1.028	1.044	1.025
Vavuniya	1.068	1.087	0.982	1.055	1.03
Mean	1.077	1.052	1.023	1.027	1.024

As a further extension, time periods were clustered into two slots and an investigation made of the trends of TFP, EC and TC and, more importantly, according to Sri Lanka's agro-ecological zones (see Table 6). According to the results, the mean TFP of the dry zone has increased considerably over the time period. In 2005–2011 the mean TFP was 6.5% which can be attributed to gains in EC of 0.5% and 6.1% in TC. In 2012–2017, the mean TFP was 9.0% attributable to gains in EC of 7.8% and 1.2% in TC. Over time, the main contributing factor to TFP growth in the dry zone changed from EC to TC. Both wet and intermediate zones show decreasing mean TFP with the contributing factors being EC and TC. In 2005–12 growth of mean TFP of the wet zone was 4.9% attributed to gains in EC of 2.7% and 2.2% in TC. However, in 2012–17 (see Table 7) mean TFP change regressed by 6%—attributed to a regression in EC of 3.6% and 0.6% in TC. Similarly, results of the intermediate zone show a regression in mean TFP in the period 2012–17 compared to the

Table 6 TFP decomposition in inland fisheries: 2005–2011

District	TFP Change	Efficiency	Technical	Pure Eff	Scale Eff
2005–2011		Change	Change	Change	Change
Dry zone					
Ampara	1.209	1.073	1.126	1.100	0.976
Anuradhapura	1.147	1.023	1.122	1.042	0.982
Batticaloa	0.920	0.856	1.074	0.893	0.958
Hambantota	1.153	1.029	1.121	1.089	0.945
Mannar	0.962	1.089	0.883	1.039	1.049
Monaragala	1.080	0.951	1.136	1.008	0.943
Polonnaruwa	1.127	0.985	1.145	1.063	0.926
Puttalam	0.935	0.970	0.964	1.000	0.970
Trincomalee	1.111	1.017	1.093	1.003	1.014
Vavuniya	1.002	1.057	0.948	1.003	1.054
Mean	1.065	1.005	1.061	1.024	0.982
Wet zone					
Colombo	1.053	1.050	1.003	1.000	1.050
Galle	1.116	1.186	0.941	1.079	1.099
Gampaha	0.898	0.998	0.899	1.000	0.998
Kalutara	0.802	0.787	1.019	0.790	0.997
Kandy	1.013	0.892	1.135	0.862	1.035
Matara	1.253	1.178	1.064	1.000	1.178
Nuwara Eliya	0.983	1.048	0.938	0.864	1.213
Ratnapura	1.270	1.080	1.177	1.048	1.030
Mean	1.049	1.027	1.022	0.955	1.075
Intermediate zone					
Badulla	1.073	0.964	1.114	0.912	1.056
Matale	1.161	1.018	1.140	0.989	1.029
Kurunegala	1.206	1.100	1.097	1.165	0.944
Mean	1.147	1.027	1.117	1.022	1.010

period 2005–12. These results further indicate inland fisheries in the dry zone have substantially contributed to the overall rise in TFP.

3.3.2 TFP Change, EC and TC in Coastal Fisheries: 2014–2016

As previously stated, for the purpose of investigative efficiency in the measurement of TFP in coastal fisheries, this study uses data from 15 coastal districts for the years 2000 to 2017.

Table 7 TFP decomposition results of 2002–2017 period

District	TFP Change	Efficiency	Technical	Pure Eff	Scale Eff
2012–2017		Change	Change	Change	Change
Dry zone					
Ampara	1.012	0.976	1.037	0.999	0.977
Anuradhapura	0.998	1.020	0.979	1.000	1.020
Batticaloa	1.393	1.414	0.985	1.360	1.040
Hambantota	1.227	1.102	1.114	1.087	1.014
Mannar	1.150	1.146	1.003	1.125	1.019
Monaragala	0.975	1.003	0.972	0.974	1.030
Polonnaruwa	0.915	0.915	1.000	0.946	0.968
Puttalam	1.017	1.000	1.017	1.000	1.000
Trincomalee	1.171	1.179	0.993	1.164	1.013
Vavuniya	1.042	1.022	1.019	1.025	0.997
Mean	1.090	1.078	1.012	1.068	1.008
Wet zone					
Colombo	0.827	0.858	0.964	1.000	0.858
Galle	1.245	1.202	1.036	1.270	0.947
Gampaha	0.761	0.902	0.844	1.000	0.902
Kalutara	1.176	1.144	1.028	1.104	1.037
Kandy	1.055	0.993	1.062	1.000	0.993
Matara	0.961	0.914	1.051	1.000	0.914
Nuwara Eliya	0.679	0.702	0.967	0.718	0.979
Ratnapura	0.994	0.996	0.999	1.000	0.996
Mean	0.962	0.964	0.994	1.012	0.953
Intermediate zone					
Badulla	0.876	0.835	1.049	0.860	0.970
Matale	0.909	0.866	1.049	0.890	0.973
Kurunegala	1.034	1.048	0.986	1.020	1.028
Mean	0.940	0.916	1.028	0.923	0.990

Table 8 illustrates the productivity change, TC and EC of the Sri Lankan coastal fisheries sector for the period 2000–2017. It shows that mean TFP is 1.5% attributed to gains in EC of 0.4% and 1.1% in TC—suggesting productivity growth mainly came from TC. Further, by decomposing efficiency we find that PTE regressed by 0.4% while SE contributed to growth in efficiency of 0.8%. For the years 2002 to 2006 there are noticeable productivity losses mainly due to regressing TC. In the 2004–2005 period—which included the year the Indian Ocean tsunami destroyed almost all the coastal villages in Sri Lanka—a massive loss of productivity was recorded due to a strong regression in both EC and TC. In the following period 2005–2006, the growth in TFP was strongly driven by the high level of TC—the highest recorded (2.081) for the past 18 years. This can largely be attributed to international aid and contributions following the tsunami. Previous literature has also observed the positive economic impacts following natural disasters [3] and [60]. Such authors have noted the occurrence of 'creative destructions' in the aftermath of disasters which produce positive economic growth. Some other studies have criticized this view [31], and the debate on this subject remains inconclusive. After the 2004 tsunami, 3 years of efficiency regress (2005–2008) was recorded indicating its negative effect on the use of resources and best-practice management. Thus, over the years 2016–17 TFP regression was experienced due to the regression of both TC and EC.

Due to the heterogeneous nature of different DMUs and to a better understanding of TFP and its decomposition, Table 9 presents the growth rate of each component by different districts involved in inland fisheries. Overall, almost all districts show, on average, an improvement of TFP. The exceptions are Chilaw, Mannar, Matara and Negombo. From the fifteen districts only two show technical regression while efficiency regression is posted in six districts. Of the fifteen costal districts twelve posted positive TFP growth—mainly attributed to improved TC—indicating productivity growth was largely derived from 'frontier shift' and not from 'catch up'.

Table 8 A decomposition of TFP, EC and TC: 2000–2017

Year	TFP change	Efficiency Change (EC)	Technical Change (TC)	Pure Tec. eff (PTE)	Scale eff (SE)
2001–2002	1.208	1.070	1.129	1.024	1.045
2002–2003	0.896	1.238	0.724	0.984	1.259
2003–2004	0.802	1.004	0.799	1.041	0.964
2004–2005	0.952	1.004	0.948	1.025	0.980
2005–2006	0.453	0.802	0.565	0.956	0.839
2006–2007	1.337	0.643	2.081	0.816	0.787
2007–2008	0.937	0.650	1.440	0.872	0.746
2008–2009	1.147	1.005	1.142	1.112	0.903
2009–2010	1.099	1.376	0.799	1.204	1.143
2010–2011	1.331	1.280	1.040	1.047	1.223
2011–2012	1.273	1.128	1.128	1.017	1.109
2012–2013	1.127	1.077	1.047	1.048	1.027
2013–2014	1.111	1.167	0.952	1.037	1.126
2014–2015	1.102	1.241	0.889	0.925	1.341
2015–2016	0.930	0.797	1.167	0.881	0.905
2016–2017	1.038	1.000	1.039	1.066	0.937
2017–2018	0.961	0.966	0.995	0.950	1.017
Mean	1.015	1.004	1.011	0.996	1.008

Table 9 A decomposition of TFP, efficiency and technical change in coastal districts

DMU	TFP change	Efficiency Change (EC)	Technical Change (TC)	Pure Tec eff. (PTE)	Scale eff (SE)
Batticaloa	1.019	1.001	1.018	1.007	0.994
Chilaw	0.994	0.985	1.009	0.984	1.001
Colombo	1.024	1.014	1.010	1.000	1.014
Galle	1.023	1.014	1.009	1.008	1.006
Jaffna	1.035	1.017	1.017	1.039	0.979
Kalmunai	1.022	0.995	1.027	1.004	0.991
Kalutara	1.020	1.000	1.020	1.000	1.000
Kilinochchi	1.110	1.114	0.997	1.107	1.006
Mannar	0.981	0.970	1.011	0.963	1.008
Matara	0.977	0.975	1.002	0.986	0.989
Mullaitivu	1.017	1.034	0.984	0.911	1.135
Negombo	0.968	0.949	1.020	0.963	0.986
Puttalam	1.000	0.995	1.005	0.973	1.023
Tangalle	1.019	1.000	1.019	1.000	1.000
Trincomalee	1.021	1.007	1.013	1.011	0.997
Mean	1.015	1.004	1.011	0.996	1.008

As a further exercise, we clustered the time period into three time slots in order to investigate trends in TFP, EC and TC, and—more importantly—in relation to the impact of tsunami destruction (see Table 10). The results show that the mean TFP in the 2000–2004 period (before the 26 December 2004 tsunami) was minus 4.7%—mainly attributed to an 11.3% technological regression. Overall, all the districts experienced technological regress before the intervention of the tsunami. Efficiency shows growth in that period of 7.5%. Only three districts out of fifteen show regression of efficiency. In the period 2005–2009[2] which is the period soon after the tsunami, TFP shows a 12.1% growth mainly attributed to technological improvement of 28%. However, efficiency regressed by 12.8% during the period. Eleven districts out of fifteen showed technical regress while all the districts showed technical improvement. Further, both scale efficiency and pure TE regressed during the period. According to the 2010–2017 analysis, it is clear that the sector returns to a more or less normal status with some improvement in efficiency which indicates a 'catch up'. Overall TFP growth is 7.2% for this period due to both EC (4.4%) and TC (2.7%). Most districts regressed in terms of efficiency during the previous period (2000–2004) but achieved improvement in efficiency and high growth in TC during 2006–2013).

[2] PLEASE see Appendix Table 1 for all three periods.

Table 10 Temporal decomposition of productivity in coastal fisheries: TFP, efficiency and technical change

District	2000–2004					2010–2017				
	TFPC	EC	TC	PECH	SECH	TFPC	EC	TC	PECH	SECH
Batticaloa	1.092	1.266	0.862	1.136	1.115	0.911	0.881	1.034	0.870	1.013
Chilaw	0.876	0.977	0.897	0.916	1.067	1.037	1.015	1.021	0.982	1.034
Colombo	0.817	0.927	0.881	1.000	0.927	1.149	1.147	1.001	1.102	1.041
Galle	0.918	1.051	0.873	0.990	1.061	1.150	1.116	1.030	1.089	1.025
Jaffna	1.091	1.222	0.893	1.311	0.932	1.101	1.080	1.019	1.044	1.034
Kalmunai	1.099	1.238	0.888	1.238	1.000	1.077	1.008	1.069	0.978	1.030
Kalutara	0.905	1.000	0.905	1.000	1.000	1.026	1.000	1.026	1.000	1.000
Kilinochchi	1.091	1.264	0.863	1.130	1.119	1.418	1.397	1.015	1.000	1.397
Mannar	0.939	1.049	0.896	1.004	1.045	1.153	1.123	1.027	1.034	1.086
Matara	0.913	1.004	0.910	1.000	1.004	0.967	0.939	1.031	0.966	0.972
Mullaitivu	0.947	1.097	0.863	0.714	1.536	0.948	0.963	0.984	0.797	1.209
Negombo	0.914	1.036	0.882	0.985	1.052	1.000	0.978	1.022	0.968	1.011
Puttalam	0.867	0.958	0.905	0.899	1.066	1.025	0.984	1.041	0.988	0.996
Tangalle	0.899	1.000	0.899	1.000	1.000	1.207	1.149	1.051	1.116	1.030
Trincomalee	0.998	1.120	0.891	1.095	1.023	1.014	0.978	1.036	0.927	1.056
Mean	0.953	1.075	0.887	1.018	1.056	1.072	1.044	1.027	0.987	1.058

3.3.3 TFP Change, EC and TC in Deep-Sea Fisheries: 2014–2016

As mentioned in the data section, our data set for Indian Ocean tuna fisheries is limited to three years. Therefore, we investigated monthly TFP change, EC and TC according to three different gear types. One of the main observations is that offshore gillnets produce higher productivity than the other two gear types. Also, efficiency contributed more to the existing productivity level than to TC. In examining different gear types, it is revealed that, where the same gear type is used overtime, there is technological regression because of depreciation of the gear/equipment which is consistent with expectations. Further, this study investigated the seasonal variation and monsoonal effect on TFP changes, given deep-sea fisheries are highly dependent on climatic conditions. In Sri Lanka, the northeast monsoon falls from September to March, while the southwest monsoon occurs during May to end of August. Accordingly, the southwest coast, which is well known for fisheries, is more accessible during September to April. As the results for September to April show, there is productivity growth in all three gear types, but a regression in TFP in the period from May to September, indicating the impact of climatic conditions on the performance of deep-sea fisheries (Appendix Table 14).

Table 11 Impact of different disasters and climatic conditions on efficiency of different fisheries sectors (SFA analysis)[3]

Dependent variable—efficiency			
	Inland	Coastal	IOT
Floods	↓	↓	↓
Droughts	↓	↑	↓
Strong winds	↑	↓	↓
All disasters	↓	↓	↓
Tsunami		↓	
Ocean pressure			↑
Wind dipole			↓

*** $p < 0.01$, ** $p < 0.05$, * $p < 0.1$

Table 12 Sectoral vulnerability and efficiency in fisheries: impacts of climate and disasters (bias-corrected bootstrap DEA analysis)

Dependent variable—efficiency			
	Inland	Coastal	IOT
Floods	↓	↓	↓
Droughts	↑	↑	↓
Strong winds	↓***	↓***	↑*
Combined disasters	↓	↓	↓
Tsunami		↓	
Ocean pressure			↑**
Wind dipole			↓

*** $p < 0.01$, ** $p < 0.05$, * $p < 0.1$

3.4 Determinants of Inefficiency

This study used the disaster-related variables only for the inefficiency model given adding more control variables that are endogenous to the weather variation could be counterproductive and partially offset part of the true weather effect [43]. Moreover, adding more variables in weather-related studies can cause problems such as over-control or poor control [19]. The study carried out the analysis individually for all disaster and weather-related variables given most types of disasters do not occur simultaneously. However, the study analysed the combined impact through the 'all disasters' variable which is a combination of the floods, drought and strong wind indices (Tables 11 and 12).

Many studies have found that climate change has a severe impact on fishing. However, most authors examine the fish catch/production in different types of fishery environments. Ahamad [1] argues that the main reason for declining catches is directly related to climatic variables, including cyclone, salinity, sea-level rise, water temperature, floods, rainfall and droughts. He goes on to state that all of these climatic

[3] ↓ indicates a negative impact on efficiency; ↑indicates a positive impact on efficiency.

variables also affect the physical operation of fishing practices by debilitating the efforts of fishers. In this study, the flood variable shows a consistent negative relationship with the efficiency of fish production in all three sectors throughout the study. Similarly, a number of studies [28] have found fish abundance is significantly negatively correlated with rainfall in coastal rivers and marine environments in California during and post-flood occurrences.

On the other hand, the literature shows that temperature and wind are an important trigger for fish migration, growth and reproduction patterns [5, 33]. However, according to the results of this study, there is an inconclusive effect of both the drought and wind variables. The drought variable shows a mostly positive relationship with efficiency but is negative for the IOT fishery analysis and is consistently so in both SFA and DEA analyses. This indicates that drought is not disrupting the Sri Lankan fisheries community as do other extreme events such as high rainfall and wind-related extreme events.

Strong winds exhibit a generally negative, though inconclusive, relationship with efficiency across different fisheries sectors under both DEA and alternative analytical frameworks. Specifically, DEA analysis indicates a negative association with efficiency in coastal and inland fisheries, while a positive relationship is observed in Indian Ocean Tuna (IOT) fisheries. These findings align with previous literature, which suggests that the impact of wind varies significantly across fisheries sectors and geographic regions [35, 45, 54].

The study further investigates the comparative effect of ocean pressure and wind dipole indicators on efficiency in IOT fisheries. The results indicate that ocean pressure is positive in both DEA and SFA models. The wind dipole is negative in both models.

The IOT fishery data consists of tuna and tuna-like species. [11] observes that the two tuna species skipjack (*Katsuwonus pelamis*) and yellowfin (*Thunnus albacares*) are among the top predators of the tropical pelagic ecosystem and that they have higher recruitments during El Niño events. El Nino and ocean currents are closely related to ocean pressure and oceanic atmosphere. Doner et al. (2017) expand on previous analyses of environmental factors related to productivity of Chinook salmon (*Oncorhynchus tshawytscha*) populations confirming that ocean currents can have a positive impact on their productivity.

4 Conclusions

Given agricultural productivity and efficiency are key drivers of the wellbeing of many developing countries, this study investigated whether and to what extent natural disasters have an impact on the TC of the fisheries sector based on Sri Lankan data. Previous global studies have highlighted the importance of considering different sectors' agro-climatic conditions and regions where responses may differ from the global trend in climate change studies [4, 50]. This study offers a new perspective from previous studies by investigating TE of three different fisheries sectors under different geo-climatic conditions.

The central finding of this study is that natural disasters have a negative impact on the TE of fishery production. That impact varies according to the different geo-climatic factors, such as agro-ecological zones and different seasons and different time periods. Accordingly, the results confirm that DMUs in different agro-climatic regions respond differently to different disasters. Therefore, heterogeneous disasters occurring under different geo-climatic conditions and seasons do not necessarily have the same impact. As indicated by the results in relation to inland fisheries, dry zones are more efficient than the intermediate and wet zones, while in coastal fisheries, the west coast is more efficient than the east coast. Further, in relation to IOT fisheries, it is confirmed that during April to September—which is the period when the southwest coast is accessible to fishing activities—greater efficiency is evident when compared to that of May to September, which is the period that the east coast of the country engages in fisheries. It is important to note that our study mainly focuses on the impact of disasters. Hence, the investigation of their role as determinants of inefficiency is a major concern. Further, the results reveal that strong winds have a significant negative relationship with efficiency in both coastal and inland fisheries, but a positive relationship with the IOT fisheries sector. Floods, the combined disaster variable and the wind dipole data variable used in IOT efficiency analysis all show a consistent negative relationship with efficiency. The drought variable shows mostly a positive impact. Ocean pressure shows a consistent positive relationship with efficiency in both the analyses. These results suggest that not all disasters and weather-related variables react in the same way on efficiency in fisheries.

This study makes a significant contribution by addressing a gap in the existing literature, which has largely overlooked technical efficiency and productivity across all three fisheries sectors in Sri Lanka. Furthermore, this study incorporates data from all districts, diverse agro-ecological zones, coastal regions and seasonal variations, offering a comprehensive national perspective. To the authors' knowledge, it is also the first study to investigate the effects of extreme weather events and natural disasters on both productivity and technical efficiency in the fisheries sector—within Sri Lanka or in comparable international contexts.

There are limitations to this study. Sample size and data quality are two of these. As well, collecting reliable input data in all three sectors is a major challenge. Even though the IOTC collects comprehensive data on deep-sea fisheries, identifying data according to different vessels is particularly complicated and incomplete. Further, disaster and weather data typically suffer from measurement errors, while the unpredictable nature of disaster variables limits our analysis in several ways. Some studies forecast the impact of climate change based on weather variables. However, disaster variables can be of a different nature and less predictable.

Lastly, the findings of this study hold significant relevance for policymakers in disaster-prone agricultural economies. They offer valuable insights for developing informed strategies to mitigate the adverse effects of weather-related disasters on fisheries efficiency, thereby helping to reduce associated economic costs. Moreover, the results can support efforts to enhance fisheries production performance and guide the formulation of more effective pre- and post-disaster management plans aimed at minimizing efficiency losses.

Acknowledgements We are thankful to the Ministry of Fisheries of Sri Lanka and the Indian Ocean Tuna Commission (website) for data. We also thank participants of the Western Economics Association International conference (2018) held in Tokyo, Japan, for their helpful comments.

Disclosure of Interests The authors have no competing interests to declare that are relevant to the content of this article.

Appendix 1

See Table 13

Table 13 Total Factor Productivity, efficiency, and technical change across different time periods: in coastal fisheries

2000–2004						2005–2009					2010–2017				
District	TFPC	EC	TC	PECH	SECH	TFPC	EC	TC	PECH	SECH	TFPC	EC	TC	PECH	SECH
Batticaloa	1.092	1.266	0.862	1.136	1.115	1.246	0.936	1.330	1.055	0.887	0.911	0.881	1.034	0.870	1.013
Chilaw	0.876	0.977	0.897	0.916	1.067	1.230	0.961	1.280	0.976	0.985	1.037	1.015	1.021	0.982	1.034
Colombo	0.817	0.927	0.881	1.000	0.927	1.130	0.860	1.314	1.000	0.860	1.149	1.147	1.001	1.102	1.041
Galle	0.918	1.051	0.873	0.990	1.061	1.124	0.881	1.276	0.915	0.962	1.150	1.116	1.030	1.089	1.025
Jaffna	1.091	1.222	0.893	1.311	0.932	1.023	0.776	1.318	0.793	0.978	1.101	1.080	1.019	1.044	1.034
Kalmunai	1.099	1.238	0.888	1.238	1.000	1.071	0.830	1.290	0.943	0.880	1.077	1.008	1.069	0.978	1.030
Kalutara	0.905	1.000	0.905	1.000	1.000	1.300	1.000	1.300	1.000	1.000	1.026	1.000	1.026	1.000	1.000
Kilinochchi	1.091	1.264	0.863	1.130	1.119	0.868	0.692	1.253	1.182	0.586	1.418	1.397	1.015	1.000	1.397
Mannar	0.939	1.049	0.896	1.004	1.045	0.981	0.775	1.266	0.812	0.954	1.153	1.123	1.027	1.034	1.086
Matara	0.913	1.004	0.910	1.000	1.004	1.276	1.000	1.276	1.000	1.000	0.967	0.939	1.031	0.966	0.972
Mullaitivu	0.947	1.097	0.863	0.714	1.536	0.882	0.689	1.280	1.229	0.560	0.948	0.963	0.984	0.797	1.209
Negombo	0.914	1.036	0.882	0.985	1.052	1.049	0.816	1.284	0.960	0.851	1.000	0.978	1.022	0.968	1.011
Puttalam	0.867	0.958	0.905	0.899	1.066	1.184	0.943	1.256	0.902	1.045	1.025	0.984	1.041	0.988	0.996
Tangalle	0.899	1.000	0.899	1.000	1.000	1.296	1.023	1.266	1.035	0.988	1.207	1.149	1.051	1.116	1.030
Trincomalee	0.998	1.120	0.891	1.095	1.023	1.300	1.000	1.301	1.121	0.892	1.014	0.978	1.036	0.927	1.056
Mean	0.953	1.075	0.887	1.018	1.056	1.121	0.872	1.286	0.988	0.882	1.072	1.044	1.027	0.987	1.058

Table 14 TFP change, EC and TC in IOT fisheries

	Gill nets			Offshore-Gillnets(GIOF)			Ring nets		
	effch	techch	tfpch	effch	techch	tfpch	effch	techch	tfpch
Jan-Feb	1.472	1.437	2.115	1.015	1.217	1.236	1.372	0.509	0.698
Feb-Mar	0.69	1.267	0.875	0.983	0.968	0.952	0.887	1.639	1.454
Mar-Apr	0.826	0.862	0.712	1.052	0.893	0.94	1.128	0.904	1.019
Apr-May	**1.635**	**0.488**	**0.798**	**0.978**	**0.795**	**0.777**	**0.827**	**0.51**	**0.422**
May-Jun	**0.99**	**1.134**	**1.123**	**0.911**	**0.926**	**0.843**	**0.683**	**1.325**	**0.905**
Jun-Jul	**0.964**	**1.209**	**1.165**	**1.008**	**1.043**	**1.052**	**0.762**	**1.497**	**1.14**
Jul-Aug	**1.087**	**0.876**	**0.952**	**1.046**	**1.325**	**1.386**	**2.215**	**0.58**	**1.285**
Aug-Sep	**0.986**	**0.636**	**0.627**	**1.058**	**0.811**	**0.858**	**0.787**	**1.029**	**0.81**
Sep-Oct	0.791	1.165	0.921	1.008	0.787	0.793	1.093	0.877	0.958
Oct-Nov	1.03	0.772	0.796	1.002	1.185	1.188	0.976	1.069	1.043
Nov-Dec	1.12	0.802	0.898	1.005	1.16	1.166	1.298	0.893	1.159
Avg.	*1.023*	*0.925*	*0.946*	*1.005*	*0.994*	*1*	*1.033*	*0.916*	*0.946*

References

1. Ahmed, N., Occhipinti-Ambrogi, A., Muir, J.F.: The impact of climate change on prawn post-larvae fishing in coastal Bangladesh: socioeconomic and ecological perspectives. Mar. Policy **39**, 224–233 (2013)
2. Aigner, D., Lovell, C.K., Schmidt, P.: Formulation and estimation of stochastic frontier production function models. J. Econ. **6**(1), 21–37 (1977)
3. Albala-Bertrand, J.-M.: Natural disaster situations and growth: a macroeconomic model for sudden disaster impacts. World Dev. **21**(9), 1417–1434 (1993)
4. Arnell, N.W.: Climate change and global water resources: SRES emissions and socio-economic scenarios. Glob. Environ. Chang. **14**(1), 31–52 (2004)
5. Attrill, M.J., Power, M.: Climatic influence on a marine fish assemblage. Nature **417**(6886), 275 (2002)
6. BOM.: Indian Ocean influence in climate change. Bureau of Meteorology, Australia (2018). http://www.bom.gov.au/climate/iod/
7. Badjeck, M.C., Allison, E.H., Halls, A.S., Dulvy, N.K.: Impacts of climate variability and change on fishery-based livelihoods. Mar. Policy **34**(3), 375–383 (2010)
8. Barange, M., Bahri, T., Beveridge, M.C., Cochrane, K.L., Funge-Smith, S., Poulain, F.: Impacts of climate change on fisheries and aquaculture. United Nations' Food Agric. Org. **12**(4), 628–635 (2018)
9. Battese, G.E., Coelli, T.J.: Frontier production functions, technical efficiency and panel data: with application to paddy farmers in India. J. Prod. Anal. **3**(1–2), 153–169 (1992)
10. Battese, G.E., Coelli, T.J.: A model for technical inefficiency effects in a stochastic frontier production function for panel data. Empir Econ **20**(2), 325–332 (1995)
11. Brander, K.: Impacts of climate change on marine ecosystems and fisheries. J. Mar. Biol. Assoc. India **51**(1), 1–13 (2009)
12. Brander, K.M.: The role of growth changes in the decline and recovery of North Atlantic cod stocks since 1970. ICES J. Mar. Sci. J. du Conseil **64**(2) (2007)
13. Charnes, A., Cooper, W.W., Rhodes, E.: Measuring the efficiency of decision making units. Eur. J. Oper. Res. **2**(6), 429–444 (1978)

14. Cheung, W.W., Lam, V.W., Sarmiento, J.L., Kearney, K., Watson, R.E.G., Zeller, D., Pauly, D.: Large-scale redistribution of maximum fisheries catch potential in the global ocean under climate change. Glob. Change Biol. **16**(1), 24–35 (2010)
15. Cheung, W.W., Lam, V.W., Sarmiento, J.L., Kearney, K., Watson, R.E.G., Zeller, D., Pauly, D.: Large-scale redistribution of maximum fisheries catch potential in the global ocean under climate change. Glob. Chang. Biol. **16**(1), 24–35 (2010)
16. Coelli, T.J., Prasada Rao, D.S., O'Donnell, C.J., Battese, G.E.: Additional topics on data envelopment analysis. An introduction to efficiency and productivity analysis, 183–208 (2005)
17. Coglan, L., Pascoe, S., Harris, R.I.D.: Measuring efficiency in demersal trawlers using a frontier production function approach. In: Xth Annual meeting of the European Association of Fisheries Economics. The Hague, Netherlands (1998)
18. Collier, T.C., Mamula, A., Ruggiero, J.: Estimation of multi-output production functions in commercial fisheries. Omega **42**(1), 157–165 (2014)
19. Dell, M., Jones, B.F., Olken, B.A.: What do we learn from the weather? The new climate-economy literature. J. Econ. Lit. **52**(3), 740–798 (2014)
20. Dorner, B., Catalano, M.J., Peterman, R.M.: Spatial and temporal patterns of covariation in productivity of Chinook salmon populations of the northeastern Pacific Ocean. Can. J. Fish. Aquat. Sci. **75**(7), 1082–1095 (2017)
21. Drinkwater, K.F.: The response of Atlantic cod (Gadusmorhua) to future climate change. ICES J. Mar. Sci. **62**(7), 1327–1337 (2005)
22. FAO.: FAO report: global fisheries and aquaculture production reaches a new record. High. Newsroom (2024). https://www.fao.org/newsroom/detail/fao-report-global-fisheries-and-aquaculture-production-reaches-a-new-record-high/en
23. Fare, R., Grosskopf, S., Norris, M., Zhang, Z.: Productivity growth, technical progress, and efficiency change in industrialized countries. Am. Econ. Rev. **84**(1), 66–83 (1994)
24. Farrell, M.J.: The measurement of productive efficiency. J. Royal Stat. Soc. Ser. A (General) **120**(3), 253–281 (1957)
25. Fomby, T., Ikeda, Y., Loayza, N.V.: The growth aftermath of natural disasters. J. Appl. Economet. **28**(3), 412–434 (2013)
26. Fousekis, P., Klonaris, S.: Technical efficiency determinants for fisheries: a study of trammel netters in Greece. Fish. Res. **63**(1), 85–95 (2003)
27. Färe, R., Grosskopf, S.: *New Directions: Efficiency and Productivity* (Vol. 3). Springer Science & Business Media (2005)
28. Garcia, A.M., Vieira, J.P., Winemiller, K.O.: Dynamics of the shallow-water fish assemblage of the Patos Lagoon estuary (Brazil) during cold and warm ENSO episodes. J. Fish Biol. **59**(5), 1218–1238 (2001)
29. Griffin, W.L., Woodward, R.T.: Determining policy-efficient management strategies in fisheries using data envelopment analysis (DEA). Mar. Policy **35**(4), 496–507 (2011)
30. Gulev, S.K.: Global climate change and the oceans. Stud. Russ. Econ. Dev. **34**(6), 738–745 (2023)
31. Hallegatte, S., Dumas, P.: Can natural disasters have positive consequences? Investigating the role of embodied technical change. Ecol. Econ. **68**(3), 777–786 (2009)
32. Halpern, B.S., Walbridge, S., Selkoe, K.A., Kappel, C.V., Micheli, F., D'agrosa, C., Fujita, R.: A global map of human impact on marine ecosystems. Science **319**(5865), 948–952 (2008)
33. Henderson, P.A., Seaby, R.M.: The role of climate in determining the temporal variation in abundance, recruitment and growth of sole Soleasolea in the Bristol Channel. J. Mar. Biol. Assoc. U.K. **85**(1), 197–204 (2005)
34. Herrero, I., Pascoe, S.: Value versus volume in the catch of the Spanish South-Atlantic trawl fishery. J. Agric. Econ. **54**(2), 325–341 (2003)
35. Hill, D.K., Magnuson, J.J.: Potential effects of global climate warming on the growth and prey consumption of Great Lakes fish. Trans. Am. Fish. Soc. **119**(2), 265–275 (1990)
36. Jackson, J.B., Kirby, M.X., Berger, W.H., Bjorndal, K.A., Botsford, L.W., Bourque, B.J., Bradbury, R.H., Cooke, R., Erlandson, J., Estes, J.A., Hughes, T.P.: Historical overfishing and recent collapse of coastal ecosystems. Science **293**(5330), 629–637 (2001)

37. Keerthiratne, S., Tol, R.S.: Impact of natural disasters on financial development. Econ. Disasters Clim. Change **1**(1), 33–54 (2017)
38. Kirkley, J., Paul, C.J.M., Squires, D.: Deterministic and stochastic capacity estimation for fishery capacity reduction. Mar. Resour. Econ. **19**(3), 271–294 (2004)
39. Kirkley, J.E., Squires, D., Strand, I.E.: Assessing technical efficiency in commercial fisheries: the mid-Atlantic sea scallop fishery. Am. J. Agr. Econ. **77**(3), 686–697 (1995)
40. Kovats, R.S., Bouma, M.J., Hajat, S., Worrall, E. and Haines, A.: El Nino and health. Lancet **362**(9394), 1481–1489 (2003)
41. Kularatne, M.G., Pascoe, S., Wilson, C., Robinson, T.: Efficiency of culture-based fisheries production in village irrigation systems of Sri Lanka. Aquac. Econ. Manag. 1–21 (2018)
42. Lehodey, P., Alheit, J., Barange, M., Baumgartner, T., Beaugrand, G., Drinkwater, K., Roy, C.: Climate variability, fish, and fisheries. J. Clim. **19**(20), 5009–5030 (2006)
43. Letta, M., Tol, R.S.: Weather, climate and total factor productivity. Environ. Resour. Econ. 1–23 (2018)
44. Link, P.M., Tol, R.S.: Economic impacts of changes in the population dynamics of fish on the fisheries of the Barents Sea. ICES J. Mar. Sci. **63**(4), 611–625 (2006)
45. Loneragan, N.R., Bunn, S.E.: River flow and estuarine food-webs: implications for the production of coastal fisheries with an example from the Logan River, southeast Queensland. Aust. J. Ecol. **24**, 431–440 (1999)
46. Maravelias, C.D., Tsitsika, E.V.: Economic efficiency analysis and fleet capacity assessment in Mediterranean fisheries. Fish. Res. **93**(1–2), 85–91 (2008)
47. Meeusen, W., & van Den Broeck, J.: Efficiency estimation from Cobb-Douglas production functions with composed error. Int. Econ. Rev. 435–444 (1977)
48. Oliveira, M.M., Camanho, A.S., Gaspar, M.B.: Technical and economic efficiency analysis of the Portuguese artisanal dredge fleet. ICES J. Mar. Sci. **67**(8), 1811–1821 (2010)
49. Parry, M.L., Canziani, O.F., Palutikof, J.P., Van Der Linden, P.J., Hanson, C.E.: IPCC, 2007: climate change 2007: impacts, adaptation and vulnerability. Contribution of working group II to the fourth assessment report of the intergovernmental panel on climate change. Cambridge University Press, Cambridge, UK (2007)
50. Parry, M.L., et al. (Eds.) Climate change 2007: impacts, adaptation and vulnerabilitiy. IPCC Report, Cambridge University Press (2007)
51. Pascoe, S., Coglan, L.: The contribution of unmeasurable inputs to fisheries production: an analysis of technical efficiency of fishing vessels in the English Channel. Am. J. Agr. Econ. **84**(3), 585–597 (2002)
52. Pascoe, S., Hutton, T., Coglan, L., Nguyen, V.Q.: Implications of efficiency and productivity change over the season for setting MEY-based trigger targets. Australian J. Agric. Resour. Econ. **62**(2), 199–216 (2018)
53. Pörtner, H.O., Peck, M.A.: Climate change effects on fishes and fisheries: towards a cause-and-effect understanding. J. Fish Biol. **77**(8), 1745–1779 (2010)
54. Qiu, Y., Lin, Z., Wang, Y.: Responses of fish production to fishing and climate variability in the northern South China Sea. Prog. Oceanogr. **85**(3–4), 197–212 (2010)
55. Rouyer, T., Fromentin, J.M., Ménard, F., Cazelles, B., Briand, K., Pianet, R., Stenseth, N.C.: Complex interplays among population dynamics, environmental forcing, and exploitation in fisheries. Proc. Natl. Acad. Sci. **105**(14), 5420–5425 (2008)
56. Sesabo, J.K., Tol, R.S.: Technical efficiency of small-scale fishing households in Tanzanian coastal villages: an empirical analysis. Afr. J. Aquat. Sci. **32**(1), 51–61 (2007)
57. Shephard, R.W.: Cost and Production Functions. Princeton University Press, Princeton (1953)
58. Shephard, R.W.: Theory of coast and production functions. Princeton University Press (1970)
59. Simar, L., Wilson, P.W.: Estimation and inference in two-stage, semi-parametric models of production processes. J. Econ. **136**(1), 31–64 (2007)
60. Skidmore, M., Toya, H.: Do natural disasters promote long-run growth? Econ. Inq. **40**(4), 664–687 (2002)
61. Sumaila, U.R., Cheung, W., Dyck, A., Gueye, K., Huang, L., Lam, V., Zeller, D.: Benefits of rebuilding global marine fisheries outweigh costs. PloS one **7**(7), e40542 (2012)

62. Tingley, D., Pascoe, S., Coglan, L.: Factors affecting technical efficiency in fisheries: stochastic production frontier versus data envelopment analysis approaches. Fish. Res. **73**(3), 363–376 (2005)
63. Van Hoof, L., De Wilde, J.W.: Capacity assessment of the Dutch beam-trawler fleet using data envelopment analysis (DEA). Mar. Resour. Econ. **20**(4), 327–334 (2005)
64. Vestergaard, N., Squires, D., Kirkley, J.E.: Measuring capacity and capacity utilization in fisheries: the case of the Danish gillnet fleet. Fisheries. Resources. **60**(2–3), 357–368 (2003)
65. Vázquez-Rowe, I., Tyedmers, P.: Identifying the importance of the "skipper effect" within sources of measured inefficiency in fisheries through data envelopment analysis (DEA). Mar. Policy **38**, 387–396 (2013)

The Cascading Effects among Natural Hazards: Interaction Maps as Tool for Risk Analysis and Management

George Halkos and Argyro Zisiadou

Abstract Natural hazards are increasingly characterized not as isolated events but as dynamic, interconnected processes that trigger cascading effects across ecological, infrastructural, and social systems. This study examines the systemic nature of cascading environmental hazards, highlighting how anthropogenic climate change, rapid urbanization, and environmental degradation amplify their frequency, intensity, and complexity. Using more than a century of data from the EM-DAT International Disaster Database (1900–2024), we quantify inter-hazard interactions and visualize their linkages through a Cascading Environmental Hazard Sankey Diagram. The results reveal dominant pathways—such as tropical cyclones, floods, and earthquakes—that frequently propagate secondary and tertiary disasters, including landslides, infrastructure failures, and technological accidents. Through case analysis, including the 2011 Great East Japan Earthquake and Fukushima nuclear disaster, we illustrate the systemic vulnerabilities that transform singular hazards into multi-layered crises. Our findings underscore the inadequacy of siloed risk management approaches and call for integrated frameworks that combine cross-sectoral risk analysis, resilience planning, and nature-based solutions. By bridging empirical evidence with governance strategies, this research contributes to advancing disaster risk reduction policies aligned with the Sendai Framework and emphasizes the need for systemic, anticipatory approaches to address the compounding risks of the twenty-first century.

Keywords Cascading Environmental Hazards · Interdependence · Climate Change · Environmental Degradation · Disaster Risk Reduction

JEL Codes D81 · H84 · O44 · Q54 · Q56 · Q58 · R11

G. Halkos (✉) · A. Zisiadou
Laboratory of Operations Research, Department of Economics, University of Thessaly, Volos, Greece
e-mail: halkos@uth.gr

G. Halkos et al. (eds.), *Environmental Risk Analysis*, Springer Proceedings in Mathematics & Statistics 540, https://doi.org/10.1007/978-3-032-24155-9_2

1 Introduction

Anthropogenic climate change refers to alterations in the Earth's climate that are primarily driven by human activities, particularly the emission of greenhouse gases (GHGs) such as carbon dioxide (CO_2), methane (CH_4), and nitrous oxide (N_2O). These emissions originate predominantly from the burning of fossil fuels for energy and transportation, deforestation, industrial processes, and agricultural practices. The scientific consensus, as established by the Intergovernmental Panel on Climate Change (IPCC), affirms that human influence is the dominant cause of observed global warming since the mid-twentieth century [1] The mechanism behind anthropogenic climate change is rooted in the enhanced greenhouse effect. GHGs trap longwave radiation emitted from the Earth's surface, thereby increasing atmospheric and surface temperatures. Pre-industrial levels of CO_2 were approximately 280 parts per million (ppm), whereas current levels exceed 420 ppm—an increase directly linked to industrialization and the exponential growth of carbon-intensive human activities [2].

Furthermore, anthropogenic climate change has profound ecological, economic, and social consequences. Ecosystems are experiencing shifts in species distribution and the timing of phenological events, such as earlier flowering and migrations, which disrupt ecological interactions and threaten biodiversity [3, 4]. Concurrently, human societies face heightened risks including food and water insecurity, climate-induced displacement, and adverse health outcomes, particularly in vulnerable regions such as Sub-Saharan Africa, South Asia, and Small Island Developing States [5, 6]. Addressing this global challenge necessitates coordinated mitigation efforts, including the transition to low-carbon energy systems, carbon pricing mechanisms, and reduction of greenhouse gas emissions [7]. Simultaneously, robust adaptation strategies—such as climate-resilient infrastructure, early warning systems, and ecosystem-based approaches—are essential to enhance resilience to both current and future climate impacts [5, 8]. The IPCC Sixth Assessment Report (AR6) underscores the increasing frequency and intensity of natural hazards as a direct consequence of anthropogenic climate change, highlighting the compounding and cascading risks these pose to ecosystems, infrastructure, and human well-being. Extreme events such as heatwaves, droughts, heavy precipitation, and tropical cyclones are not only becoming more severe but are also interacting across sectors and regions, leading to systemic risks [1]. The report emphasizes that these hazards are often interconnected, triggering compound and cascading impacts, which are particularly acute in regions with high socio-ecological [3]. For instance, the concurrence of drought and heat extremes can simultaneously disrupt agriculture, water supply, and energy systems, creating complex, multi-layered crises [9].

Given this context, the IPCC strongly advocates for transformative risk mitigation and adaptation strategies, emphasizing the urgency of integrated approaches that address not just individual hazards but their interdependencies. The report stresses that traditional siloed disaster risk management approaches are increasingly inadequate in the face of cascading hazards exacerbated by climate change. Instead, it calls for multi-risk frameworks, strengthened early warning systems, and adaptive governance that incorporate climate projections, local knowledge, and equity considerations [5]. Without ambitious and coordinated mitigation efforts—particularly limiting warming to 1.5 °C—and investment in climate-resilient infrastructure, many regions are projected to face irreversible losses and non-linear impacts [7]. Thus, understanding the dynamic interplay between climate change and natural hazards is central to advancing global climate resilience.

The frequency, intensity, and complexity of natural hazards are increasing in the context of a changing climate and growing human vulnerability. Among these, cascading natural hazards—where one hazard triggers or amplifies another—pose a particularly grave threat to communities, infrastructure, and ecosystems. Traditional hazard assessments often consider events in isolation, neglecting how interconnected processes can escalate risk. However, recent disasters, such as the 2011 Great East Japan Earthquake, which triggered a tsunami and a nuclear crisis, have underscored the urgent need to adopt integrated approaches. This paper argues that studying cascading natural hazards is essential for improving risk assessment, strengthening early warning systems, and enhancing the resilience of socio-ecological systems in the face of compound and interconnected threats. For that reason, Sect. 2 reviews the existing literature on cascading environmental hazards, providing a conceptual foundation for the analysis. Section 3 outlines a typology of environmental hazards and presents illustrative examples, while Sect. 4 examines the key drivers, including climate change, urbanization, and environmental degradation. Section 5 discusses policy and governance implications with a focus on disaster risk reduction and resilience planning, and Sect. 6 explores the interdependence of hazards through empirical data and case analysis. Finally, Section 7 synthesizes the findings and offers concluding reflections on advancing systemic approaches to risk reduction.

2 Literature Review

The following literature review is structured to provide a foundational understanding of the evolving nature of environmental hazards in the context of both climate change and systemic interconnectivity. Section 2.1, From Anthropogenic Climate Change to Natural Hazards, explores the growing body of research linking human-induced climate change to the increased frequency, intensity, and unpredictability of natural hazards such as floods, droughts, and extreme weather events. This section sets the stage for understanding how global environmental shifts contribute to hazard emergence. Section 2.2, Defining Cascading Hazards, delves into the concept of cascading hazards—events where one hazard triggers another in a chain of impacts—and reviews current theoretical and empirical frameworks used to study these phenomena. Together, these sections aim to contextualize the data presented in this study and establish the relevance of examining hazards through a multi-dimensional and interconnected lens.

2.1 From Anthropogenic Climate Change to Natural Hazards

The scientific consensus is unequivocal: anthropogenic, or human-induced, climate change is accelerating due to the substantial increase in greenhouse gas (GHG) emissions resulting from industrial activities, fossil fuel combustion, deforestation, and unsustainable land-use practices [7]. This phenomenon is contributing to rising global temperatures, sea-level rise, biodiversity loss, and more frequent and intense weather-related disasters. In response, various metrics and indices have been developed to evaluate environmental performance and climate vulnerability, including the Environmental Performance Index (EPI), the Climate Change Performance Index (CCPI), and the Climate Risk Index (CRI). These indices not only provide comparative analyses of countries' environmental and climate policies but also highlight the global inequality inherent in climate change impacts and responses.

Human activities have drastically altered the Earth's atmosphere since the onset of the Industrial Revolution. The concentration of atmospheric carbon dioxide (CO_2) has increased from approximately 280 parts per million (ppm) in pre-industrial times to over 420 ppm in 2023, primarily due to the burning of fossil fuels and deforestation [10]. Methane (CH_4) and nitrous oxide (N_2O) emissions from agriculture and industry further exacerbate the greenhouse effect. The Intergovernmental Panel on Climate Change [7] states with high confidence that more than 95% of the observed warming since 1950 is attributable to human activities.

The impacts of anthropogenic climate change are far-reaching. Global average surface temperatures have increased by approximately 1.1 °C above pre-industrial levels [7]. This warming has led to widespread glacier retreat, ocean acidification, increased frequency of heatwaves, floods, and wildfires, and severe threats to global food and water security [11]. These effects are unevenly distributed, disproportionately affecting low-income countries and marginalized communities.

The Environmental Performance Index (EPI), developed by Yale University and Columbia University in collaboration with the World Economic Forum, provides a data-driven evaluation of a country's environmental health and ecosystem vitality [12, 13]. The EPI ranks countries based on several indicators grouped under climate change mitigation, air quality, water and sanitation, and biodiversity conservation. The EPI includes climate-related metrics such as projected GHG emissions per capita and progress toward net-zero targets. High-scoring nations, such as Denmark and Sweden, are often characterized by robust environmental regulations, renewable energy deployment, and sustainable urban planning. However, the EPI also reveals gaps in implementation, particularly among rapidly developing economies and major fossil fuel producers [12].

The *Climate Change Performance Index* (CCPI), published annually by Germanwatch, the NewClimate Institute, and Climate Action Network International, evaluates the climate protection performance of 60 countries that are collectively responsible for over 90% of global GHG emissions [14, 15]. The CCPI is based on four key dimensions: GHG emissions (40%), renewable energy (20%), energy use (20%), and climate policy (20%). The CCPI is notable for its methodological rigor and critical stance: the top three positions are often left blank if no country meets the stringent criteria for a "very high" rating. Recent editions have highlighted the relative progress of countries like India and Morocco, which have expanded renewable energy and improved policy frameworks. In contrast, large emitters such as the United States, Russia, and China have often received poor evaluations for inadequate policy efforts or rising emissions [14].

The *Climate Risk Index* (CRI), also developed by Germanwatch, assesses the extent to which countries have been affected by extreme weather events. The CRI is based on data from the Munich Re NatCatSERVICE and includes indicators such as death toll, economic losses, and loss events per 100,000 inhabitants [15, 16].

The CRI underscores the disproportionate impact of climate change on vulnerable nations, especially in the Global South. Countries such as Haiti, the Philippines, and Pakistan have frequently ranked among the most affected due to their high exposure to tropical storms, floods, and limited adaptive capacities. These findings support the argument for climate justice and the urgent need for financial and technological support to build resilience in low-income nations [16].

There is a growing body of evidence that anthropogenic climate change is increasing the frequency, severity, and unpredictability of natural hazards. Rising global temperatures are intensifying the hydrological cycle, leading to more extreme precipitation events, while warming oceans contribute to stronger hurricanes and cyclones. Melting glaciers and thermal expansion are causing sea levels to rise, exacerbating coastal flooding and erosion. Prolonged heatwaves and droughts, linked to disrupted atmospheric circulation, are becoming more common across many regions [1, 17, 18]. The CRI captures these phenomena, showing that climate-vulnerable countries—particularly in the Global South—suffer disproportionately from climate-related disasters, often despite contributing minimally to global emissions. For instance, countries like Mozambique, the Philippines, and Bangladesh consistently rank high on the CRI due to recurrent tropical storms, floods, and heatwaves [16]. These events, while natural in origin, are amplified by human-induced climate change, making them more destructive and frequent. Thus, the CRI serves as a crucial empirical bridge linking anthropogenic climate change with real-world, observable impacts on human societies and economies.

Together, the EPI, CCPI, and CRI offer a multidimensional understanding of global climate governance and vulnerability. While the EPI emphasizes environmental sustainability and long-term policy outcomes, the CCPI focuses specifically on climate mitigation performance. In contrast, the CRI reveals real-world impacts and risks faced by nations regardless of their contribution to global emissions. Conjunctively, these indices show a clear chain of cause and effect: poor environmental and climate performance (low EPI and CCPI scores) often correlates with increased exposure to and vulnerability from natural disasters (high CRI scores), especially in regions already facing socio-economic challenges. These indices provide complementary insights that are invaluable for researchers, policymakers, and civil society actors working to design effective and equitable climate strategies.

2.2 Defining Cascading Hazards

Cascading natural hazards refer to sequences of events where an initial hazard triggers secondary or tertiary hazards, often across spatial, temporal, or sectoral boundaries. These can include obvious chains—such as earthquakes triggering landslides or tsunamis—as well as more complex interactions involving technological systems, social infrastructure, or climate feedback loops [19]. Cascading hazards are characterized by non-linear impacts, uncertainty, and systemic risks, making them difficult to predict and manage using conventional disaster risk models. Natural hazards can often trigger cascading effects, where one hazard initiates another. For instance, earthquakes are known to cause secondary natural hazards such as tsunamis and landslides. A historical example is the 2011 Tōhoku earthquake in Japan, which generated a massive tsunami that devastated coastal areas, caused over 15,000 deaths, and triggered numerous landslides [20]. This illustrates the interconnected nature of geophysical processes, where the initial seismic event set off a chain reaction of other natural disasters, compounding the overall impact.

Moreover, natural hazards can lead to technological (or "natech") disasters [21]. The same 2011 Tōhoku earthquake and tsunami also caused catastrophic failures at the Fukushima Daiichi Nuclear Power Plant. The tsunami disabled the plant's cooling systems, resulting in nuclear meltdowns, hydrogen explosions, and widespread radioactive contamination. This event highlights the vulnerability of critical infrastructure to natural forces and demonstrates how inadequate hazard resilience in technological systems can exacerbate disaster impacts [22].

Conversely, technological hazards can also trigger natural hazards, although such cases are less frequent. One notable example is the 1962 underground nuclear test at the Soviet Union's Novaya Zemlya site, which reportedly caused a small-scale seismic event and permafrost destabilization in the region. Additionally, large-scale industrial activities such as reservoir-induced seismicity—where the weight of artificial reservoirs can trigger earthquakes—are documented, as in the case of the 1967 Koyna Dam earthquake in India [23]. These examples illustrate that human-engineered systems can significantly alter natural systems, sometimes leading to unexpected and hazardous outcomes.

Cascading natural environmental hazards represent an increasingly recognized dimension of disaster risk in an interconnected and rapidly changing world. These phenomena occur when a primary hazard triggers secondary, often heterogeneous events, resulting in compounded, non-linear impacts across environmental, social, and infrastructural systems. Understanding cascading hazards is critical for advancing comprehensive risk assessment, policy integration, and resilience building.

According to Pescaroli and Alexander [19], cascading disasters are characterized by a sequence of events in which the initial disruption causes subsequent hazards, often amplifying the scale and complexity of the overall impact. In the environmental context, cascading natural hazards specifically involve sequential or interrelated geophysical, meteorological, hydrological, or biological events.

Climate change is intensifying the conditions that lead to cascading hazards. For example, prolonged droughts can weaken vegetation, making landscapes more susceptible to wildfires; subsequent rainfall on fire-scarred terrain can then result in flash floods or landslides. The Intergovernmental Panel on Climate Change [3] warns that global warming is increasing the likelihood of compound events, including extreme heat combined with drought, sea-level rise combined with storm surges, and cyclones followed by inland flooding. Without a systemic understanding of these linkages, policymakers and emergency planners risk being unprepared for the full spectrum of cascading consequences.

Conventional hazard assessment tools typically focus on singular hazards within defined geographic or temporal frames. These models, while useful for understanding isolated events, often fail to account for interdependencies across systems or time lags between primary and secondary effects [24]. For instance, while a model may predict the likelihood of flooding in a river basin, it may not account for how upstream landslides triggered by an earthquake could worsen downstream flood severity days later. This analytical gap can lead to inadequate risk mitigation measures, poorly informed land-use planning, and suboptimal allocation of emergency resources.

Cascading hazards are most effectively conceptualized through a systems thinking framework, which emphasizes the interconnectedness and interdependence of physical, ecological, and human systems. Unlike linear hazard models that isolate events, systems thinking recognizes that natural hazards rarely occur in isolation; instead, they interact dynamically with ecological processes, built infrastructure, and socio-economic structures, often triggering non-linear and emergent consequences [19, 25]. For instance, a primary geophysical event such as an earthquake can instigate secondary hazards like landslides or tsunamis, which in turn may damage critical infrastructure, disrupt supply chains, and compromise public health systems—producing impacts far beyond the initial shock [26].

The vulnerabilities and adaptive capacities embedded within these interlinked systems play a pivotal role in shaping how hazards cascade and amplify. Social systems characterized by poverty, inequality, or weak governance are often more susceptible to cascading impacts, as they lack the redundancy and flexibility needed to absorb multiple, interacting stressors [2]. Moreover, critical infrastructure interdependencies—such as between energy, water, and transportation systems—can act as conduits for cascading failures, particularly in densely urbanized or climate-sensitive regions [27, 28]. Therefore, understanding hazard propagation through a systems perspective is not only essential for risk assessment but also for designing integrated and anticipatory risk reduction strategies that address systemic vulnerabilities rather than isolated threats.

3 Typologies and Examples

Cascading hazards often emerge from the dynamic interplay among geophysical, hydrometeorological, biophysical, and ecological processes, where one disturbance can amplify or trigger subsequent hazards across domains. For instance, a geophysical event such as an earthquake may induce landslides that dam rivers, subsequently causing hydrological flooding downstream [29]. Similarly, hydrometeorological extremes—such as cyclones or intense rainfall—can erode soils, heightening the risk of landslides, while also facilitating outbreaks of vector-borne diseases, demonstrating biophysical consequences [30]. Ecological degradation, including deforestation or wetland loss, further exacerbates these interactions by reducing natural buffers, allowing relatively moderate hazards to escalate into systemic crises [31]. Understanding these linkages requires moving beyond hazard-specific approaches towards integrated risk frameworks that account for multi-hazard interdependencies, where environmental, physical, and biological processes converge to produce complex disaster cascades (see Table 1).

3.1 *Geophysical Cascades*

Geophysical cascades exemplify the sequential nature of natural hazard interactions, where one geophysical event initiates subsequent processes that amplify overall risk. A typical example is the chain reaction of earthquake-induced landslides leading to dam failure and subsequent flooding, which is especially prevalent in seismically active and mountainous regions [32, 33]. Earthquakes destabilize slopes by reducing shear strength and triggering mass movements; these landslides can block river channels, forming landslide-dammed lakes that pose significant downstream flood risks if the natural dam subsequently fails [34]. The probability and magnitude of such cascades are further influenced by precipitation regimes, as heavy rainfall accelerates slope instability and increases the likelihood of rapid overtopping or breaching of landslide dams [35]. The 2008 Wenchuan Earthquake in China provides a salient case study, where thousands of coseismic landslides led to the formation of over 800 landslide dams, many of which presented severe flood hazards to downstream communities [36]. This illustrates how geophysical cascades operate as multi-hazard processes, requiring integrated monitoring and early warning systems to mitigate their compound impacts on human settlements and infrastructure.

3.2 *Hydrometeorological Cascades*

Hydrometeorological cascades often unfold across both temporal and spatial scales, with hurricanes serving as a prominent driver of multi-hazard sequences. For example, intense tropical cyclones can cause severe coastal erosion, which not only reshapes shorelines but also compromises natural barriers such as dunes and mangroves [37]. The loss of these protective features, coupled with storm surge and seawater intrusion, leads to the salinization of freshwater aquifers and soils, undermining local water quality and long-term agricultural productivity [38, 39]. Such processes have been documented in deltaic and low-lying island systems, where repeated hurricane activity has increased soil and groundwater salinity, forcing shifts in crop suitability and threatening food security [40, 41]. The cascading nature of these hazards illustrates how a single extreme climatic event can propagate through ecological and socio-economic systems, generating enduring impacts on livelihoods, ecosystem services, and regional resilience.

3.3 Biophysical and Ecological Cascades

Wildfires often act as catalysts for cascading geomorphological hazards, where the initial destruction of vegetation precipitates subsequent soil erosion and hydrological instability. The removal of vegetation cover reduces interception, root reinforcement, and soil cohesion, leaving slopes highly susceptible to erosion and mass wasting [42]. In addition, combustion processes can create hydrophobic soil layers that significantly reduce infiltration capacity, thereby increasing surface runoff during rainfall events [43]. These altered conditions frequently lead to flash flooding and debris flows in post-fire landscapes, with impacts often more severe than the direct fire damage itself [44]. For example, studies in Mediterranean and western North American ecosystems show that intense precipitation following wildfire events can mobilize large volumes of sediment, altering river morphology, threatening downstream infrastructure, and endangering human lives [45]. This sequence underscores how changes in land cover fundamentally transform the geomorphological and hydrological response of landscapes to precipitation, amplifying the risks of cascading natural hazards in fire-prone regions.

4 Drivers of Cascading Hazards

Cascading hazards are increasingly shaped by a convergence of global and local drivers that amplify the frequency, intensity, and interconnectedness of risks. Climate change acts as a primary driver by intensifying hydrometeorological extremes such as floods, droughts, and storms, thereby increasing the likelihood of secondary and compounding impacts [3]. At the same time, rapid and often unplanned urbanization concentrates populations and assets in hazard-prone areas, while the interdependence of critical infrastructures—such as transport, energy, and water systems—creates conditions where failures in one sector propagate rapidly across others [46]. Additionally, environmental degradation, including deforestation, wetland loss, and soil erosion, undermines natural buffers that traditionally mitigate hazard impacts, increasing the probability that one hazard will trigger another [31]. Together, these drivers interact to produce highly complex risk landscapes in which cascading effects are no longer exceptional events but systemic features of contemporary hazard dynamics.

4.1 Climate Change

Warming temperatures caused by anthropogenic greenhouse gas emissions are unequivocally linked to the increased frequency, duration, and intensity of primary natural hazards such as storms, droughts, heatwaves, and wildfires. The thermodynamic properties of a warmer atmosphere drive these changes: rising global mean surface temperatures increase the capacity of the air to hold water vapor, thereby intensifying extreme precipitation events and storm surges, while elevated ocean surface temperatures provide more energy for tropical cyclones, increasing their destructive potential [47, 48]. Similarly, higher temperatures accelerate evapotranspiration and soil moisture depletion, exacerbating drought severity in arid and semi-arid regions [49, 50]. The IPCC Sixth Assessment Report (AR6) confirms with high confidence that the frequency and magnitude of concurrent climate extremes, such as simultaneous droughts and heatwaves, have increased since the mid-twentieth century and are projected to escalate further under higher warming scenarios [1, 9] Such extremes are no longer isolated anomalies but part of a coherent global pattern of intensification, challenging historical baselines of hazard frequency and severity.

These climate-driven hazards rarely act in isolation; instead, they interact with systemic vulnerabilities within ecological, infrastructural, and socio-economic systems, thereby producing cascading and compound risks. For example, persistent droughts in California have been shown to dry vegetation and create conditions conducive to megafires, which then degrade air quality, damage infrastructure, and increase public health risks far beyond the fire zones [51, 52]. Similarly, hurricanes and cyclones can trigger multi-hazard cascades, combining coastal storm surge with extreme rainfall to produce compound flooding, which disrupts energy networks, transport corridors, and water systems simultaneously [53, 54]. Vulnerability to such cascading effects is exacerbated by human drivers, including rapid urbanization, coastal settlement expansion, ecosystem degradation, and socio-economic inequalities, which reduce adaptive capacity and increase exposure to interacting hazards [3, 28]. These dynamics illustrate that global warming amplifies not only the intensity of primary hazards but also the interconnectedness of risks, reinforcing the urgent need for multi-hazard risk assessment frameworks and systemic resilience strategies. Without integrated approaches, societies face a growing likelihood of compound and cascading disasters that exceed traditional disaster management capacities.

4.2 Urbanization and Infrastructure Interdependence

Highly networked infrastructure systems, encompassing energy, transport, water, telecommunications, and health services, form the backbone of modern societies but also represent a source of systemic fragility. Their high degree of technical and functional interdependence means that disruptions in one network can cascade rapidly into others, magnifying disaster impacts [55, 56]. For example, electricity grids are central to multiple critical services, from water pumping and treatment to digital communication and hospital operations. A power failure during a hazard event can therefore precipitate a breakdown in multiple dependent infrastructures, creating compounding challenges for emergency response and recovery [57]. These risks are exacerbated under climate change, as more frequent and severe hazards—such as heatwaves stressing power grids, or floods damaging transport corridors—test the resilience of interdependent systems beyond their design thresholds [5, 58]. This interconnectivity produces non-linear risk propagation, where small initial disruptions can escalate into disproportionate socio-economic consequences that far exceed the scale of the triggering event [27, 28].

The cascading consequences of such interdependencies are increasingly visible in global disasters. Hurricane Sandy (2012) illustrates how the inundation of electrical substations in New York City caused widespread power outages, which in turn disabled fuel distribution systems, paralyzed subway networks, and forced hospital evacuations [59]. Similarly, during the 2003 European heatwave, electricity demand surged while river levels dropped, reducing hydropower capacity and cooling water availability for thermal plants, leading to rolling blackouts across multiple countries [60]. The 2011 Tōhoku earthquake and tsunami further exemplifies cascading infrastructure failures, as the loss of power and inundation of backup generators at the Fukushima Daiichi nuclear plant led to a nuclear accident, amplifying an already catastrophic event [61]. These cases highlight that disruptions in one sector can compound humanitarian crises, economic losses, and environmental damage when interdependent systems are simultaneously stressed. Addressing such risks requires integrated resilience strategies, including diversification of critical networks, investment in redundancies, cross-sectoral contingency planning, and governance frameworks that move beyond siloed infrastructure management [28, 62]. Without systemic approaches, highly networked infrastructure will remain a vulnerability multiplier under intensifying climate and hazard regimes.

The vulnerability of highly networked infrastructure to cascading failures is expected to grow under accelerating climate change, as multiple environmental hazards increasingly occur simultaneously or in close succession. Rising global temperatures intensify compound events—such as coastal flooding driven by the co-occurrence of storm surge and heavy precipitation, or simultaneous droughts and heatwaves—which place unprecedented stress on interdependent infrastructures [9, 53]. For instance, extreme rainfall linked to climate change has been shown to compromise transport systems through landslides and flooding, while at the same time overwhelming drainage and power infrastructure, leading to knock-on failures in emergency services and public health systems [5, 58]. The systemic risks arising from such multi-hazard interactions can trap societies in feedback loops where environmental shocks degrade infrastructure, which in turn reduces adaptive capacity to withstand subsequent hazards, thereby magnifying the likelihood of cascading disasters [28, 63]. Anticipating and mitigating these interconnected risks demands a paradigm shift towards systemic resilience planning, integrating climate projections, cross-sectoral risk assessments, and adaptive governance to reduce vulnerabilities and prevent cascading infrastructure crises under worsening climatic extremes.

4.3 Environmental Degradation

Deforestation, wetland degradation, and unsustainable land use planning dismantle natural protective systems that historically served as buffers against environmental hazards, creating conditions where one hazard can easily trigger another. Forest ecosystems play a central role in regulating hydrological processes, enhancing infiltration, and stabilizing soils; when removed, the risk of erosion, landslides, and flash floods rises dramatically following heavy rainfall events [64, 65]. Similarly, wetlands act as natural retention basins capable of absorbing floodwaters, filtering pollutants, and attenuating storm surges. The widespread drainage of wetlands for agriculture and urban development has substantially reduced this buffering capacity, exacerbating downstream flooding and intensifying the consequences of extreme precipitation [66, 67]. Poor land use planning amplifies these vulnerabilities: the expansion of settlements into floodplains, coastal zones, and steep slopes increases exposure while simultaneously degrading ecosystems that once mediated hazard intensity [68, 69]. As a result, the removal of these ecological safeguards not only increases the magnitude of primary hazards but also heightens the probability of cascading effects, whereby one disturbance sets off secondary hazards and systemic failures.

The interaction of climate change with land degradation further amplifies cascading risks, creating feedback loops between environmental hazards and human vulnerability. In the Amazon, extensive deforestation has reduced evapotranspiration, weakening regional precipitation cycles and contributing to recurrent droughts, which in turn elevate wildfire risk and accelerate further forest dieback [70, 71]. In coastal regions, the conversion of mangroves and wetlands to agricultural and urban land has undermined storm surge protection, as illustrated during Cyclone Nargis (2008) in Myanmar, where the destruction of mangrove forests significantly increased mortality and damages [72]. Likewise, the 2005 Hurricane Katrina disaster was exacerbated by decades of wetland loss in the Mississippi Delta, which had previously functioned as a natural buffer against storm surges [73]. More recently, the 2022 Pakistan floods were magnified by upstream deforestation and poorly regulated land use, which accelerated surface runoff and worsened inundation in downstream plains [74]. These cases reveal how anthropogenic degradation of natural buffers transforms otherwise localized hazards into system-wide crises, underscoring the urgent need to integrate ecosystem-based approaches—such as forest conservation, wetland restoration, and nature-based flood defenses—into both disaster risk reduction (DRR) and climate adaptation frameworks [31, 75].

5 Policy and Governance Implications

Addressing cascading hazards requires a fundamental rethinking of policy and governance frameworks, with particular emphasis on Disaster Risk Reduction (DRR) and resilience planning. Traditional hazard management approaches, which often focus on single-event responses, are inadequate in the face of systemic and interconnected risks that transcend sectoral and jurisdictional boundaries [76]. The Sendai Framework for Disaster Risk Reduction (2015–2030) underscores the need for multi-hazard, multi-sectoral governance that integrates risk knowledge into development planning and prioritizes proactive rather than reactive strategies. Resilience planning, in turn, shifts the focus from restoring systems to their pre-disaster state toward building adaptive capacities that enable societies, economies, and ecosystems to absorb, adapt, and transform in the face of cascading disruptions [77]. Effective governance therefore depends on coordination across scales of government, collaboration with civil society and the private sector, and the mainstreaming of DRR principles into urban planning, climate adaptation, and environmental management.

5.1 Disaster Risk Reduction (DRR)

The Sendai Framework for Disaster Risk Reduction (2015–2030) represents a significant evolution in global disaster governance by emphasizing the need to understand disaster risk in "all its dimensions," explicitly recognizing hazard interactions and systemic risks as central challenges [76]. This shift reflects a growing acknowledgment in academic and policy circles that disasters are not merely the outcome of single hazards, but rather the product of interconnected processes across social, ecological, and infrastructural systems [28, 78]. By moving beyond event-based understandings of disasters, the Sendai Framework underscores the importance of systemic analysis that incorporates multi-hazard interactions, exposure, and vulnerabilities across scales, thereby aligning with contemporary systems thinking in disaster risk science [27, 79].

Despite its conceptual advances, the operationalization of systemic risk thinking within the Sendai Framework remains limited. One persistent gap lies in the translation of systemic risk concepts into actionable national and local policies. Many disaster risk reduction (DRR) strategies continue to prioritize siloed, hazard-specific planning—such as separate frameworks for floods, earthquakes, or cyclones—rather than integrated multi-hazard approaches [80, 81]. This disconnect is further reinforced by institutional fragmentation, where responsibilities for risk governance are divided across ministries or sectors, hindering coordination [82]. Moreover, while the Sendai Framework emphasizes the role of science and risk knowledge, it provides limited guidance on methodologies for assessing systemic and cascading risks, particularly in low- and middle-income countries with constrained technical and financial capacities [83].

Another critical gap lies in the Framework's integration with other global governance agendas. Although the Sendai Framework is positioned as complementary to the Paris Agreement on climate change and the Sustainable Development Goals (SDGs), in practice these regimes often operate in parallel rather than in synergy [84]. For instance, climate change adaptation and disaster risk reduction initiatives are frequently managed by separate agencies and financed through distinct streams, despite overlapping goals and systemic drivers of risk. This lack of coherence limits the capacity of states to address cascading risks that cut across environmental, social, and economic domains. The COVID-19 pandemic further highlighted this challenge: despite the Sendai Framework's recognition of biological hazards, most countries were underprepared for systemic disruptions to health systems, supply chains, and economies [85]. The pandemic revealed how systemic risks manifest globally and propagate across borders, underscoring the need to move beyond localized, hazard-specific risk governance to approaches that address globally networked risks [79, 82].

Finally, scholars argue that the Sendai Framework insufficiently addresses the political economy of risk creation, focusing more on risk management than on the structural drivers of vulnerability [86, 87]. Land use practices, economic inequality, urbanization, and environmental degradation are central to the production of systemic risks, yet the Framework tends to frame these issues in technocratic rather than political terms. For example, rapid urban expansion into floodplains or deforestation in hazard-prone regions are often discussed as "exposures" rather than as outcomes of governance failures and development choices. This technocratic framing risks obscuring the socio-political processes that generate and reinforce vulnerability, limiting the potential of the Framework to drive transformative change. Addressing this gap requires not only advancing risk assessment methodologies but also embedding justice, equity, and governance reform into the core of systemic risk reduction strategies [88, 89].

5.2 Resilience Planning

Resilience planning has become a cornerstone of contemporary disaster risk reduction (DRR) and climate adaptation, shifting the focus from reactive responses to anticipatory, adaptive, and transformative strategies. It emphasizes that systems—whether ecological, infrastructural, or social—should not only withstand shocks but also adapt and evolve in response to disturbances [90, 91]. The essence of resilience thinking lies in its rejection of linear, command-and-control approaches to hazard management, instead recognizing the dynamic, uncertain, and interconnected nature of risks. In this view, resilience is not merely about "bouncing back" to a pre-disaster condition but about bouncing forward into more sustainable, equitable, and adaptive states [77, 92]. This conceptual shift has profound implications for policy, requiring governance frameworks that prioritize flexibility, inclusivity, and long-term system robustness over short-term efficiency.

One of the most critical applications of resilience thinking is the call for cross-sectoral risk analysis, which identifies vulnerabilities and interdependencies across energy, transport, water, food, and health systems. Highly networked societies mean that a disruption in one sector can propagate rapidly, creating cascading impacts that overwhelm local coping capacities [27, 46]. For example, during the 2003 European heatwave, high energy demand for cooling stressed electricity grids, while drought conditions reduced hydropower availability and disrupted river transport, amplifying socio-economic damages [93]. By integrating knowledge across disciplines and sectors, cross-sectoral risk analysis enables policymakers to anticipate such interactions and prevent disproportionate consequences. Importantly, this requires moving beyond siloed approaches in planning institutions, embedding systemic perspectives into urban development, infrastructure investment, and national adaptation strategies [79, 94].

Equally essential to resilience-oriented governance are robust, people-centered early warning systems (EWS). These systems extend beyond technical hazard monitoring to incorporate risk knowledge, communication strategies, and social preparedness, ensuring that warnings translate into timely and effective action [95, 96]. Empirical studies show that effective EWS can reduce disaster mortality by up to 30% when integrated with community-level engagement and education [63]. Recent innovations, such as forecast-based financing, further link early warnings to pre-emptive humanitarian action, thereby reducing losses before hazards manifest [97]. However, the effectiveness of EWS is conditioned by social trust, governance capacity, and equity of access, with marginalized communities often excluded from warning dissemination or lacking the resources to act upon it [94]. Hence, resilient EWS are not purely technological solutions but require participatory design, inclusive governance, and sensitivity to cultural and socio-economic contexts.

Resilience thinking also emphasizes the importance of nature-based solutions (NbS), which harness ecosystems to mitigate risks while delivering co-benefits for climate mitigation, biodiversity, and livelihoods. Forest restoration reduces landslide and flood risk by stabilizing soils and regulating hydrological flows; wetlands store excess water and filter pollutants; and mangroves dissipate wave energy, shielding coastal settlements from cyclones and storm surges [31, 98, 99]. Compared with hard infrastructure, NbS are adaptive, self-regenerating, and cost-effective, particularly under conditions of climate uncertainty [100]. Beyond physical protection, NbS enhance social resilience by providing sustainable livelihoods and cultural ecosystem services that strengthen community cohesion [101]. Nonetheless, implementation faces challenges: institutional silos often separate environmental management from disaster governance, while the benefits of NbS are dervalued in cost–benefit analyses compared with engineered alternatives [102]. Effective integration requires policy coherence, robust governance mechanisms, and long-term investment strategies that acknowledge the multifunctionality of ecosystems.

Despite its conceptual appeal, resilience thinking is not without limitations. Scholars caution that resilience can be used as a political or normative construct, shifting responsibility for risk management onto local communities without addressing structural drivers of vulnerability such as poverty, inequality, or weak governance [103, 104]. For instance, encouraging communities to adapt to climate-induced flooding may inadvertently normalize exposure rather than incentivize systemic changes like relocation, land-use regulation, or emission reduction. Similarly, cross-sectoral risk analysis and NbS, while widely endorsed, often encounter barriers in practice due to fragmented governance, short political cycles, and competing economic interests [80]. There is also a risk that resilience discourses prioritize system continuity over transformative change, thereby sustaining unsustainable socio-economic pathways [105]. Thus, while resilience thinking provides a valuable framework for integrating cross-sectoral analysis, early warning, and ecosystem-based approaches, its effectiveness depends on ensuring equity, accountability, and transformative governance that address root causes of vulnerability rather than simply managing symptoms of systemic risk.

6 The Interdependence of Hazards

To systematically examine the occurrence and interactions of cascading hazards, we employed the EM-DAT International Disaster Database, which provides standardized records of disaster events globally from 1900 to 2024. EM-DAT categorizes disasters into main types (e.g., geophysical, meteorological, hydrological, climatological, biological, etc.) and further into subtypes (e.g., earthquakes, floods, droughts, storms, etc.). Using this dataset, we extracted all disaster observations within the defined timeframe and disaggregated them by subtype. This allowed us to establish a comprehensive baseline of event frequencies while also identifying instances where one disaster subtype triggered or contributed to another, thus enabling a systematic mapping of cascading relationships across hazards.

Following data extraction, we quantified causal linkages between hazard subtypes. For each recorded event, we identified whether it directly or indirectly caused another disaster subtype to occur (for example, an earthquake leading to a tsunami, or floods leading to disease outbreaks). The number of such causative interactions was then compared against the total number of events recorded for that specific subtype. By dividing the frequency of causation by the total occurrences per subtype, we derived probabilistic frequencies of cascading interactions, which provided a normalized measure of how likely each hazard was to trigger subsequent events.

These frequency estimates were then used to construct a Cascading Environmental Hazard Sankey Diagram (Fig. 1), a visual representation that captures both the direction and magnitude of inter-hazard linkages. In the diagram, the width of each flow corresponds to the relative frequency of cascading interactions, thereby illustrating not only the dominant initiating hazards but also the most common secondary impacts. This approach offers a holistic perspective of hazard interdependencies across more than a century of disaster data, facilitating a systems-level understanding of cascading risks that can inform disaster risk reduction strategies and resilience planning.

While the EM-DAT database provides a robust foundation for analyzing global disaster patterns, several methodological limitations must be acknowledged. First, the completeness and reliability of records vary across regions and time periods, with earlier decades (particularly pre-1950) often suffering from underreporting and inconsistent classification [106]. This introduces potential bias in long-term trend analyses and may underestimate cascading relationships in less-documented contexts. Second, while EM-DAT records secondary impacts in many instances, identifying causality between hazards remains challenging, as disaster chains are complex and sometimes influenced by underlying vulnerabilities or governance failures rather than direct physical triggers. Third, the classification of hazards into discrete subtypes, although analytically useful, may oversimplify the multifaceted and overlapping nature of certain events, such as storm surges, compound floods, or ecosystem-mediated disasters. Consequently, the frequencies we derived should be interpreted as indicative probabilities rather than deterministic measures of cascading dynamics. Despite these constraints, the use of EM-DAT remains valuable for systematically exploring hazard interactions at the global scale, and the Sankey visualization provides a meaningful tool for communicating systemic risk pathways to both academic and policy audiences.

Since 1900, a total of 26,835 environmental hazard occurrences have been recorded across all known subtypes. These hazards encompass a wide range of events, including natural disasters, biological outbreaks, and technological accidents. However, not all of these incidents led to further consequences beyond their immediate impact. Certain subtypes, such as bacterial, viral, and parasitic events, while significant in isolation, often did not result in cascading effects that extended beyond their initial domain. These non-cascading events typically remained confined in scope, suggesting a more contained interaction with surrounding systems.

In contrast, a substantial portion of the recorded hazards—23,353 occurrences—fall under subtypes that provide clear evidence of interdependence between systems (Table I, Appendix). These events illustrate how a disturbance in one domain, such as the environment or infrastructure, can ripple across multiple sectors, triggering a chain of consequences. Out of these interdependent events, 7565 led to cascading effects, highlighting the complex and often fragile nature of interconnected systems. This data emphasizes the importance of understanding systemic interlinkages in hazard analysis and disaster preparedness, as the potential for cascading impacts can significantly amplify the overall consequences of an initial event.

The data illustrates a detailed and interconnected web of environmental hazards and the cascading events they often trigger. Notably, some hazard types, such as Tropical Cyclones (2589 recorded occurrences), show extensive multi-hazard linkages. A single cyclone event can simultaneously or sequentially lead to floods (26.19 frequency), landslides or other ground movements (12.01), and storm surges (2.39).

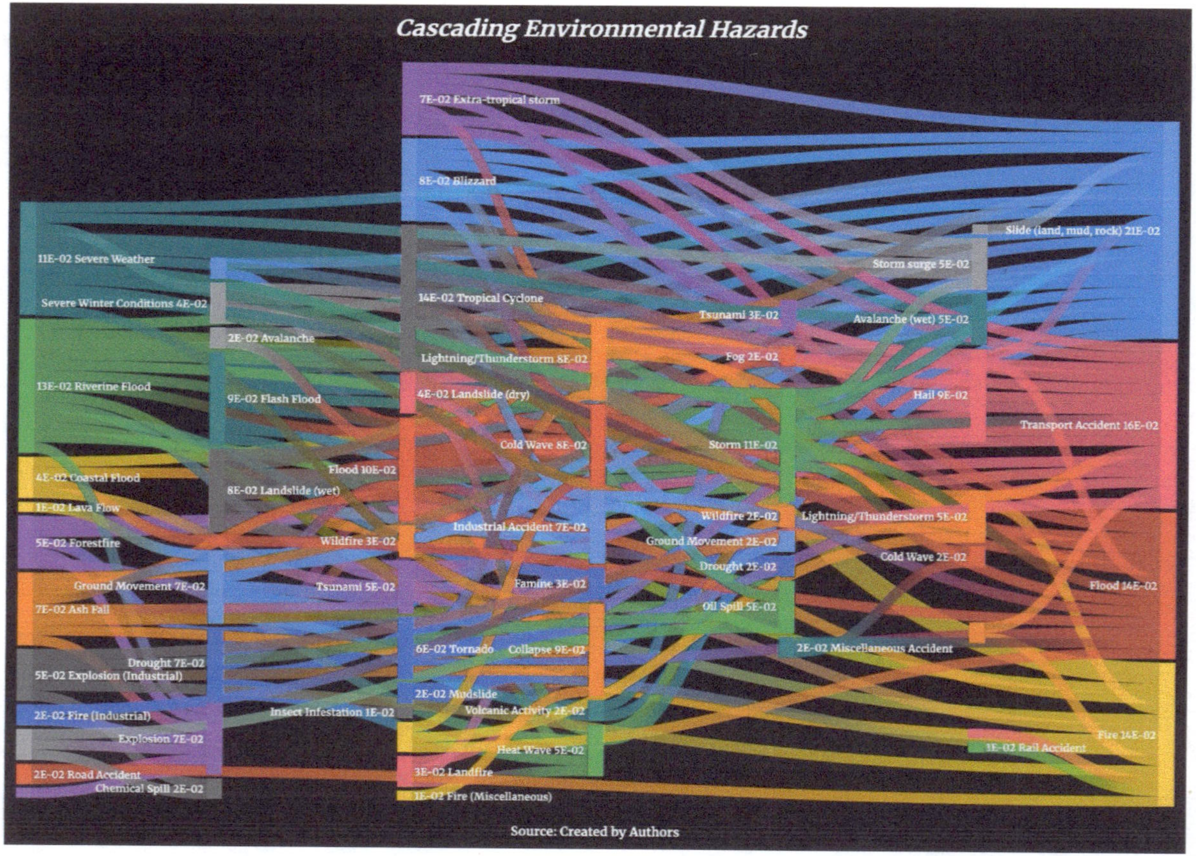

Fig. 1 Cascading environmental Hazards Sankey Diagram

This clearly illustrates how one primary hazard can destabilize multiple systems—hydrological, geological, and atmospheric. Similarly, Riverine Floods (2645 events) frequently lead to landslides (20.57), lightning/thunderstorms, and even tsunamis in certain contexts, highlighting their capacity to cause both localized destruction and widespread secondary impacts. Ground Movement events (1561) also serve as a central node in cascading hazard patterns, often associated with floods, landslides, and tsunamis. In another example, Tsunamis (56 occurrences) themselves result in high-impact consequences such as landslides (5.36) and industrial accidents (1.78)—both of which can escalate the event from a natural disaster to a technological and infrastructural crisis.

In contrast, some hazards—while dangerous—tend to generate more isolated impacts. For example, Cold Waves (305 occurrences) show relatively few cascading links, most notably to floods, storms, and fog, each at a frequency of around 0.65 to 0.98, suggesting these are generally more contained events unless overlapped with other hazard contexts. Similarly, Heat Waves (312)—although responsible for a strong direct connection to droughts (7.37) and wildfires (2.88)—do not exhibit a broad spread of cascading consequences across diverse hazard domains. However, even in these cases, the risks can intensify when one hazard primes the system for another. For instance, a drought may reduce soil moisture and vegetation health, increasing the susceptibility to wildfires, which in turn can produce land degradation and even localized flooding when heavy rains follow.

The dataset also includes examples of technological and anthropogenic hazards that integrate into the cascade chain, particularly in industrial settings. Events like Industrial Explosions (790 occurrences) are associated with fires (1.65), chemical spills, and oil spills, adding a hazardous materials dimension to environmental risk. Transport accidents, which frequently appear as outcomes across multiple hazard types (e.g., blizzards, storms, and floods), demonstrate how extreme weather can disrupt infrastructure and human systems. The presence of compound events like Glacial Lake Outburst Floods, which show a 100% frequency of flooding (60) and 20 for landslides, emphasizes how geological and hydrological risks converge in high-altitude environments. Ultimately, the dataset underlines a crucial insight for risk assessment and disaster management: while some hazards are episodic and self-contained, many more interact dynamically within complex systems. Understanding these cascading relationships is essential for developing more resilient infrastructure, early-warning systems, and cross-sectoral preparedness strategies.

The Sankey diagram constructed from this dataset visually represents the cascading relationships between primary environmental hazard sources and their resulting outcomes. By mapping the frequency and intensity of these interactions, the diagram highlights how certain hazards—such as tropical cyclones, riverine floods, and ground movements—serve as central nodes that trigger a wide array of secondary events. The width of each flow corresponds to the frequency of occurrence, allowing for an intuitive understanding of which hazard pathways are most prominent. This visualization not only reveals the complexity and interdependence of environmental systems but also underscores the importance of integrated disaster risk management, where a single initiating event can have widespread and compounding effects across sectors and geographies.

The Sankey diagram below visualizes the complex interconnections among cascading environmental hazards, offering a systemic perspective on how different hazard subtypes interact across temporal and spatial scales. Drawing on more than a century of disaster records from the EM-DAT International Disaster Database (1900–2024), the diagram illustrates not only the occurrence of hazards but also the directional flows of causality, whereby one hazard subtype triggers another. The thickness of each connecting band represents the relative frequency of cascading interactions, providing a quantitative dimension to what is otherwise often described qualitatively in disaster studies. In this way, the diagram transforms vast datasets into an accessible yet rigorous representation of systemic risk.

A closer examination of the flows highlights the multifaceted pathways of cascading hazards. Geophysical events, such as earthquakes, often initiate chains leading to tsunamis, landslides, and secondary technological failures, while hydrometeorological hazards, such as tropical cyclones and floods, commonly generate secondary crises including industrial accidents, transport disruptions, and infrastructure collapses. Similarly, droughts and heatwaves can trigger wildfires, which in turn may escalate into industrial explosions, loss of biodiversity, or longer-term ecological degradation. These interactions demonstrate that hazards are rarely discrete, bounded events; rather, they operate as part of dynamic, interacting systems, where environmental processes, technological vulnerabilities, and socio-economic dependencies converge to produce escalating risks.

The diagram also underscores the role of infrastructure interdependencies and environmental degradation in amplifying cascading effects. For example, storms not only create immediate physical hazards, such as storm surges and flooding, but also disrupt energy supply chains, transportation networks, and emergency response systems, compounding their impacts. Similarly, the loss of natural buffers like forests and wetlands exacerbates the probability of one hazard triggering another, such as flooding leading to landslides or coastal erosion. By incorporating such interlinked pathways, the visualization reflects a systems-thinking approach that aligns with recent academic calls for integrated, multi-hazard risk assessments [29, 46].

Beyond its analytical value, the Sankey diagram holds significant policy relevance. It illustrates vividly why disaster risk reduction cannot remain siloed within individual hazard domains, but must instead account for cascading and compounding effects across sectors and scales. The Sendai Framework for Disaster Risk Reduction (2015–2030) explicitly emphasizes the importance of understanding systemic risk, and visualizations such as this provide a practical means to operationalize that principle [76]. By revealing dominant cascading pathways—such as the frequent transition from hydrometeorological hazards to technological and infrastructural failures—the diagram highlights where governance interventions, such as early warning systems, nature-based solutions, and resilience planning, can have the greatest preventive impact.

Finally, this diagram reflects the potential of data-driven visualization tools for bridging the gap between science and policy. By translating empirical frequencies of cascading interactions into a coherent visual narrative, it enables stakeholders to grasp the magnitude and complexity of systemic risks more intuitively than raw data alone. At the same time, it also exposes methodological challenges: under-reporting in earlier decades, limitations in attributing causality, and simplifications inherent in hazard categorization. Nevertheless, the Sankey approach provides a valuable step toward systemic risk communication, supporting both academic inquiry and decision-making. In sum, the figure encapsulates the urgent need for holistic, cross-sectoral, and forward-looking approaches to disaster risk reduction in an era of climate change, rapid urbanization, and global interconnectedness.

The cascading dynamics illustrated in the Sankey diagram find a concrete manifestation in the 2011 Great East Japan Earthquake and Fukushima Daiichi nuclear disaster. The event exemplifies how a single geophysical shock—a magnitude 9.0 earthquake—can propagate through interconnected systems to generate a cascade of secondary and tertiary hazards, including a devastating tsunami, infrastructural collapse, and ultimately a nuclear accident. This mirrors the pathways highlighted in the diagram, where geophysical hazards frequently trigger hydrometeorological and technological failures, amplifying both humanitarian and environmental consequences. The Tōhoku case underscores the systemic vulnerabilities revealed in the global hazard flows: insufficient protective infrastructure, energy system dependence, and the compounding of natural and technological risks. By linking global patterns to a specific catastrophe, the diagram and the case study together highlight the universality of cascading risk processes while also emphasizing the critical importance of localized resilience planning and governance.

6.1 Case Study: 2011 Great East Japan Earthquake

The 2011 Great East Japan Earthquake (also referred to as the Tōhoku earthquake) provides a paradigmatic example of cascading hazards and highlights the vulnerabilities inherent in highly interconnected socio-technical systems. The magnitude 9.0 undersea earthquake off the Pacific coast of Japan triggered one of the most powerful tsunamis ever recorded, which inundated more than 500 km^2 of land, devastated coastal settlements, and caused massive loss of life and displacement [20]. Beyond the immediate geophysical and hydrological destruction, the tsunami crippled essential infrastructure, including transport networks, power supply systems, and communication facilities, revealing the profound interdependencies between natural hazards and technological systems [26].

Among the most critical cascading impacts was the failure of the Fukushima Daiichi Nuclear Power Plant. When the tsunami inundated the facility, it disabled backup diesel generators, preventing cooling of the nuclear reactors and ultimately leading to core meltdowns and the release of radioactive materials [61]. This nuclear accident transformed an already catastrophic natural disaster into a complex technological crisis, compounding humanitarian suffering, necessitating the long-term evacuation of over 100,000 people, and contaminating surrounding ecosystems. The

Fukushima incident illustrates how initial geophysical hazards can propagate through infrastructural fragilities, producing multi-hazard crises with global implications for energy security, environmental safety, and nuclear policy [107].

The cascading nature of the Tōhoku disaster also exposed institutional and systemic vulnerabilities. Inadequate risk assessments of tsunami heights, insufficient protective seawalls, and overreliance on centralized energy systems amplified the scale of the impacts [19]. Moreover, the disruption of transportation networks and energy supply chains hampered emergency response efforts and prolonged recovery. These systemic failures demonstrate how insufficient integration of disaster risk governance across geophysical, infrastructural, and socio-economic domains can exacerbate cascading effects and undermine resilience [108].

Beyond its national consequences, the disaster had profound international implications. Global supply chains were disrupted, particularly in the automotive and electronics sectors, revealing the extent to which local disasters can reverberate through interconnected economic systems [22]. The nuclear accident also reignited global debates over nuclear safety and energy transitions, prompting policy reviews in countries such as Germany, which accelerated its nuclear phase-out in response [109]. These cascading transboundary impacts underscore the importance of embedding multi-level governance and cross-sectoral risk assessments into disaster risk reduction strategies.

Overall, the 2011 Great East Japan Earthquake illustrates how a primary geophysical hazard—the earthquake—can evolve into secondary hazards such as tsunamis and infrastructure collapse, and ultimately into tertiary hazards like nuclear disaster, each amplifying the scale and complexity of the crisis. This event demonstrates the need for integrated, systemic disaster risk reduction approaches that account for hazard interactions, infrastructural interdependencies, and socio-economic vulnerabilities. It highlights the importance of resilience planning that transcends sectoral boundaries and emphasizes flexibility, redundancy, and adaptive governance to better prepare societies for future cascading disasters.

7 Conclusion

Cascading natural environmental hazards represent complex, systemic challenges that transcend traditional single-hazard boundaries. Historical and contemporary evidence demonstrates that hazards do not occur in isolation but instead interact dynamically across geophysical, hydrometeorological, biophysical, and ecological domains. Events such as the 2011 Great East Japan Earthquake, tsunami, and Fukushima nuclear disaster exemplify how initial geophysical shocks can propagate through interdependent technological and socio-economic systems, producing multi-hazard crises with long-term environmental and humanitarian consequences [26, 61]. These dynamics are becoming increasingly pronounced under anthropogenic climate change, which is intensifying the frequency and severity of primary hazards such as storms, droughts, floods, and heatwaves, thereby increasing the likelihood of secondary and tertiary cascading effects across already stressed systems [3].

The systemic propagation of risks is further shaped by the interdependencies of modern infrastructure, rapid urbanization, and environmental degradation. Highly networked infrastructures, including transport, power, and water systems, create pathways for disruptions to cascade across multiple sectors, amplifying direct damage and complicating recovery [46]. At the same time, the erosion of natural buffers through deforestation, wetland loss, and poor land-use practices increases susceptibility to hazard interactions, as illustrated by floods triggering landslides or droughts exacerbating wildfire risk [31]. These drivers interact to create a highly interconnected risk landscape in which vulnerabilities extend beyond local contexts and may escalate into transboundary or global crises, as evidenced by supply chain disruptions following the Tōhoku disaster [22].

Within this landscape, policy and governance frameworks play a critical role in shaping outcomes. The Sendai Framework for Disaster Risk Reduction (2015–2030) emphasizes the importance of understanding risk in all its dimensions, explicitly recognizing systemic and cascading interactions [76]. However, gaps remain in translating these principles into practice. Risk governance continues to be fragmented, with responsibilities dispersed across sectors and scales, while hazard datasets often lack the causal detail necessary to model cascading relationships effectively. Moreover, resilience planning is unevenly mainstreamed into national development agendas, with a tendency toward reactive approaches rather than proactive, adaptive strategies [108]. Addressing these gaps requires embedding cross-sectoral coordination, inclusive governance, and long-term sustainability planning into disaster risk reduction policies.

Resilience thinking offers a compelling paradigm for addressing cascading hazards by fostering proactive, flexible systems that can absorb shocks, adapt, and transform. Integrating cross-sectoral risk analysis, developing robust early warning systems, and implementing nature-based solutions such as reforestation and wetland restoration represent tangible strategies for reducing systemic vulnerabilities while simultaneously enhancing ecological resilience [77]. At the same time, advancing systemic resilience requires interdisciplinary research that combines physical sciences, systems modeling, and social sciences to capture the dynamic and interconnected nature of cascading risks. Only through such integrative approaches can societies anticipate, prepare for, and respond to the accelerating hazard interactions driven by climate change, urbanization, and environmental degradation.

Reflecting on this study, our analysis using the EM-DAT dataset (1900–2024) underscores both the opportunities and challenges of quantifying cascading hazards at the global scale. By calculating the frequency with which one hazard subtype triggers another, and visualizing these relationships through a Cascading Environmental Hazard Sankey Diagram, we provided an innovative lens for identifying dominant pathways of systemic risk. While the results highlight the potential of global datasets for advancing multi-hazard analysis, they also reveal inherent limitations, including underreporting in earlier decades and difficulties in attributing causality. Nevertheless, the Sankey approach demonstrates the value of data-driven visualization tools for bridging science and policy, offering decision-makers an accessible representation of hazard interdependencies. By linking empirical evidence to governance frameworks and resilience strategies, this study reinforces the urgent need for systemic, multi-disciplinary approaches to disaster risk reduction in an era of unprecedented global change.

Looking forward, future research should focus on integrating global disaster databases with emerging technologies, including remote sensing, artificial intelligence, and big-data analytics, to capture cascading dynamics in real time and improve predictive modeling. There is also a critical need to couple hazard interaction data with socio-economic vulnerability assessments to better understand how systemic risks translate into unequal impacts across populations. On the policy side, aligning cascading hazard analysis with climate adaptation strategies and national DRR plans could enhance coherence between global frameworks such as the Sendai Framework, the Paris Agreement, and the Sustainable Development Goals. Ultimately, building resilience to cascading hazards will depend not only on technological innovation and scientific modeling but also on inclusive governance, community engagement, and long-term sustainability planning that recognizes the interconnected nature of risk in the twenty-first century.

Disclosure of Interests The authors declare that they have no known competing financial interests or personal relationships that could have appeared to influence the work reported in this paper.

Appendix

See Table 1

Table 1 Source and outcome of cascading environmental hazards

Source	Outcome	Frequency
Drought (795)	Insect infestation	0.1257862
	Wildfire	2.1383648
	Flood	0.1257862
	Heat wave	3.6477987
	Cold wave	0.5031447
	Storm	0.2515723
	Famine	8.9308176
Glacial lake outburst flood (5)	Flood	60
	Slide (land, mud, rock)	20
Forestfire (35)	Drought	0.6349206
	Heat wave	6.031746
	Storm	0.952381
	Explosion	0.3174603
	Transport accident	0.3174603
Landfire (92)	Heat wave	6.5217391
	Storm	1.0869565
	Industrial accident	1.0869565
Wildfire (80)	Drought	1.25
	Heat wave	13.75
	Transport accident	1.25
Ash fall (215)	Flood	1.1952191
	Slide (land, mud, rock)	1.9920319
	Ground movement	1.1952191
	Tsunami	0.7968127
	Oil spill	0.3984064
	Explosion	0.3984064
	Fire	0.3984064
Lava flow (11)	Ground movement	27.2727273
Volcanic activity (11)	Ground movement	9.0909091
	Tsunami	9.0909091
Ground movement (1561)	Flood	0.384369
	Avalanche (wet)	0.0640615
	Slide (land, mud, rock)	11.5310698
	Tsunami	8.9686099
	Volcanic activity	0.3203075
	Collapse	0.128123
	Fire	1.024984

(continued)

Table 1 (continued)

Source	Outcome	Frequency
Tsunami (56)	Slide (land, mud, rock)	5.3571429
	Volcanic activity	1.7857143
	Industrial accident	1.7857143
	Fire	5.3571429
Landslide–wet (644)	Flood	7.1428571
	Slide (land, mud, rock)	1.242236
	Tsunami	0.1552795
	Lightning / thunderstorm	0.310559
	Storm	0.1552795
	Industrial accident	0.4658385
	Collapse	1.242236
	Transport accident	0.7763975
Landslide–dry (30)	Slide (land, mud, rock)	3.3333333
	Heat wave	20
	Storm	3.3333333
	Industrial accident	10
Avalanche - wet (122)	Flood	0.8196721
	Slide (land, mud, rock)	2.4590164
Mudslide (79)	Slide (land, mud, rock)	10.1265823
	Collapse	3.7974684
Flood (2440)	Slide (land, mud, rock)	15.3278689
	Cold wave	0.1229508
	Hail	0.3688525
	Lightning / thunderstorm	1.4344262
	Storm	3.1967213
	Industrial accident	0.0819672
	Collapse	0.3688525
	Fire	0.0409836
	Transport accident	0.1229508
	Famine	0.0409836
Coastal flood (84)	Flood	1.1904762
	Slide (land, mud, rock)	7.1428571
	Tsunami	1.1904762
	Storm surge	2.3809524
Flash flood (864)	Flood	0.3472222
	Slide (land, mud, rock)	24.0740741
	Cold wave	0.1157407

(continued)

Table 1 (continued)

Source	Outcome	Frequency
	Hail	0.3472222
	Lightning / thunderstorm	0.3472222
	Storm	2.1990741
	Industrial accident	0.1157407
	Collapse	0.3472222
	Transport accident	0.2314815
Riverine flood (2645)	Wildfire	0.1134216
	Flood	0.1512287
	Slide (land, mud, rock)	20.5671078
	Tsunami	0.1134216
	Cold wave	0.1512287
	Hail	0.1890359
	Lightning / thunderstorm	0.5293006
	Storm	0.4914934
	Collapse	0.0378072
	Explosion	0.0756144
	Fire	0.0756144
	Transport accident	0.3024575
	Famine	0.0378072
Heat wave (312)	Drought	7.3717949
	Wildfire	2.8846154
Cold wave (305)	Flood	0.9836066
	Storm	0.6557377
	Fog	0.6557377
Severe winter conditions (80)	Flood	1.25
	Cold wave	11.25
	Fog	1.25
	Transport accident	1.25
Tropical cyclone (2589)	Flood	26.1877173
	Slide (land, mud, rock)	12.01236
	Ground movement	0.038625
	Tsunami	0.0772499
	Cold wave	0.038625
	Hail	0.0772499
	Lightning / thunderstorm	0.1158749
	Storm	0.038625
	Storm surge	2.394747

(continued)

Table 1 (continued)

Source	Outcome	Frequency
	Fire	0.038625
	Industrial accident	0.0772499
	Oil spill	0.0772499
	Fire	0.038625
	Transport accident	0.6179992
Tornado (300)	Flood	12.6666667
	Slide (land, mud, rock)	0.6666667
	Hail	12.6666667
	Lightning / thunderstorm	1.3333333
	Fire	0.3333333
	Transport accident	0.3333333
Blizzard (231)	Flood	7.7922078
	Avalanche (wet)	4.3290043
	Slide (land, mud, rock)	2.1645022
	Cold wave	23.8095238
	Hail	0.8658009
	Lightning / thunderstorm	0.4329004
	Storm surge	0.8658009
	Transport accident	5.6277056
Extra-tropical storm (164)	Flood	14.0243902
	Avalanche (wet)	1.2195122
	Slide (land, mud, rock)	4.8780488
	Tsunami	0.6097561
	Lightning / thunderstorm	0.6097561
	Storm surge	0.6097561
	Transport accident	3.6585366
Hail (113)	Flood	11.5044248
	Slide (land, mud, rock)	0.8849558
Lightning/Thunderstorm (191)	Flood	32.460733
	Slide (land, mud, rock)	6.2827225
	Hail	25.6544503
	Lightning / thunderstorm	1.5706806
	Fire	0.5235602
	Transport accident	0.5235602
Sand/Dust storm (23)	Flood	13.0434783
	Cold wave	4.3478261
	Lightning / thunderstorm	4.3478261

(continued)

Table 1 (continued)

Source	Outcome	Frequency
Severe weather (304)	Wildfire	0.3289474
	Flood	37.8289474
	Avalanche (wet)	0.3289474
	Slide (land, mud, rock)	10.8552632
	Severe winter conditions	0.3289474
	Hail	20.7236842
	Lightning / thunderstorm	14.4736842
	Storm	0.6578947
	Storm surge	1.9736842
	Collapse	0.3289474
	Transport accident	0.9868421
Storm (915)	Flood	10.3825137
	Avalanche (wet)	0.2185792
	Slide (land, mud, rock)	3.2786885
	Cold wave	1.6393443
	Hail	4.0437158
	Lightning / thunderstorm	0.6557377
	Storm surge	0.4371585
	Transport accident	0.7650273
Storm surge (10)	Flood	40
Collapse–industrial (188)	Slide (land, mud, rock)	1.5957447
Explosion–industrial (790)	Chemical spill	0.3797468
	Collapse	0.1265823
	Fire	1.6455696
	Oil spill	0.1265823
	Transport accident	0.1265823
Fire–industrial (221)	Lightning / thunderstorm	0.4524887
	Explosion	2.7149321
Industrial accident (126)	Flood	0.7936508
	Oil spill	0.7936508
Radiation (9)	Explosion	11.1111111
Collapse–miscellaneous (310)	Fire	0.3225806
Explosion–miscellaneous (225)	Fire	2.6666667
	Transport accident	0.4444444
Fire–miscellaneous (821)	Collapse	0.1218027
Miscellaneous accident (281)	Flood	0.3558719
	Cold wave	0.3558719

(continued)

Table 1 (continued)

Source	Outcome	Frequency
Rail accident (647)	Fire	0.1545595
Road accident (2936)	Explosion	0.1702997
	Fire	0.1702997
Marine accident (1674)	Chemical spill	0.0597372
	Oil spill	0.0597372
	Explosion	0.0597372

References

1. IPCC: Climate change 2021: the physical science basis. In: Contribution of Working Group I to the Sixth Assessment Report of the Intergovernmental Panel on Climate Change (2021)
2. NOAA: Trends in Atmospheric Carbon Dioxide. National Oceanic and Atmospheric Administration. (2023). https://www.esrl.noaa.gov/gmd/ccgg/trends/
3. IPCC: Climate change 2022: impacts, adaptation and vulnerability. In: Contribution of Working Group II to the Sixth Assessment Report of the Intergovernmental Panel on Climate Change. Cambridge University Press (2022)
4. Parmesan, C., Yohe, G.: A globally coherent fingerprint of climate change impacts across natural systems. Nature. **421**(6918), 37–42 (2003)
5. IPCC: Summary for policymakers. In Climate Change 2022: Impacts, Adaptation and Vulnerability. WGII AR6 (2022)
6. Watts, N., et al.: The 2021 report of the lancet countdown on health and climate change: code red for a healthy future. Lancet. **398**(10311), 1619–1662 (2021)
7. IPCC: Climate Change 2023: Synthesis Report. Contribution of Working Groups I, II and III to the Sixth Assessment Report of the Intergovernmental Panel on Climate Change, IPCC (2023) https://www.ipcc.ch/report/ar6/syr/
8. Hallegatte, S., Rentschler, J., Rozenberg, J.: Shock Waves: Managing the Impacts of Climate Change on Poverty. World Bank (2016)
9. Seneviratne, S.I., et al.: Weather and climate extreme events in a changing climate. Rev. Geophys. **59**(3), e2020RG000727 (2021)
10. NASA: Climate Change: Vital Signs of the Planet (2024). https://climate.nasa.gov/vital-signs/carbon-dioxide/
11. World Bank: Climate change overview (2023). https://www.worldbank.org/en/topic/climatechange/overview
12. Wendling, Z.A., Emerson, J.W., de Sherbinin, A., Esty, D.C., et al.: 2022 environmental performance index. Yale Cent. Environ. Law & Policy. (2022) https://epi.yale.edu
13. Halkos, G., Zisiadou, A.: Environmental performance in transition: an empirical examination of Greece through the EPI framework. MPRA. **No. 125315** (2025)
14. Burck, J., Hagen, U., Höhne, N., Nascimento, L., Bals, C.: Climate change performance index 2024. In: Germanwatch, NewClimate Institute, and Climate Action Network (2023) https://ccpi.org
15. Halkos, G., Zisiadou, A.: The effects of climate change to weather-related environmental hazards: interlinkages of economic factors and climate risk. J. Risk Financ. Manag. **16**(5), 264 (2023)
16. Eckstein, D., Künzel, V., Schäfer, L., Winges, M.: Global climate risk index 2021. Germanwatch. (2021) https://germanwatch.org/en/cri
17. Halkos, G., Zisiadou, A.: An overview of the technological environmental hazards over the last century. Econ. Disasters Clim. Chang. **4**(2), 411–428 (2020a)

18. Halkos, G., Zisiadou, A.: Examining the natural environmental hazards over the last century. Econ. Disasters Clim. Chang. **3**(2), 119–150 (2019)
19. Pescaroli, G., Alexander, D.: A definition of cascading disasters and cascading effects: going beyond the "toppling dominos" metaphor. Planet@Risk. **3**(1) (2015)
20. Mimura, N., Yasuhara, K., Kawagoe, S., Yokoki, H., Kazama, M.: Damage from the great East Japan earthquake and tsunami–a quick report. Mitig. Adapt. Strateg. Glob. Chang. **16**, 803–818 (2011)
21. Halkos, G., Zisiadou, A.: Is investors' psychology affected due to a potential unexpected environmental disaster? J. Risk Financ. Manag. **13**(7), 151 (2020b)
22. Park, J., Seager, T.P., Rao, P.S.C., Convertino, M., Linkov, I.: Integrating risk and resilience approaches to catastrophe management in engineering systems. Risk Anal. **33**(3), 356–367 (2013)
23. Gupta, H.K.: A review of recent studies of triggered earthquakes by artificial water reservoirs with special emphasis on earthquakes in Koyna, India. *Earth-Sci. Rev.* **58**(3–4), 279–310 (2002)
24. Kappes, M.S., Keiler, M., von Elverfeldt, K., Glade, T.: Challenges of analyzing multi-hazard risk: a review. Nat. Hazards. **64**(2), 1925–1958 (2012). https://doi.org/10.1007/s11069-012-0294-2
25. Cutter, S.L.: Compound, cascading, or complex disasters: what's in a name? Environ. Sci. Policy Sustain. Dev. **60**(6), 16–25 (2018)
26. Gill, J.C., Malamud, B.D.: Reviewing and visualizing the interactions of natural hazards. Rev. Geophys. **52**(4), 680–722 (2014)
27. Helbing, D.: Globally networked risks and how to respond. Nature. **497**(7447), 51–59 (2013)
28. Pescaroli, G., Alexander, D.: Understanding compound, interconnected, interacting, and cascading risks: a holistic framework. Risk Anal. **38**(11), 2245–2257 (2018)
29. Gill, J.C., Malamud, B.D.: Hazard interactions and interaction networks (cascades) within multi-hazard methodologies. Earth Syst. Dynam. **7**(3), 659–679 (2016). https://doi.org/10.5194/esd-7-659-2016
30. Ebi, K.L., Bowen, K.: Extreme events as sources of health vulnerability: drought, floods, and storms. Clim. Chang. **134**(1–2), 61–73 (2016). https://doi.org/10.1007/s10584-013-0889-7
31. Renaud, F.G., Sudmeier-Rieux, K., Estrella, M. (eds.): The Role of Ecosystems in Disaster Risk Reduction. UNU Press (2013)
32. Keefer, D.K.: Investigating landslides caused by earthquakes–a historical review. Surv. Geophys. **23**(6), 473–510 (2002)
33. Fan, X., Scaringi, G., Xu, Q., Peng, D., Dai, L., Huang, R.: Coseismic landslides triggered by the 2008 Wenchuan earthquake, China: understanding the long term impacts. Eng. Geol. **250**, 53–69 (2019)
34. Schuster, R.L., Highland, L.M.: Impact of landslides and innovative landslide-mitigation measures on the natural environment. Environ. Manag. **30**(5), 653–665 (2003)
35. Tang, C., Zhu, J., Li, W.L., Liang, J.T.: Rainfall-triggered debris flows following the Wenchuan earthquake. Bull. Eng. Geol. Environ. **70**(4), 625–631 (2011)
36. Fan, X., Xu, Q., Scaringi, G., Dai, L., Li, W., Dong, X., Pei, X.: Characteristics of landslide dams induced by the 2008 Wenchuan earthquake, Sichuan, China. Eng. Geol. **129–130**, 95–110 (2012)
37. Dillen, Y., Tolhurst, T., Watson, G.: Impacts of extreme storms on coastal erosion and morphology. Earth Surf. Process. Landf. **45**(12), 2740–2754 (2020)
38. Werner, A.D., et al.: Seawater intrusion processes, investigation and management: recent advances and future challenges. Adv. Water Resour. **51**, 3–26 (2013)
39. Dasgupta, S., Huq, M., Khan, Z.H., Ahmed, M.M.Z., Mukherjee, N., Khan, M.F., Pandey, K.: Cyclones in a changing climate: the case of Bangladesh. Clim. Dev. **11**(10), 876–885 (2018)
40. Nicholls, R.J., Wong, P.P., Burkett, V., Codignotto, J., Hay, J., McLean, R., Ragoonaden, S., Woodroffe, C.: Coastal systems and low-lying areas. In: *Climate Change 2007: Impacts, Adaptation and Vulnerability*. Contribution of Working Group II to the IPCC Fourth Assessment Report. Cambridge University Press (2007)

41. Alam, M., Shamsudduha, M., Ahmed, N., Islam, M., Clarke, D.: Salinity intrusion in interior coast of Bangladesh: challenges to agriculture in south-central coastal zone. J. Coast. Conserv. **21**(5), 563–574 (2017)
42. Shakesby, R.A., Doerr, S.H.: Wildfire as a hydrological and geomorphological agent. Earth Sci. Rev. **74**(3–4), 269–307 (2006)
43. Moody, J.A., Shakesby, R.A., Robichaud, P.R., Cannon, S.H., Martin, D.A.: Current research issues related to post-wildfire runoff and erosion processes. Earth Sci. Rev. **122**, 10–37 (2013)
44. Cannon, S.H., Gartner, J.E., Rupert, M.G., Michael, J.A., Rea, A.H., Parrett, C.: Predicting the probability and volume of postwildfire debris flows in the intermountain western United States. Bulletin. **122**(1–2), 127–144 (2010)
45. Ebel, B.A., Moody, J.A., Martin, D.A.: Hydrologic conditions controlling runoff generation immediately after wildfire. Water Resour. Res. **48**(3) (2012)
46. Pescaroli, G., Alexander, D.: Critical infrastructure, panarchies and the vulnerability paths of cascading disasters. Nat. Hazards. **82**(1), 175–192 (2016)
47. Knutson, T.R., et al.: Tropical cyclones and climate change assessment: part II. Projected response to anthropogenic warming. Bull. Am. Meteorol. Soc. **101**(3), E303–E322 (2020)
48. Trenberth, K.E., et al.: Attribution of extreme rainfall from hurricane Harvey, august 2017. Environ. Res. Lett. **13**(1), 014501 (2018)
49. Dai, A.: Increasing drought under global warming in observations and models. Nat. Clim. Chang. **3**(1), 52–58 (2013)
50. Cook, B.I., Mankin, J.S., Anchukaitis, K.J.: Climate change and drought: from past to future. Curr. Clim. Chang. Rep. **4**(2), 164–179 (2018)
51. Abatzoglou, J. T., & Williams, A. P. (2016). Impact of anthropogenic climate change on wildfire across western US forests. Proceedings of the National Academy of Sciences, 113(42), 11770–11775
52. Xu, R., et al.: Wildfires, global climate change, and human health. New Engl. J. Med. **383**(22), 2173–2181 (2020)
53. Zscheischler, J., et al.: Future climate risk from compound events. Nat. Clim. Chang. **8**(6), 469–477 (2018)
54. Wahl, T., Jain, S., Bender, J., Meyers, S.D., Luther, M.E.: Increasing risk of compound flooding from storm surge and rainfall for major US cities. Nat. Clim. Chang. **5**(12), 1093–1097 (2015)
55. Rinaldi, S.M., Peerenboom, J.P., Kelly, T.K.: Identifying, understanding, and analyzing critical infrastructure interdependencies. IEEE Control. Syst. Mag. **21**(6), 11–25 (2001)
56. Buldyrev, S.V., Parshani, R., Paul, G., Stanley, H.E., Havlin, S.: Catastrophic cascade of failures in interdependent networks. Nature. **464**(7291), 1025–1028 (2010)
57. Ouyang, M.: Review on modeling and simulation of interdependent critical infrastructure systems. Reliab. Eng. Syst. Saf. **121**, 43–60 (2014)
58. Pant, R., Thacker, S., Hall, J.W., Alderson, D., Barr, S.: Critical infrastructure impact assessment due to flood exposure. J. Flood Risk Manag. **11**(1), 22–33 (2018)
59. Zimmerman, R., Restrepo, C.E., Culpen, D.: Multi-sector infrastructure impacts from hurricane Sandy and climate adaptation implications. J. Extrem. Events. **6**(01), 1950003 (2019)
60. Forzieri, G., Cescatti, A., e Silva, F. B., & Feyen, L.: Increasing risk over time of weather-related hazards to the European population: a data-driven prognostic study. Lancet Planet. Health. **2**(5), e200–e208 (2018)
61. Funabashi, Y., Kitazawa, K.: Fukushima in review: a complex disaster, a disastrous response. Bull. Atomic Sci. **68**(2), 9–21 (2012)
62. Oughton, E.J., Usher, W., Tyler, P.: Infrastructure as a complex adaptive system. Complexity. **2018**, 1–11 (2018)
63. Liu, W., Tang, Q., Voisin, N.: Cascading impacts of global warming on natural and human systems. Nat. Clim. Chang. **8**(5), 357–363 (2018)
64. Sidle, R.C., Ziegler, A.D., Negishi, J.N., et al.: Erosion processes in steep terrain—truths, myths, and uncertainties related to forest management in Southeast Asia. For. Ecol. Manag. **224**(1–2), 199–225 (2006)

65. Alimohammadlou, Y., Najafi, A., Yalcin, A.: Landslide susceptibility mapping using GIS and fuzzy logic in the Gorgan–Rud watershed. Iran. Environ. Earth Sci. **68**(6), 1875–1888 (2013)
66. Davidson, N.C.: How much wetland has the world lost? Long-term and recent trends in global wetland area. Mar. Freshw. Res. **65**(10), 934–941 (2014)
67. Junk, W.J., et al.: Current state of knowledge regarding the world's wetlands and their future under global climate change: a synthesis. Aquat. Sci. **75**(1), 151–167 (2013)
68. Hallegatte, S., Green, C., Nicholls, R.J., Corfee-Morlot, J.: Future flood losses in major coastal cities. Nat. Clim. Chang. **3**(9), 802–806 (2013)
69. Koks, E.E., Rozenberg, J., Zorn, C., Tariverdi, M., Vousdoukas, M., Fraser, S.A., Hall, J.W., Hallegatte, S.: A global multi-hazard risk analysis of road and railway infrastructure assets. Nat. Commun. **10**(1), 2677 (2019)
70. Brando, P. M., et al. (2014). Abrupt increases in Amazonian tree mortality due to drought–fire interactions. Proceedings of the National Academy of Sciences, 111(17), 6347–6352
71. Lovejoy, T.E., Nobre, C.: Amazon tipping point. Sci. Adv. **4**(2), eaat2340 (2018)
72. Das, S., Vincent, J.R.: Mangroves protected villages and reduced death toll during Indian super cyclone. PNAS. **106**(18), 7357–7360 (2009)
73. Day, J.W., et al.: Restoration of the Mississippi Delta: lessons from hurricanes Katrina and Rita. Science. **315**(5819), 1679–1684 (2007)
74. Webster, P.J., Jian, J., Singh, O.P.: Compound flooding in Pakistan: land use, climate change, and vulnerability. Nat. Clim. Chang. **12**, 983–985 (2022)
75. Spalding, M.D., McIvor, A.L., Beck, M.W., et al.: Coastal ecosystems: a critical element of risk reduction. Conserv. Lett. **7**(3), 293–301 (2014)
76. UNDRR (United Nations Office for Disaster Risk Reduction): Sendai Framework for Disaster Risk Reduction 2015–2030. UNDRR, Geneva (2015)
77. Meerow, S., Newell, J.P., Stults, M.: Defining urban resilience: a review. Landsc. Urban Plan. **147**, 38–49 (2016)
78. Zscheischler, J., Westra, S., van den Hurk, B.J., et al.: Future climate risk from compound events. Nat. Clim. Chang. **10**(6), 469–477 (2020)
79. Renn, O.: Systemic risks: the new kid on the block. Environ. Sci. Policy Sustain. Dev. **63**(1), 26–36 (2021)
80. Oliver-Smith, A., Alcántara-Ayala, I., Burton, I., Lavell, A.: Forensic investigations of disasters (FORIN): a conceptual framework and guide to research. In: Integrated Research on Disaster Risk (IRDR) (2017)
81. Weichselgartner, J., Pigeon, P.: The role of knowledge in disaster risk reduction. Int. J. Disaster Risk Sci. **6**(2), 107–116 (2015)
82. Galaz, V., Tallberg, J., Boin, A., Ituarte-Lima, C., Hey, E., Olsson, P., Westley, F.: Global governance dimensions of globally networked risks: the state of the art in social science research. Risk, Hazards Cris. Pub. Policy. **8**(1), 4–27 (2017)
83. Aitsi-Selmi, A., Egawa, S., Sasaki, H., Wannous, C., Murray, V.: The Sendai framework for disaster risk reduction: renewing the global commitment to people's resilience, health, and Well-being. Int. J. Disaster Risk Sci. **7**(2), 164–176 (2016)
84. Kelman, I., Glantz, M.H.: Early warning systems defined. In: Reducing Disaster: Early Warning Systems for Climate Change, pp. 89–108. Springer Netherlands, Dordrecht (2014)
85. Phillips, C.A., Caldas, A., Cleetus, R., Dahl, K.A., Declet-Barreto, J., Licker, R., Merner, L.D., Ortiz-Partida, J.P., Phelan, A.L., Carlson, C.J.: Compound climate risks in the COVID-19 pandemic. Nat. Clim. Chang. **10**(7), 586–588 (2020)
86. Gaillard, J.C.: Vulnerability, capacity and resilience: perspectives for climate and development policy. J. Int. Dev. **22**(2), 218–232 (2010)
87. Wisner, B.: Five years beyond Sendai—can we get beyond frameworks? Int. J. Disaster Risk Sci. **11**, 239–249 (2020)
88. Pelling, M., O'Brien, K., Matyas, D.: Adaptation and transformation. Clim. Chang. **133**(1), 113–127 (2015)
89. Kelman, I.: Disaster by Choice: how our Actions Turn Natural Hazards into Catastrophes. Oxford University Press (2020)

90. Folke, C., Carpenter, S.R., Walker, B., Scheffer, M., Chapin, T., Rockström, J.: Resilience thinking: integrating resilience, adaptability and transformability. Ecol. Soc. **15**(4), 20 (2010)
91. Holling, C.S.: Resilience and stability of ecological systems. Annu. Rev. Ecol. Syst. **4**(1), 1–23 (1973)
92. Walker, B., Salt, D.: Resilience Practice: Building Capacity to Absorb Disturbance and Maintain Function. Island Press (2012)
93. Kovats, R.S., Kristie, L.E.: Heatwaves and public health in Europe. Eur. J. Pub. Health. **16**(6), 592–599 (2006)
94. UNDRR: Global Assessment Report on Disaster Risk Reduction 2022. UNDRR, Geneva (2022)
95. Basher, R.: Global early warning systems for natural hazards: systematic and people-centred. Philos. Trans. R. Soc. A. **364**(1845), 2167–2182 (2006)
96. Kelman, I.: Linking disaster risk reduction, climate change, and the sustainable development goals. Disaster Prev Manag. **26**(3), 254–258 (2017)
97. Coughlan de Perez, E., van den Hurk, B., van Aalst, M.K., Amuron, I., Bamanya, D., Hauser, T., Jongma, B., Lopez, A., Mason, S., Mendler de Suarez, J., Pappenberger, F., Zsoter, E.: Action-based flood forecasting for triggering humanitarian action. Hydrol. Earth Syst. Sci. **20**(9), 3549–3560 (2016)
98. Temmerman, S., Meire, P., Bouma, T.J., Herman, P.M., Ysebaert, T., De Vriend, H.J.: Ecosystem-based coastal defence in the face of global change. Nature. **504**(7478), 79–83 (2013)
99. Seddon, N., Smith, A., Smith, P., Key, I., Chausson, A., Girardin, C., House, J., Srivastava, S., Turner, B.: Getting the message right on nature-based solutions to climate change. Glob. Chang. Biol. **27**(8), 1518–1546 (2020)
100. Kabisch, N., Frantzeskaki, N., Pauleit, S., Naumann, S., Davis, M., Artmann, M., Haase, D., Knapp, S., Korn, H., Stadler, J., Zaunberger, K., Bonn, A.: Nature-based solutions to climate change mitigation and adaptation in urban areas. Ecol. Soc. **22**(2), 39 (2017)
101. Raymond, C.M., Frantzeskaki, N., Kabisch, N., Berry, P., Breil, M., Nita, M.R., Geneletti, D., Calfapietra, C.: A framework for assessing and implementing the co-benefits of nature-based solutions in urban areas. Environ. Sci. Pol. **77**, 15–24 (2017)
102. Frantzeskaki, N., McPhearson, T., Collier, M.J., Kendal, D., Bulkeley, H., Dumitru, A., Walsh, C., Noble, K., Van Wyk, E., Ordóñez, C., Oke, C., Elmqvist, T.: Nature-based solutions for urban climate change adaptation: linking science, policy, and practice communities for evidence-based decision- making. Bioscience. **69**(6), 455–466 (2019)
103. Davoudi, S., Shaw, K., Haider, L.J., Quinlan, A.E., Peterson, G.D., Wilkinson, C., Fünfgeld, H., McEvoy, D., Porter, L.: Resilience: a bridging concept or a dead end? Plan. Theory Pract. **13**(2), 299–333 (2012)
104. Joseph, J.: Varieties of Resilience: Studies in Governmentality. Cambridge University Press (2018)
105. Pelling, M.: Adaptation to Climate Change: from Resilience to Transformation. Routledge (2011)
106. Guha-Sapir, D., Hoyois, P., Below, R.: EM-DAT: the CRED/OFDA International Disaster Database—Data Documentation v11. Université catholique de Louvain, Brussels, Belgium (2016) https://emdat.be
107. Hasegawa, R.: Disaster evacuation from Japan's 2011 tsunami disaster and the Fukushima nuclear accident. Asian J. Soc. Sci. **41**(3), 312–331 (2013). https://doi.org/10.1163/15685314-12341297
108. Aven, T., Renn, O.: An evaluation of the treatment of risk and uncertainties in the IPCC reports on climate change. Risk Anal. **35**(4), 701–712 (2015). https://doi.org/10.1111/risa.12298
109. Schneider, M., Froggatt, A., Thomas, S.: The world nuclear industry status report 2010–2011: nuclear power in a post-Fukushima world. In: Worldwatch Institute Report (2011)

Analysis of the Environmental Tax in EU. A Study of its Equity through Gini Index and Lorenz Curve

Antonio Seijas-Macias, Amílcar Oliveira, and Teresa A. Oliveira

Abstract This study utilises the Gini Index to analyse the discrepancy between environmental tax revenues and greenhouse gas emissions within the European Union (EU). The objective of the research is to quantify the concentration of tax revenues among countries based on their pollution levels. The methodology employed in this study is an analysis of tax revenues and greenhouse gas emissions data from 2008 to 2023. The study's key findings indicate a high degree of equity between tax revenues and carbon dioxide (CO_2) emissions, but a lower degree of equity for methane (CH_4) and nitrous oxide (N_2O) emissions. The findings imply a necessity for enhanced congruence between environmental taxes and EU climate objectives, along with the promotion of clean technologies. This research contributes to the understanding of the effectiveness of environmental taxation in reducing emissions and promoting sustainability in the EU.

Keywords Environmental taxes · Air pollution · Gini index · Lorenz curve

1 Introduction

In the European Union (EU), environmental taxation constitutes a pivotal policy instrument aimed at curbing greenhouse gas emissions and fostering sustainable development. This approach involves the imposition of taxes on activities that

A. Seijas-Macias (✉)
Department of Economics, Universidade da Coruña, Coruña, Spain
e-mail: antonio.smacias@udc.gal

A. Seijas-Macias · A. Oliveira · T. A. Oliveira
CEAUL, Faculdade de Ciências, Universidade de Lisboa, Lisbon, Portugal
e-mail: amilcar.oliveira@uab.pt

T. A. Oliveira
e-mail: teresa.oliveira@uab.pt

A. Oliveira · T. A. Oliveira
Department of Science and Technology, Universidade Aberta, Lisbon, Portugal

G. Halkos et al. (eds.), *Environmental Risk Analysis*, Springer Proceedings in Mathematics & Statistics 540, https://doi.org/10.1007/978-3-032-24155-9_3

contribute to environmental degradation, with the objective of incentivising cleaner practices and technologies.

The "Polluter Pays Principle" (PPP) is a foundational concept in environmental law and policy, which asserts that those who cause environmental damage should bear the costs associated with it. This principle is recognised and implemented in various forms across the globe, with the aim of allocating the financial responsibility of pollution to the polluters themselves rather than to the public.

We need to use a statistical measure relating to inequality. We must use a statistical concept that allows analysts to quantify the degree of inequality in each distribution (e.g. environmental taxes and gas emissions). This concept is like entropy in information theory, which measures uncertainty or disorder; both aim to quantify diversity or concentration in a system. The analysis of divergence between two elements is a common theme in scientific literature. One example of this can be found in politics, where the challenge lies in devising proportional methods to mitigate the unavoidable distortion that arises from translating votes into seats [1]. Several measures have been used to analyse the distribution of votes and seats, but more recently, some authors have expressed a preference for using the Lorenz Curve and Gini Index [2] or α-divergence methods [3]. Another example is the analysis of equity in income distribution within a country. The Gini Index, a prevalent metric of such distortions, has been used to compare situations in multiple countries. Utilisation of the Gini Index is associated with the Lorenz Curve [4], which provides a graphical representation of the divergence between two distributions, such as population and income [5].

As demonstrated in the seminal work [6], the Gini Index can be applied to environmental data. In this study, the authors analyse the distribution of household carbon footprints and consumption across 27 European Union (EU) countries.

The present paper analyses the divergence between environmental tax revenues and greenhouse gas emissions in the EU. The Gini Index is utilised to quantify the divergence in tax revenues among countries based on their respective levels of pollution.

The remainder of the paper is divided into three sections. The subsequent section comprises a review of the extant literature on the Gini Index and environmental expenditure. The third section of the paper is concerned with the calculation of the Gini Index and the Lorenz Curve, and the distortions of these values when there are small differences between distributions, in particular, tail differences. We propose the use of alternative approaches to avoid or overcome this problem. The fourth section of the paper considers the concentration of environmental tax revenues in EU countries in basis to level of greenhouse effect gas air emissions. In our study we have considered three basic gases: carbon dioxide (CO_2), methane (CH_4) and nitrous oxide (N_2O). The analysis covers the period from 2008 to 2023. It is evident that as the calculated Gini Index approaches 0, there is a corresponding enhancement observed in the distribution of income in accordance with the polluter pays principle. The polluter pays principle (PPP) is a fundamental tenet of environmental taxation, under which the onus is on the polluting entity to bear the financial responsibility for the environmental consequences of their actions.

The conclusions of this study are outlined in the final section.

2 The Environmental Taxes on Emissions in the EU

The correlation between increasing greenhouse gas emissions and climate change has necessitated the reduction of such emissions to mitigate the repercussions of global climate change. In lieu of the imposition of radical, prohibitive measures to curtail greenhouse gas emissions and reduce pollution, a course of action that has been adopted is the penalisation of these emissions through the imposition of so-called environmental taxes on companies.

The "Polluter Pays Principle" (PPP) is a foundational concept in environmental law and policy, which asserts that those who cause environmental damage should bear the costs associated with it. This principle is recognised and implemented in various forms across the globe, with the aim of allocating the financial responsibility of pollution to the polluters themselves rather than the public.

The PPP can be defined as an economic rule of cost allocation, originating from the theory of externalities [7, 8]. This theory requires polluters to take responsibility for the external costs arising from their pollution. This principle has been refined and disseminated through various international organisations, including the OECD and the EU, exerting influence on national legislation and policies [7, 9]. The principle serves multiple functions, including redistributive, preventive, and curative roles, and is considered a key element in environmental law [8].

In developing countries, the PPP has been adapted to include state obligations to compensate victims when polluters are unidentified or insolvent, thereby incentivising local governments to monitor environmentally risky activities [10].

Despite its widespread adoption, the PPP faces challenges in its application. A significant challenge pertains to the difficulty in defining who the polluter is and the extent of their financial responsibility [11]. The European Court of Justice has sought to address these issues by interpreting the principle broadly, yet challenges remain in establishing liability, especially in cases of accidental pollution [11]. Moreover, the principle's prioritisation of economic considerations over environmental protection objectives has the potential to create a discordance between these two fundamental values [12].

The impact of environmental taxes on emissions can be observed in three distinct ways:

- Reductions in CO_2 Emissions: Numerous studies have demonstrated that environmental taxes have been shown to significantly reduce CO_2 emissions in EU countries. Studies confirm that higher levels of environmental taxation correlate with a decrease in emissions, particularly when the tax levels exceed certain thresholds [13–15]. For instance, a €20/tonne CO_2 tax has been shown to lower emissions by approximately 1% on average [16].

- Nonlinear Effects: The relationship between environmental taxes and emissions is non-linear. Once certain tax thresholds are surpassed, the impact of these taxes on reducing emissions becomes significantly negative, indicating a stronger reduction in emissions [14].
- Sector-Specific Impacts: Carbon taxes, particularly in the energy sector, have been effective in reducing greenhouse gas emissions. The effectiveness of these taxes is enhanced when combined with other environmental policy tools, such as renewable energy incentives [17].

The efficacy of such measures in achieving the objective of reducing emissions is ambiguous and can be impacted by several factors.

- Fiscal versus Regulatory Role: While environmental taxes are intended to regulate emissions, their fiscal component often dominates. This results in a significant revenue generation, although this is not always reinvested into environmental protection measures [13, 18].
- Variability Across Member States: The impact of environmental taxes varies across EU member states due to differences in economic structures, energy consumption patterns, and existing environmental policies. Countries with higher GDP per capita tend to have higher emissions, which can affect the uniformity of tax impacts [16, 19].
- Policy Alignment and EU Law: The harmonisation of environmental taxes across member states has been both facilitated and hindered by EU law. While some EU laws encourage the adoption of environmentally driven tax measures, others limit the ability of member states to implement such taxes effectively [20].

The effectiveness of the PPP is influenced by economic and political factors. For instance, public acceptance of environmental taxes is influenced by political trust and the perceived integrity of public officials [21]. Furthermore, variations in economic conditions, such as elevated interest rates and protracted judicial delays, have the potential to influence the desirability and implementation of the PPP in different regions [10].

In the European Union, environmental taxation plays a significant role in reducing emissions. However, the effectiveness of this policy is influenced by various factors, including tax thresholds, economic conditions, and the integration of policy measures. Future efforts should focus on harmonising these taxes with broader environmental goals and supporting the development of clean technologies:

- Integration with Clean Technologies: There is a need for a more integrated approach that combines environmental taxes with the promotion of clean technologies. This could involve the implementation of differentiated tax rates to support various stages of technological development [18].
- Harmonization and Redistribution: To enhance the effectiveness of environmental taxes, there is a call for better alignment of these taxes with EU climate goals. The redistribution of tax revenues could also help balance the economic impacts across different countries [16, 20].

3 The Gini Index for Analysis of the Difference between Two Distributions

The Gini Index [22] is a common measure of income inequality, which can be used to compare it between different regions [23]. While the index was initially developed to assess income distribution, it has since been employed in various scientific disciplines, including astrophysics, ecology, econophysics, engineering, finance, geography, medicine, biology, and transport [24]. The interpretation of the Gini Index is uncomplicated: the closer to zero, the more equitable the distribution. The maximum value of the index is 1, denoting a situation in which the entire welfare is possessed by a single agent. This maximum value represents the extent of inequality at its zenith.

The calculation of the Gini Index is derived from the original approach of Gini. The following alternative approaches are noteworthy: The absolute mean difference [25] and the relative mean difference approach [26] produce identical values for the index. However, the Geometric Approach [27] and the Gini Coefficient [28] yield different values, with the latter being employed to enhance the accuracy of the index.

The Gini Index can be derived from the Lorenz Curve [29]. The Lorenz Curve is a graphical representation of wealth distribution, which plots the cumulative proportion of the population on the x-axis and the cumulative proportion of income on the y-axis, both from the lowest to the highest. The Gini Index is calculated as the ratio of the area between the equity line and the Lorenz Curve, divided by the total area between the equity line and the x-axis. To derive the Gini Index, it is necessary to consider the cumulative distribution function (CDF) of a continuous random variable X:

$$L(p) = \frac{\int_0^p F^{-1}(y)dy}{\int_0^1 F^{-1}(y)dy}, 0 \leq p \leq 1, \tag{1}$$

where $F^{-1}(y) = \sup \{x: F(x) \leq y\}$, is the Lorenz Curve (LC) associated with F(x). Lorenz Curve satisfies the following conditions: $L(0) = 0, L(1) = 1$. The LC function $L(p)$ is an increasing and convex function for $p \in (0, 1)$. Several authors consider the use of parametric models to calculate LC, the most used approaches consider the use of one parameter [30], two parameters [31] and three parameters [32]. Results indicate that the more parameters the more accuracy [33]. Some authors (see [5]) consider a new interpretation of the Gini Index from the Lorenz Curve because the Lorenz Curve is more sensitive to changes in the tails of distributions.

Let the cumulative distribution function (CDF) of a continuous random variable X, following a distribution with parameter α, be denoted by F(X| α). The Lorenz Curve is then defined as follows.

$$L(x, \alpha) = \frac{1}{\mu}\int_0^x F^{-1}(y \mid \alpha)dy, \qquad x \in [0, 1] \tag{2}$$

where μ is the mean of F and $F^{-1} = inf \{z: F(z \mid \alpha) \geq y\}$.

Using this geometrical approach, the Gini Index is:

$$G = 1 - \frac{2}{\mu}\int_{0}^{+\infty} z\,F(z \mid \alpha)f(z \mid \alpha)dy = 1 - 2\int_{0}^{1} L(y \mid \alpha)\,dy \tag{3}$$

To enhance the efficacy of the Lorenz Curve, it is possible to utilise a range of distributions and to augment the number of parameters under consideration.

In accordance with the methodology outlined by [33], the Lorenz Curve can be defined in several ways, depending on the underlying distribution selected.

1. Lorenz Curve with 1 parameter. Pareto Distribution

$$L_{CP}(x \mid\alpha) = 1 - (1 - x)^{\frac{\alpha-1}{\alpha}}, \qquad \alpha > 1 \tag{4}$$

2. Lorenz Curve with 1 parameter [30].

$$L_1(x \mid\alpha) = \frac{\exp(\alpha\, x) - 1}{\exp(\alpha) - 1}, \qquad \alpha > 0. \tag{5}$$

3. Lorenz Curve with 2 parameters [31].

$$L_2(x \mid\alpha, \beta) = x^{\alpha}\big(1 - (1 - x)^{\beta}\big), \quad \alpha \geq 0,\ 0 \leq \beta \leq 1. \tag{6}$$

4. Lorenz Curve with 3 parameters [32].

$$L_3(x \mid\alpha, \beta, \gamma) = x^{\alpha}\big(1 - (1 - x)^{\beta}\big)^{\gamma}, \ \alpha \geq 0, 0 \leq \beta \leq 1, \gamma \geq 1. \tag{7}$$

The percentile method is utilised to estimate the parameters and fit the Lorenz curves to observed data [33]. The findings suggest that the proposed methods demonstrate a level of performance that is comparable to, if not superior to, the currently utilised methods in numerous instances [33]. Recent developments in modelling the distribution of products of normal variables [34, 35], offer enhanced tools for capturing tail behaviour and skewness in equity analysis, which are particularly useful when applying Lorenz Curve methodologies to environmental data.

Alternatively, other approaches [36] consider estimating the Lorenz Curve assuming a particular distribution: Dirichlet Distribution [37]. "Maximum likelihood estimates under the Dirichlet Distribution assumption provide better fitting Lorenz curves than nonlinear least squares and another estimation technique that has appeared in the literature" [36]. An alternative approach [38] uses the Bayesian model averaging to estimate the parameters of the Lorenz Curve. A novel Bayesian approach to estimating the Gini coefficient and the Lorenz curves is in [39].

3.1 Gini Index and Environmental Sciences

The Gini Index has been a pivotal component in the study of income distribution among the population since its inception. Its utilisation as a metric for evaluating the degree of equity among nations has been demonstrated in numerous studies, and it is among the most frequently employed reference statistics. However, its application is not limited to this, since it is an index that allows us to study the similarity or difference in the distribution of two variables. In the field of environmental studies, it has been employed in the analysis of diverse economic dimensions, including regional development, resource distribution and environmental efficiency.

In [40], the authors present a Gini Index adapted to assess disparities in regional economic output, energy consumption and pollution emissions in China. In [41], a multidimensional Gini Index is presented as a means of investigating inequalities in sustainable economic development.

In accordance with the study previously conducted [42], the authors employ the Gini Index to analyse CO_2 emissions, thereby facilitating a comparison of the degree of environmental efficiency between nations. The findings indicate that while developed countries exhibit higher levels of pollution concentration, they demonstrate a superior capacity for effective emission management.

In another line of study, the Gini Index has also been used to establish the distribution of resources in cities [43].

To the best of our knowledge, this study is the first to utilise the Gini Index to analyse the similarity between the distribution of revenue obtained from environmental taxes and the reduction of the main greenhouse gases (CO_2, CH_4 and N2O) in the European Union as a whole. The objective of this study is to analyse the efficacy of taxation as a regulatory instrument in promoting the adoption of environmentally sustainable systems, with a view to reducing emissions.

A plethora of studies have previously been conducted on these two variables, with a particular emphasis on the analysis of their negative correlation [44], in addition to aspects such as regional variations [45] and causality [46].

The Gini Index can be utilised as a reference point to analyse the efficiency of environmental taxes in the EU and to detect differences between member countries. Consequently, policies appropriate to each country could be implemented based on its economic development and productive structure.

4 Environmental Taxes and Pollution in the EU

The introduction of environmental taxation in the European Union can be traced back to the establishment of the Single European Act, which marked a significant development in the region's environmental policy. The aforementioned act established the Polluter Pays Principle (PPP). Presently, the focus of taxation has been directed towards sectors exhibiting the highest levels of pollution. These include the energy and transport sectors, as well as two more generic taxes on pollution and resources. Most of the tax revenue is derived from the energy sector, with a smaller contribution from the transport sector.

Nevertheless, the fundamental principle remains that of the 'polluter pays' principle. The overarching objective of environmental taxation should be to curtail pollution, with a particular emphasis on the reduction of greenhouse gas emissions. The EU has identified three categories of gases for the purpose of its regulatory agenda: carbon dioxide (CO_2), methane (CH_4) and nitrous oxide (N_2O).

4.1 Relationship between Environmental Taxes and Pollution

The rise in environmental tax revenues despite declining emission can be attributed to a combination of fiscal, policy and behavioural factors.

1. Tax Rate Increases
   (a) Higher environmental tax rates can offset declining emissions. "In 2023, total environmental tax revenue in the EU amounted to 341.5 billion, representing 2% of EU GDP and 5.1% of total EU government revenue form taxes and social contributions (TSC)".[1] During the whole period 2008–2023, the revenues have increased with two-year exceptions (2020 and 2022).
   (b) Governments may increase rates over time to strengthen climate action or maintain revenue neutrality, especially when emissions are falling. "To succeed in the EU's climate neutrality objective by 2050, taxation systems are a key policy tool. Well-designed tax reforms can stimulate economic growth, reduce greenhouse gas emissions through effective carbon pricing, and support a equitable transition."[2]
2. Expansion of the Tax Base
   (a) Establish new taxes on other sources of pollution. Until now, the EU has focused its taxes on the emission of three types of greenhouse gases. However, in the future, the harmful and polluting effects of other types of emissions linked to other industries could be considered as a source of taxation. The EU's current fiscal policy has focused on the energy sector (production), the ownership and use of motor vehicles (which use fossil fuels) and a group of various polluting effects such as the extraction of raw materials, the emission of pollutants into water and air, noise and waste management.
   (b) Creation or modification of current environmental taxes to rebalance sectors subject to higher tax pressure. In 2003, 48.3% of environmental tax revenue was borne by households, while 49.4% was attributed to productive sectors.

[1] Eurostat: environmental tax statistics. https://ec.europa.eu/eurostat/statistics-explained/index.php?title=Environmental_tax_statistics.

[2] Eurostat: environmental tax statistics. https://ec.europa.eu/eurostat/statistics-explained/index.php?title=Environmental_tax_statistics.

However, in some countries such as Ireland, the Netherlands, Slovenia and France, the contribution of households reached 55%.[3]

3. Temporal Lag Effects
   (a) The impact of environmental taxation on reducing emissions is not instantaneous. It is imperative that any such action be informed by the understanding that a period of adjustment is requisite until such time as initial, perceptible outcomes are manifest in relation to emission levels. To illustrate this point, the introduction of an environmental tax on fuels does not result in an immediate reduction in emissions. This is because vehicles do not adapt immediately; rather, they require a period of adaptation until more efficient vehicles with lower emissions become available.
   (b) Conversely, this adaptation period may result in price increases by those industries and sectors that are able to pass on these tax costs to the end consumer via price increases.

4.2 Empirical Analysis of the 2008–2023 Period

In this paper we study the evolution of the pollution and environmental taxes during the years 2008–2023, and we have considered the 27 countries of the EU. The available data used were EUROSTAT data with two specific tables: Air emissions accounts by NACE Rev. 2 activity and environmental tax revenues. We have only considered the total revenues and the total air emissions of the cited greenhouse gases.

Table 1 presents the variation of the tax revenues in the 27 countries of the EU between years 2008 and 2023. An analysis of the data indicates that, in the period under review, all countries have substantially increased the revenue attributed to environmental taxes. For the European Union (EU) as a whole, the increase would be 30%, but while countries such as Romania increased by 193%, Bulgaria by 160%, and Estonia by 150%. It is noteworthy that only one EU member state has demonstrated a decline in its environmental tax revenue: Denmark, with a 23% decrease. From an environmental perspective, this rise in tax revenue can be regarded as a negative development, as it contradicts the Polluter Pays Principle. This is because if revenue is increasing, it should be because of an increase in pollution.

Tables 2, 3 and 4 illustrate the evolution of greenhouse gas emissions within the EU from 2008 to 2023. Contrary to the anticipated increase in greenhouse gas emissions, associated with the rise in tax revenues, the data reveals a general decrease across most EU countries.

Specifically, the EU's carbon dioxide (CO2) emissions experienced a 32% decline during the examined period. However, it is noteworthy that certain member countries, namely Ireland, Lithuania and Malta, have demonstrated an increase in their

[3] Eurostat: environmental tax statistics. https://ec.europa.eu/eurostat/statistics-explained/index.php?title=Environmental_tax_statistics.

Table 1 Environmental tax revenues in EU countries (Mill Euros)

Country	2008	2023	Abs. variation	%
Austria (AT)	6995.98	9533.28	2537.3	36.27%
Belgium (BE)	8081.1	13,820.15	5739.05	71.02%
Bulgaria (BG)	1221.74	3176.26	1954.52	159.98%
Cyprus (CY)	625.78	661.05	35.27	5.64%
Czechia (CZ)	3643.79	4957.42	1313.63	36.05%
Germany (DE)	64,140	71,407	7267	11.33%
Denmark (DK)	10,077.75	7716.67	−2361.08	−23.43%
Estonia (EE)	396.75	995.14	598.39	150.82%
Greece (EL)	4972	9257	4285	86.18%
Spain (ES)	18,179	24,640	6461	35.54%
Finland (FI)	5028	6182	1154	22.95%
France (FR)	36,455	49,859	13,404	36.77%
Croatia (HR)	1823.36	2575.87	752.51	41.27%
Hungary (HU)	2907.43	4249.89	1342.46	46.17%
Ireland (IE)	4314.14	4909.09	594.95	13.79%
Italy (IT)	42,749	54,224	11,475	26.84%
Lithuania (LT)	533.52	1188.66	655.14	122.80%
Luxembourg (LU)	985.22	1016.39	31.17	3.16%
Latvia (LV)	506.87	949.88	443.01	87.40%
Malta (MT)	199.88	315.55	115.67	57.87%
Netherlands (NL)	24,453	29,983	5530	22.61%
Poland (PL)	10,215.83	19,181.2	8965.37	87.76%
Portugal (PT)	4483.84	5378.36	894.52	19.95%
Romania (RO)	2487.18	7294.16	4806.98	193.27%
Sweden (SE)	9364.76	10,856.27	1491.51	15.93%
Slovenia (SI)	1110.79	1791.44	680.65	61.28%
Slovakia (SK)	1363.74	2456.04	1092.3	80.10%

respective emissions. The remaining countries have reduced their emissions, albeit by percentages ranging from 7.5% in Poland to 53.8% in Estonia. Conversely, the statistics pertaining to methane (CH_4) emissions are less favourable. Despite the EU experiencing a 13% reduction in emissions, several member states have demonstrated an increase, ranging from 7.5% in Poland to 53.8% in Estonia. About methane (CH_4) emissions, however, the figures are less favourable. Despite the EU having reduced its emissions by 13%, several countries have shown an increase in emissions: Bulgaria, Cyprus, Greece, Ireland, Luxembourg, Latvia, Malta and Portugal. Notably, Malta

Table 2 CO_2 Air emissions in EU countries (Ton)

Country	2008	2023	Abs. Variation	%
Austria (AT)	54,848,096	44,701,734.6	−10,146,361	−18.50%
Belgium (BE)	87,656,849.1	61,481,321.4	−26,175,528	−29.86%
Bulgaria (BG)	53,252,861.2	31,872,057.3	−21,380,804	−40.15%
Cyprus (CY)	7,588,141.3	5,568,842.86	−2,019,298.4	−26.61%
Czechia (CZ)	103,864,484	67,950,178.6	−35,914,306	−34.58%
Germany (DE)	733,952,060	477,174,998	−256,777,061	−34.99%
Denmark (DK)	85,911,439	60,965,641.3	−24,945,798	−29.04%
Estonia (EE)	17,951,402.7	8,391,837.16	−9,559,565.5	−53.25%
Greece (EL)	101,476,311	46,879,880	−54,596,431	−53.80%
Spain (ES)	269,884,683	168,009,695	−101,874,988	−37.75%
Finland (FI)	55,041,688.6	30,160,301	−24,881,388	−45.20%
France (FR)	312,003,867	216,764,078	−95,239,789	−30.53%
Croatia (HR)	18,149,226.2	11,610,000.2	−6,539,226	−36.03%
Hungary (HU)	46,430,545.9	32,511,269.1	−13,919,277	−29.98%
Ireland (IE)	30,938,584.2	38,287,501.3	7,348,917.11	23.75%
Italy (IT)	380,272,530	228,167,640	−152,104,890	−40.00%
Lithuania (LT)	13,282,020.2	14,807,599.2	1,525,579.02	11.49%
Luxembourg (LU)	7,391,374	6,596,821.78	−794,552.22	−10.75%
Latvia (LV)	8,678,132.93	6,668,554.46	−2,009,578.5	−23.16%
Malta (MT)	2,697,888.18	3,086,872.33	388,984.15	14.42%
Netherlands (NL)	157,690,729	111,964,055	−45,726,674	−29.00%
Poland (PL)	286,923,920	265,202,736	−21,721,184	−7.57%
Portugal (PT)	52,793,755.9	32,283,670.9	−20,510,085	−38.85%
Romania (RO)	95,346,626.3	52,411,532.2	−42,935,094	−45.03%
Sweden (SE)	45,208,606	31,720,386.5	−13,488,220	−29.84%
Slovenia (SI)	13,282,176.3	8,517,453.08	−4,764,723.2	−35.87%
Slovakia (SK)	33,901,805.1	24,900,706.8	−9,001,098.3	−26.55%

has recorded a 52% increase in methane emissions. Conversely, the remaining countries have demonstrated a decline in their emissions, with figures ranging from −3.4% in Estonia to 27.5% in Sweden.

Furthermore, nitrous oxide (N_2O) emissions have been reduced by 26% for the EU countries, although in several of them the figures show increases of varying magnitude. For instance, Bulgaria has shown a 34% increase, while Estonia has demonstrated an increase of less than 10%.

When considering the aggregate data for the EU, it is evident that 19 countries have achieved a reduction in their emissions for the three gases under consideration, while in another five countries, a reduction has been observed in two of the three

Table 3 CH_4 air emissions in EU countries (Ton)

Country	2008	2023	Abs. Variation	%
Austria (AT)	279,963.15	246,488.16	−33,474.99	−11.96%
Belgium (BE)	318,968.93	294,493.9	−24,475.03	−7.67%
Bulgaria (BG)	236,404.45	247,855.35	11,450.9	4.84%
Cyprus (CY)	34,177.72	42,848.34	8670.62	25.37%
Czechia (CZ)	551,147.03	471,461.19	−79,685.84	−14.46%
Germany (DE)	2,211,588.09	1,718,307.52	−493,280.57	−22.30%
Denmark (DK)	319,754.64	291,833.71	−27,920.93	−8.73%
Estonia (EE)	44,793.06	43,265.29	−1527.77	−3.41%
Greece (EL)	441,380.18	459,509.69	18,129.51	4.11%
Spain (ES)	1,550,285.75	1,469,769.32	−80,516.43	−5.19%
Finland (FI)	201,715.85	168,450.55	−33,265.3	−16.49%
France (FR)	2,420,291.31	1,934,566.8	−485,724.51	−20.07%
Croatia (HR)	143,083.03	123,049.44	−20,033.59	−14.00%
Hungary (HU)	339,152.29	322,204.38	−16,947.91	−5.00%
Ireland (IE)	539,894.04	564,312.85	24,418.81	4.52%
Italy (IT)	1,795,539.72	1,537,497.21	−258,042.51	−14.37%
Lithuania (LT)	144,377.99	116,751.15	−27,626.84	−19.14%
Luxembourg (LU)	23,777.55	25,659.96	1882.41	7.92%
Latvia (LV)	61,781.22	64,213.69	2432.47	3.94%
Malta (MT)	5851.39	8926.41	3075.02	52.55%
Netherlands (NL)	754,862.61	613,943.86	−140,918.75	−18.67%
Poland (PL)	1,660,008.07	1,452,585.05	−207,423.02	−12.50%
Portugal (PT)	389,216.7	396,093.26	6876.56	1.77%
Romania (RO)	1,159,007.77	965,255.77	−193,752	−16.72%
Sweden (SE)	215,235.24	155,986.61	−59,248.63	−27.53%
Slovenia (SI)	82,554.82	71,059.46	−11,495.36	−13.92%
Slovakia (SK)	164,269.15	135,429.61	−28,839.54	−17.56%

gases. It is noteworthy that three countries have demonstrated a decline in emissions with respect to only one of the three gases.

Figure 1 illustrates that the most significant emission reduction results are achieved for CO_2, followed by N_2O, with methane (CH_4) emissions demonstrating the least favourable outcomes.

To analyse the distribution of tax revenues linked to environmental taxes on the one hand, and the evolution of pollutant gas emissions in the different EU countries on the other, the Gini Index and the Lorenz Curve have been utilised. As previously mentioned, in the period analysed (2008–2023), the EU has seen an increase of almost 30% in tax revenue linked to environmental taxes. This rise in revenue has

Table 4 N2O air emissions in EU countries (Ton)

Country	2008	2023	Abs. Variation	%
Austria (AT)	11,909.26	9368.49	−2540.77	−21.33%
Belgium (BE)	22,888.08	14,254.25	−8633.83	−37.72%
Bulgaria (BG)	9770.14	13,170.12	3399.98	34.80%
Cyprus (CY)	597.13	551.3	−45.83	−7.68%
Czechia (CZ)	21,254.06	14,754.63	−6499.43	−30.58%
Germany (DE)	128,327.63	70,923.46	−57,404.17	−44.73%
Denmark (DK)	20,893.41	18,346.16	−2547.25	−12.19%
Estonia (EE)	2590.71	2844.26	253.55	9.79%
Greece (EL)	19,523.41	13,493.77	−6029.64	−30.88%
Spain (ES)	39,897.3	36,796.97	−3100.33	−7.77%
Finland (FI)	19,816.07	12,759.29	−7056.78	−35.61%
France (FR)	122,867.1	89,037.96	−33,829.14	−27.53%
Croatia (HR)	9114.08	4413.84	−4700.24	−51.57%
Hungary (HU)	13,257.16	11,549.91	−1707.25	−12.88%
Ireland (IE)	19,108.68	17,968.28	−1140.4	−5.97%
Italy (IT)	68,051.93	54,236.71	−13,815.22	−20.30%
Lithuania (LT)	17,255.61	7405.64	−9849.97	−57.08%
Luxembourg (LU)	802.18	798.77	−3.41	−0.43%
Latvia (LV)	3831.31	4420.53	589.22	15.38%
Malta (MT)	210.85	208.33	−2.52	−1.20%
Netherlands (NL)	30,059.33	24,139.87	−5919.46	−19.69%
Poland (PL)	84,821.06	64,457.52	−20,363.54	−24.01%
Portugal (PT)	13,654.4	11,107.61	−2546.79	−18.65%
Romania (RO)	37,043.76	30,689.83	−6353.93	−17.15%
Sweden (SE)	14,681.8	13,543.35	−1138.45	−7.75%
Slovenia (SI)	2217.19	1890.45	−326.74	−14.74%
Slovakia (SK)	8955.98	4445.02	−4510.96	−50.37%

been accompanied by a decline in greenhouse gas emissions, ranging from just over 13% (CH_4) to 30% of CO_2. In this study, we have analysed the concentration in environmental tax revenues and emissions of the different types of gases considered. We have recreated the Lorenz Curve for each of the pollutants under scrutiny and calculated the Gini Index for each year of the period studied, thus obtaining an evolution of this index (Table 5).

The Gini Index has remained relatively constant for the period analysed. When analysing the degree of concentration between tax revenues linked to environmental taxes and greenhouse gas emissions, it is evident that the concentration is low or very low. Except for 2019 and 2020, when the Gini Index registered at approximately

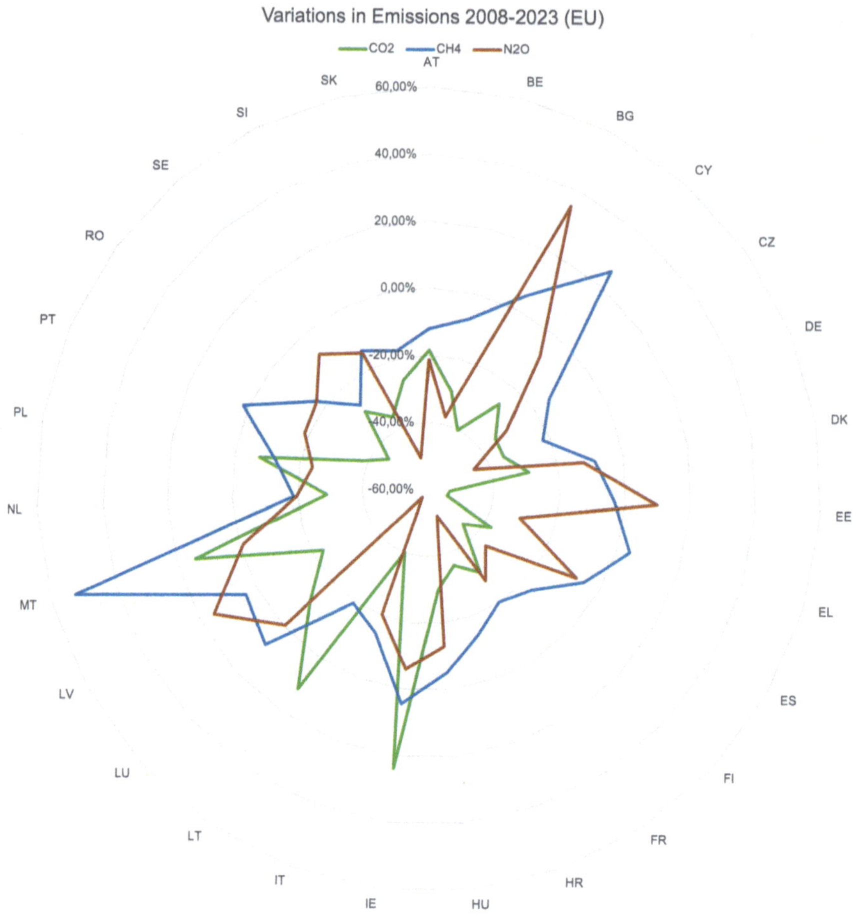

Fig. 1 Relative variation in gas air emissions EU (2008–20,023)

0.1, the figure has consistently remained at around 0.05 for the remaining years. This finding suggests a substantial degree of equity between tax revenues and CO_2 emissions. Conversely, for methane (CH_4), the concentration levels of the distribution exhibit notably higher values, approximating 0.3. A comparable trend is observed for N_2O, with an index of approximately 0.2. It is evident that in this instance, the distribution of taxes is less effective in aligning with the emission of these greenhouse gases, as evidenced by the higher concentration levels observed.

The Lorenz Curve, plotted for various years within the period under analysis (see Figs. 2, 3 and 4), substantiates the hypothesis that the degree of equity in the distributions, as gauged by the Gini Index, is indeed valid.

In the figures presented, the revenues from environmental taxes are plotted on the vertical axis, while the emission of the various greenhouse gases is plotted on the horizontal axis, for the different EU countries. The equity line indicates the scenario

Table 5 The Gini Index

Year	CO_2	CH_4	N2O
2008	0.0058	0.2574	0.1506
2009	0.0201	0.2752	0.1674
2010	0.0124	0.2488	0.1828
2011	0.0316	0.2697	0.2007
2012	0.0427	0.2617	0.1804
2013	0.0314	0.2851	0.2443
2014	0.0423	0.2949	0.2389
2015	0.0664	0.2873	0.2238
2016	0.0823	0.2906	0.2265
2017	0.0994	0.2892	0.2407
2018	0.0906	0.2849	0.2339
2019	0.1157	0.2981	0.2404
2020	0.1326	0.3227	0.2694
2021	0.0898	0.2912	0.2364
2022	0.0488	0.2586	0.1783
2023	0.0337	0.2374	0.1945

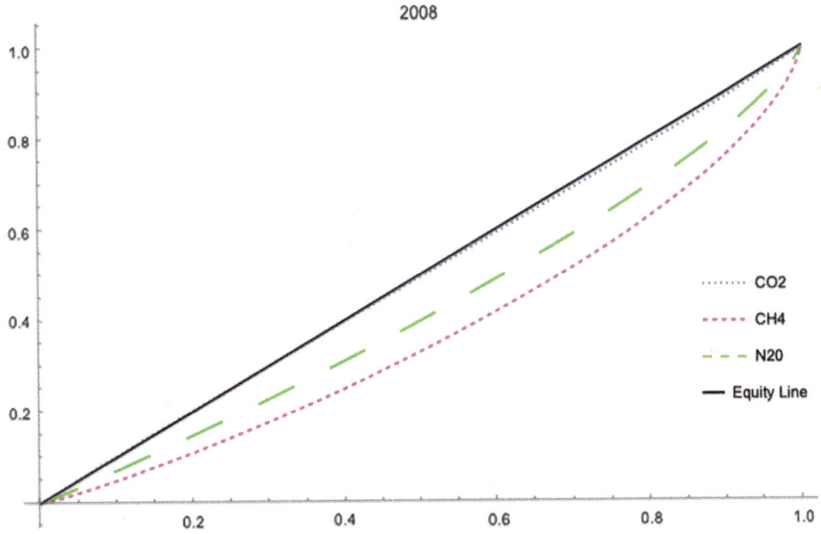

Fig. 2 Lorenz curve tax revenues versus gas emission (2008)

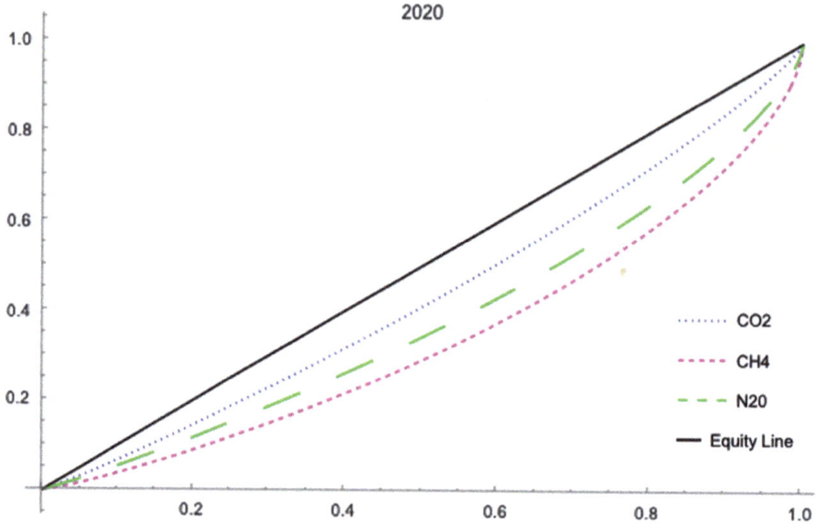

Fig. 3 Lorenz curve tax revenues versus gas emission (2020)

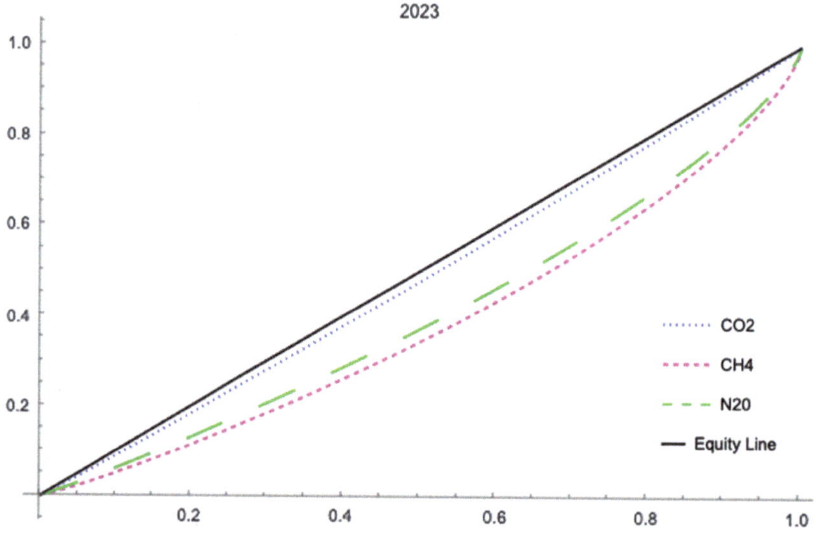

Fig. 4 Lorenz curve tax revenues versus gas emission (2023)

in which the nation's contribution to the tax revenue of the EU is equivalent to its share of greenhouse gas emissions within the broader EU context.

The analysis demonstrates that environmental tax revenues and CO_2 emissions exhibit a remarkably equitable distribution, with close to 0% variance. This phenomenon has remained virtually constant throughout the entire period under scrutiny (2008–2023), except for a negligible decline in equity observed in 2019 and 2020, resulting in Gini Index values approximating 0.1. Furthermore, an examination of the concentration of environmental tax revenues with respect to the value of emissions of the other two greenhouse gases considered (CH_4 and N_2O) reveals that the Gini Index value ranges from 0.2 to 0.3 across the various years under study. This finding suggests that the distribution is less equitable, which could indicate the presence of a less efficient taxation system for agents emitting these gases.

The investigation of the weak correlation between taxes and the reduction of CH_4 and N_2O emissions in relation to CO_2 emissions may be attributable to several factors:

1. The following investigation will address the question of the origin of emissions. In contrast, CO_2 emissions are more clearly and precisely attributable to anthropogenic activities, particularly those associated with energy production and the utilisation of fossil fuels. The emissions of CH_4 and N_2O have a more diffuse origin, including livestock, landfills, agricultural soils and fertilisers, which makes it difficult to establish clear tax schemes.
2. The pressures exerted by specific industry sectors. Conversely, certain sectors offer incentives that act as a deterrent to the implementation of taxes on these emissions. Of particular significance is the agricultural and livestock sector, where the influence of lobby groups has been demonstrated to impede the introduction of novel or more effective taxation measures. It is also necessary to consider the need to maintain a competitive product for the affected sectors in relation to production in countries where this type of tax does not exist.
3. The issue at hand is predicated on the inherent technical and institutional complexities. The sectors affected by this type of emissions encounter difficulties in accurately measuring the origin of the emissions. The magnitude of emissions can be subject to considerable variation, contingent on the characteristics of the agricultural soil or the type of feed utilised in livestock farming. These factors have been demonstrated to have a significant impact on N_2O and CH_4 emissions.
4. The following factors are to be considered in relation to the socio-political situation: While their impact may be less significant, it is crucial to acknowledge that, from the perspective of public opinion, CO_2 is widely regarded as the primary driver of climate change. A significant proportion of research and analysis aimed at addressing the issue has been directed towards this aspect. From a scientific perspective, the role of the other two gases (i.e. CH_4 and N_2O) in terms of climate change is equally detrimental. However, from a social perspective, these gases have been marginalised and as a result governments have not focused on reducing these types of emissions. Conversely, there has been a paucity of international coordination about the reduction of such gases.

4.3 Practical Implications

The Gini Index is a measure of statistical dispersion between the distributions of two variables. In the case of the index being equal to or very close to zero, this is indicative of identical or very similar distributions of the variables. Conversely, when the index approaches 1 or takes values greater than 0.3 or 0.4, it indicates a significant deviation between the two distributions.

Considering this measure, economic policymakers are equipped to act with the objective of correcting these deviations and bringing the value of the Gini Index closer to zero.

In the present study, an analysis was conducted of the distribution of environmental tax revenue and the emission of three greenhouse gases (CO_2, CH_4 and N_2O) in the countries of the current European Union in the period 2008–2023. In consideration of the data obtained and the value of the Gini Index, policy makers can act on several fronts:

1. The distribution of greenhouse gas emissions must be identified. The Gini Index has been identified as a potential tool for the analysis of disparities in greenhouse gas emissions among European Union member states. This information is of paramount importance to comprehend the varying situations of the respective states and establish the requisite targets to achieve a balance between all member states.
2. The purpose of this text is to provide a detailed explanation of the role of the individual in the decision-making process. Distinct variations in the distribution of emissions give rise to the necessity of determining whether tax rates on greenhouse gas emissions should be increased or decreased. However, it also highlights the necessity for the implementation of measures to encourage investment in industries that are more efficient and less polluting. Furthermore, it identifies those countries where the transport sector should be reformed to ensure the sustainability of transport systems. The primary objective should not be solely revenue-raising but rather should be focused on reducing greenhouse gases.
3. The impact of policies is the focus of this study. A temporal analysis of the evolution of the Gini Index facilitates an assessment of the efficacy of the policies adopted in the EU within the framework of reducing greenhouse gas emissions. Notwithstanding the primary objective of adopting the ‘polluter pays’ principle, it is imperative that this notion is complemented by the overarching goal of reducing emissions. In this regard, it is imperative that the objectives and strategies to be implemented in the various Member States are distinctly delineated, as opposed to the current approach of concentrating policy endeavours on a generic objective for the EU. In this regard, our analysis identifies the need for action in various states to promote fewer polluting means of transport and encourage the adoption of production models with lower emission rates.
4. The utilisation of benchmarking and comparisons is an integral component of the experimental process. The present analysis incorporates a temporal dimension, encompassing the evolution of the variables under scrutiny between the years

2008 and 2023. While the analysis has been focused on the initial and final years of the period, the data would permit a more in-depth analysis by following their annual evolution. This approach, informed by the knowledge of the measures adopted by the various states in each year, would facilitate the evaluation of their efficiency and effectiveness in reducing polluting gas emissions, while also allowing for a differentiation between states.

5. The analysis conducted in this study has enabled the identification of states that are particularly vulnerable, necessitating the implementation of structural measures to rectify greenhouse gas emission levels. In this regard, it is important to note that the 'polluter pays' principle is a significant component of environmental policy, but it should not be considered the primary objective of EU policies aimed at reducing emissions and pollution.

It is evident that an analysis of the Gini Index can provide valuable guidance to economic policymakers seeking to achieve their objectives of reducing pollutant emissions in EU countries. Furthermore, a more in-depth analysis could be carried out using higher levels of regional disaggregation, such as NUTS 1 and 2, to set more precise targets.

This type of analysis could be a possible line of future research. The aim would be to identify the differences not only between states, but also between the different regional divisions within each state. This would enable the proposed corrective measures to be more specific and potentially more effective.

A promising direction for future research is to conduct a comparative analysis of EU member states, examining how their distinct economic structures, energy dependence, and income levels influence the distribution and efficiency of environmental taxes in reducing greenhouse gas emissions. This work would highlight some of the potential reasons for the discrepancies observed in our study regarding the effectiveness of tax collection in reducing emissions in certain member states.

This analysis could even be carried out in more detail, not only considering differences between countries, but also differences at regional level within a country.

5 Conclusions

The Gini Index is the most widely employed index for the analysis of the divergence between two data series. Typically, the Gini Index is utilised to assess the distribution of income across the population of a nation. A comparative study between countries was performed, based on the Gini Index values.

However, the optimal version of the Gini Index remains an open problem, and it is our objective to continue our research in this area. The original Gini Index formula produces values that are insensitive to minor alterations, particularly in the tails of the distribution. Conversely, the Lorenz Curve, when utilised in conjunction with one or several parameters to define the most accurate curve, has been considered by some authors to yield a more accurate value. The Gini Index is then calculated by integrating the area between the Lorenz Curve and the equity line.

We have used the Gini Index to measure the concentration of the environmental tax revenues based on greenhouse effect gas emissions for the 27 countries of the EU, during the years 2008–2023. We have used the L_2 Lorenz Curve approach [31] to calculate the Gini Index of the different years. Parameters of the model were calculated using Maximum Likelihood Estimation and Dirichlet Distribution.

There is a very high equity (close to zero) between environmental tax revenues and CO_2 emissions. In contrast, for the other two gases analysed, the equity is much lower, and the Gini Index values are around 0.3 for CH_4 and 0.2 for N_2O. If we analyse the time evolution in the period 2008–2023, the values of the Gini Index are constant, although in the period 2018–2022 there is a slight increase in the values recorded in the Gini Index.

It is observed that between the years 2008 and 2023, there has been a consistent rise in revenue derived from environmental taxes (for the European Union as a whole). Concurrently, there has been a decline in greenhouse gas emissions. Notably, there has been a substantial decrease in CO_2 emissions, accompanied by somewhat lower levels of CH_4 and N_2O emissions. This suggests that environmental taxes may function as a regulatory mechanism for polluting gases. However, a more comprehensive analysis, considering the distinct circumstances of the various EU member states, reveals that the impact is not uniform across all of them. While most countries have achieved a reduction in all the analysed gases, some countries have demonstrated a decrease in only one or two gases.

The analysis conducted can facilitate the establishment of more efficient tax models for reducing emissions. In the case of CO_2, the distributions of emission reductions and tax revenue are very close (Gini Index close to 0). An analysis of the data on a country-by-country basis reveals that in certain states, the implemented measures have not been effective, resulting in an increase in CO_2 emissions over the period from 2008 to 2023. This phenomenon is evident in the cases of Lithuania, Malta and Ireland. In such cases, it would be necessary to analyse the existence of specific factors affecting the country and attempt to correct them through structural measures. For example, this could involve reducing dependence on coal-fired power plants in favour of more efficient ones or encouraging the construction of electric transport systems.

Regarding the emissions of methane (CH_4) and nitrous oxide (N_2O), the focus should be on the agricultural and livestock sectors, with the aim of establishing measures to promote policies aimed at reducing emissions. Such measures might include the creation of subsidies and tax incentives to encourage environmental efficiency actions.

Finally, enhanced coordination with non-EU countries is imperative, as well as the implementation of environmental tariffs on products manufactured in countries where there is an absence of emission control measures and environmental taxes are not in effect.

Acknowledgments This work is funded by national funds through FCT – Fundação para a Ciência e a Tecnologia, I.P., under CEAUL Research Unit, UIDB/0006/2020 and UIDB/00006/2025 (https://doi.org/10.54499/UID/00006/2025), and by the Laboratorio de Educação a Distância e eLearning (LE@D) through the project UID/PRR/04372/2025 and UID/PRR2/04372/20205, DOI: https://doi.org/10.54499/UID/PRR/00006/2025.

Disclosure of Interests The authors have no competing interests.

References

1. Pretolani, D.: Apportionments with minimum Gini index of disproportionality: a quadratic knapsack approach. Ann. Oper. Res. **215**, 257–267 (2014). https://doi.org/10.1007/s10479-013-1383-7
2. Mouzon, O., Thibault, L., Le Breton, M.: One Man, One Vote Part 2: Measurement of Malapportionment and Disproportionality and the Lorenz Curve. Université de Toulouse Capitole, Toulouse (2020)
3. Wada, J., Yamahara, Y.: Studying malapportionment using α-divergence. Math. Soc. Sci. **93**, 77–89 (2018) https://doi.org/10.1106/j.mathsocsci.2018.02.003
4. Kakwani, N.C., Podder, N.: On the estimation of Lorenz curves form grouped observations. Int. J. Econ. Rev. **14**(2), 278–292 (1973). https://doi.org/10.2307/2525921
5. Rogerson, P.A.: The Gini coefficient of inequality: a new interpretation. Lett. Spat. Resour. Sci. **6**(3), 109–120 (2013). https://doi.org/10.1007/s12076-013-0091-x
6. Ivanova, D., Wood, R.: The unequal distribution of household carbon footprints in Europe and its link to sustainability. Global Sustainability. **3**(18), 1–12 (2020)
7. De Sadeleer, N.: Environmental Principles: from Political Slogans to Legal Rules. Oxford University Press, Oxford (2020). https://doi.org/10.1017/sus.2020.12
8. Barthakur, A.: Polluter pays principle as the key element to environmental law. Int. J. Sci. Res. Publ. **11**(3), 274–277 (2021). https://doi.org/10.29322/IJSRP.11.03.2021.p11137
9. Israngkura, A.: The polluter-pays principle: so who pays? TDRI Q. Rev. **11**(3), 12–15 (1996)
10. Luppi, B., Parisi, F., Rajagopalan, S.: The rise and fall of the polluter-pays principle in developing countries. Int. Rev. Law Econ. **32**(1), 135–144 (2012). https://doi.org/10.1016/j.irle.2011.10.002
11. Bleeker, A.: Does the polluter pay? The polluter-pays principle in the case law of the European court of justice. Eur. Energy Environ. Law Rev. **18**(6), 289–306 (2009). https://doi.org/10.54648/eelr2009024
12. Baykal, F.: The polluter should... pay? Herit. Sustain. Dev. **3**(1), 71–77 (2021). https://doi.org/10.37868/hsd.v3i1.58
13. Koziuk, V., Dluhopolskyi, O., Voznyi, K.: Environmental taxes and emissions of pollutants in the EU: empirical analysis. MEST J. 9, 68–74 (2021) 10.12709/mest.09.09.01.09
14. Aydın, C., Esen, Ö.: Reducing CO_2 emissions in the EU member states: do environmental taxes work? J. Environ. Plan. Manag. **61**, 2396–2420 (2018). https://doi.org/10.1080/09640568.2017.1395731
15. Firtescu, B., Brînză, F., Grosu, M., Doaca, E., Siriteanu, A.: The effects of energy taxes level on greenhouse gas emissions in the environmental policy measures framework. Front. Environ. Sci. **10**, 965841. (2023) (2023). https://doi.org/10.3389/fenvs.2022.965841
16. Charlier, D., Fodha, M., Kirat, D.: Residential CO_2 emissions in Europe and carbon taxation: a country-level assessment. Energy J. **44**(5), 187–206 (2023). https://doi.org/10.5547/01956574.44.4.dcha
17. Hájek, M., Zimmermannová, J., Helman, K., Rozenský, L.: Analysis of carbon tax efficiency in energy industries of selected EU countries. Energy Policy. **134**, 110955 (2019). https://doi.org/10.1016/j.enpol.2019.110955

18. Koval, V., Laktionova, O., Udovychenko, I., Olczak, P., Palii, S., Prystupa, L.: Environmental taxation assessment on clean technologies reducing carbon emissions cost-effectively. Sustainability. **14**(21), 14044 (2022). https://doi.org/10.3390/su142114044
19. Miceikienė, A., Gesevičienė, K., Rimkuviene, D.: Assessment of the dependence of GHG emissions on the support and taxes in the EU countries. Sustainability. **13**(14), 7650 (2021). https://doi.org/10.3390/su13147650
20. Pirlot, A.: Exploring the impact of EU law on energy and environmental taxation. Social science research network. In: Panayi, C., Haslehner, W., Traversa, E. (eds.) Research Handbook on European Union Taxation Law, pp. 359–388. Edwar Elgar Publishing, Routledge (2020). https://doi.org/10.4337/9781788110846.00025
21. Fairbrother, M.: When will people pay to pollute? Environmental taxes, political trust and experimental evidence from Britain. Br. J. Polit. Sci. **49**(2), 661–682 (2017). https://doi.org/10.1017/S0007123416000727
22. Gini, C.: Variabilità e Mutuabilità. In: Contributo allo Studio delle Distribuzioni e delle Relazioni Statistiche. C. Cuppini, Bologna (1912)
23. Mahinpei, R.: A comparative analysis of the Gini index. STEM Fellowsh. J. **6**(1), 34–52 (2021). https://doi.org/10.17975/sfj-2020-010
24. Sitthiyot, T., Holasut, K.: A simple method for measuring inequality. Palgrave Communications 6(1), 1–9 (2020). https://doi.org/10.1057/s41599-020-0484-6
25. Dalton, H.: The measurement of the inequality of incomes. Econ. J. **30**(119), 348–361 (1920). https://doi.org/10.2307/2223525
26. Kendall, M.G., Stuart, A.: The Advance Theory of Statistics 1. Griffin, London (1958)
27. Lows, R.L.: Measurement of inequality: the Gini coefficient and school finance studies. J. Educ. Financ. **10**(1), 83–94 (1984)
28. Dixon, P.M., Weiner, J., Mitchell-Olds, T., Woodley, R.: Bootstrapping the Gini coefficient of inequality. Ecology. **68**(5), 1548–1551 (1987). https://doi.org/10.2307/1939238
29. Lorenz, M.O.: Methods of measuring the concentration of wealth. Publ. Am. Stat. Assoc. **9**(70), 209–219 (1905). https://doi.org/10.1080/15225437.1905.10503443
30. Chotikapanich, D.: A comparison of alternative functional forms for the Lorenz curve. Econ. Lett. **41**(2), 129–138 (1993). https://doi.org/10.1016/0165-1865(93)90186-G
31. Ortega, P., Martín, G., Fernández, A., Ladoux, M., García, A.: A new functional form for estimating Lorenz curves. The review of Income and Wealth 37(4), 447–452 (1991). https://doi.org/10.1111/j.1475-4991.1991.tb00383.x
32. Sarabia, J.M., Castillo, E., Slottje, D.J.: An ordered family of Lorenz curves. J. Econ. **91**, 43–60 (1999). https://doi.org/10.1016/S0304-4076(98)00048-7
33. Castillo, E., Hadi, A.S., Sarabia, J.M.: A method for estimating Lorenz curves. Commun. Statistics—Theory Methods. **27**(8), 2037–2063 (1998). https://doi.org/10.1080/03610929808832208
34. Seijas-Macias, A., Oliveira, A., Oliveira, T.A.: A new R-function to estimate the PDF of the product of two uncorrelated normal variables. Mathematics. **11**(16), 3515 (2023). https://doi.org/10.3390/math11163515
35. Seijas-Macias, A., Oliveira, A., Oliveira, T.A.: The skewness and kurtosis of the product of two normally distributed random variables. Commun. Statistics—Theory Methods. **52**(1), 80–93 (2021). https://doi.org/10.1080/03610926.2021.1909734
36. Chotikapanich, D., Griffiths, W.E.: Estimating Lorenz curves using a Dirichlet distribution. J. Bus. Econ. Stat. **20**(2), 290–295 (2002). https://doi.org/10.1198/073500102318352029
37. Chotikapanich, D., Griffiths, W.E.: Averaging Lorenz curves. J. Econ. Inequal. **3**, 1–19 (2005). https://doi.org/10.1007/s10888-004-5866-2
38. Minka, T. P.: Estimating a Dirichlet Distribution. https://tminka.github.io/papers/dirichlet/minka-dirichlet.pdf. Accessed 08 Jan 2025
39. Kobayashi, G., Kazuhiko, K.: Approximate Bayesian computation for Lorenz curves from grouped data. Comput. Stat. **34**, 253–279 (2019). https://doi.org/10.1007/s00180-018-0831-x
40. Shu, H., Xiong, P.: The Gini coefficient structure and its application for the evaluation of regional balance development in China. J. Clean. Prod. **199**, 668–686 (2018). https://doi.org/10.1016/j.jclepro.2018.07.224

41. Bono, F., Giacomarra, M., Giaimo, R.: Measuring sustainable economic development through a multidimensional Gini index. In: Proceedings of the 48th Scientific Meeting of the Italian Statistical Society, Salerno (2016)
42. Costa Soares, T., Aparecida Fernandes, E., Harumi Toyoshima, S.: The CO_2 emission Gini index and the environmental efficiency: an analysis for 60 leading world economies. Economia, **19**(2), 266–277 (2018) https://doi.org/10.1016/j.econ.2017.06.001
43. Martin, A.J.F., Conway, T.M.: Using the Gini index to quantify urban green inequality: a systematic review and recommended reporting standards. Landsc. Urban Plan. **254** (2025). https://doi.org/10.1016/j.landurbplan.2024.105231
44. Wolde-Rufael, Y., Mulat-Weldemeskel, E.: Effectiveness of environmental taxes and environmental stringent policies on CO_2 emissions: the European experience. Environ. Dev. Sustain. **25**, 5211–5239 (2023). https://doi.org/10.1007/s10668-022-02262-1
45. Xu, Y., Wen, S., Tao, C.: Impact of environmental tax on pollution control: a sustainable development perspective. Econ. Anal. Policy. **79**, 89–106 (2023). https://doi.org/10.1016/j.eap.2023.06.0006
46. Dogan, B., Chu, L.K., Ghosh, S., Truong, H.H.D., Balsalobre-Lorente, D.: How environmental taxes and carbon emissions are related in G7 economies? Renew. Energy. **187**, 646–656 (2022). https://doi.org/10.1016/j.renene.2022.01.077

Dynamic Environmental Subsidy Policies

George Emm. Halkos, George J. Papageorgiou, Emmanuel G. Halkos, and John G. Papageorgiou

Abstract We study a market for an extracted environmental resource, where a representative firm acts as the sole exploiter. The resource may be renewable (e.g., fish biomass) or nonrenewable (e.g., fossil fuels). We begin with a static benchmark in which a per-unit subsidy induces the monopolist to produce the socially optimal output. Building on this, we develop dynamic subsidy rules based on a performance index that summarizes the firm's extraction history. These rules require only local information and are directly comparable with the static scheme. Our main result is that the present value of subsidies under the dynamic scheme is strictly lower than under the static benchmark, making it a more efficient policy instrument. Moreover, the dynamic policy has a feedback property: the subsidy depends on the state variable representing the firm's past performance, thereby aligning incentives over time.

Keywords Optimal control · Environmental subsidies · Monopolistic firms

1 Introduction

Since Hotelling's seminal work [1], it has been well established that the monopolistic extraction path of an exhaustible resource is socially suboptimal, falling below the socially optimal level. The regulator's ability to induce socially desirable outcomes crucially depends on information about the cost structure, which determines the subsidy required for the monopolist to align extraction with the social optimum.

In this paper, starting from the fundamental problem of monopolistic extraction, we address the issue of environmental optimality by examining how a social planner, through an appropriate subsidy policy, can maximize ecological quality in a dynamic framework. Specifically, we derive a method for solving one of the fundamental optimal control problems in resource economics: how to design subsidy policies for the exploitation of natural resources so as to achieve social optimality [2].

G. Emm. Halkos (✉) · G. J. Papageorgiou · E. G. Halkos · J. G. Papageorgiou
Laboratory of Operations Research, Department of Economics, University of Thessaly, Volos, Greece
e-mail: halkos@uth.gr

G. Halkos et al. (eds.), *Environmental Risk Analysis*, Springer Proceedings in Mathematics & Statistics 540, https://doi.org/10.1007/978-3-032-24155-9_4

Following the capital-theoretic approach, we treat environmental resources as a form of natural capital and the social planner as its regulator. The planner, through the time path of subsidies, is responsible for maintaining the integrity of this capital. While the planner cannot directly determine the resource stock at each instant, he can influence its trajectory by choosing the subsidy path, thereby maximizing intertemporal social utility derived from environmental quality. Our analysis shows that in the optimal solution, any marginal gain in social utility from current subsidy policy must be exactly balanced by the marginal loss in future utility due to changes in the natural resource stock. Using the shadow price of the natural capital, we further establish that the depreciation of one unit of resource capital over a short interval equals the sum of its marginal contribution to social benefits during that interval and its marginal contribution to the future value of the capital stock.

We then extend the analysis by treating environmental subsidies as addictive goods within the framework of rational addiction theory [3]. Rational addiction models emphasize forward-looking behavior, where current utility depends not only on present consumption but also on accumulated past consumption. A good is considered addictive if higher past consumption raises the marginal utility of present consumption. Crucially, addiction may be either harmful or beneficial [4]: if greater current consumption reduces future utility, the addiction is harmful, whereas if it enhances future outcomes, it is beneficial.

In our model, environmental subsidies function as a *performance booster* for the representative resource-extracting firm. The booster index captures the accumulation of past subsidies, which influences current utility through a learning-by-doing process. In addition, environmental quality evolves not only with its intrinsic growth function but also with the performance booster of the firm. The planner's objective is to maximize intertemporal environmental quality, taking into account both the stock dynamics of natural resources and the addictive nature of subsidies.

Solving the model, we identify the conditions under which cyclical subsidy policies are feasible and derive several propositions regarding their implications. The rest of the paper is organized as follows: Sect. 2 considers the static monopolistic case; Sect. 3 examines the dynamic subsidy models; Sect. 4 introduces the subsidy addiction framework; Sect. 5 analyzes the dynamic model under concavity assumptions; and Sect. 6 concludes.

2 The Monopoly Static Case

We begin with the well-known static monopoly model. Consider a monopolistic firm extracting quantity q of an environmental resource. The inverse demand function is denoted by $p = p(q)$ while the cost function is given by $c(q)$. We impose the following standard assumptions:

$$c(0) = 0,\;\; c'(q) > 0,\;\; c'(0) < p(0),\;\; \text{and}\; p'(q) < 0 \tag{1}$$

Under these assumptions, the firm's revenue is $qp(q)$ and the marginal revenue is:

$$MR(q) = p(q) + qp'(q)$$

which is decreasing in q. The marginal cost $c'(q)$ may be increasing or decreasing. The only restriction required is that if $c'(q)$ is decreasing, its slope at the intersection with the marginal revenue curve must be less negative than the slope of the marginal revenue curve.

The monopolist's profit function is:

$$R(q) = p(q)q - c(q) \tag{2}$$

We abstract from taxes or subsidies in this static setup and assume that $R(q)$ is strictly concave, with a unique maximum at some $q_L > 0$. Clearly, profits are zero when $q = 0$.

Following Hotelling [1], social welfare is defined as:

$$W(q) = \int_0^q p(\tilde{q})d\tilde{q} - c(q)$$

The socially optimal extraction level, denoted q_{so} satisfies the first-order condition:

$$p(q_{so}) = c'(q_{so}) \tag{3}$$

We assume that Eq. (3) admits a unique solution.

The regulator's role is to induce the monopolist to extract q_{so}.. We assume that the regulator knows both the socially optimal quantity q_{so} and the slope of the demand curve at that point, but has no information about demand or cost conditions away from q_{so}. Under these assumptions, the following result holds.

Proposition 1 *Let q_{so} be the socially optimal extraction level defined by (3) Then a per-unit subsidy of*

$$s^* = -p'(q_{so})q_{so} > 0 \tag{4}$$

is sufficient to induce the monopolist to extract q_{so}.

Proof The monopolist's profit with a per-unit subsidy s is

$$\pi(q, s) = p(q)q - c(q) + sq = R(q) + sq$$

The first-order condition for profit maximization is

$$p(q) + qp'(q) - c'(q) + s = 0$$

At the social optimum q_{so} condition (3) implies $p(q_{so}) = c'(q_{so})$. Substituting into the above first-order condition maximization yields

$$q_{so}p'(q_{so}) + s^* = 0$$

Since $p'(q_{so}) < 0$ we have $s^* > 0$. Thus, setting the subsidy equal to s^* ensures that the monopolist's private optimality condition coincides with the social planner's condition, and q_{so} is implemented.

Remark 1 The per-unit subsidy rate s^* is defined as the difference between the market price at the social optimum $p(q_{so})$ and the marginal revenue at that quantity:

$$s^* = p(q_{so}) - \big(p(q_{so}) + p'(q_{so})q_{so}\big).$$

Thus, if the monopolist is induced to extract exactly at the socially optimal level q_{so}, her net profits are:

$$\pi(q_{so}) = R(q_{so}) + s^* q_{so}$$

where $R(q) = p(q)q$ is the revenue function.

At the social optimum, the first-order condition holds:

$$\frac{\partial \pi(q_{so})}{\partial q_{so}} = R'(q_{so}) + s^* = 0.$$

Finally, the assumption

$$R(q_{so}) + s^* q_{so} > 0$$

is necessary. It guarantees that at the socially optimal extraction level the monopolist's total net profit is non-negative. This ensures that the firm strictly prefers extracting q_{so} to extracting nothing, given that the subsidy s^* is in place.

We record the following conclusion.

Corollary 1 *There exists at least one subsidy rate* $\overline{s} < s^*$ *which ensures that the monopolistic firm extracts the quantity* $\overline{q} \in (0, q_{so})$ *under the assumption* $R(\overline{q}) + s^*\overline{q} > 0$.

Proof The result is easily obtained following the assumption $R(q_{so}) + s^* q_{so} > 0$ and the continuity and the strict concavity of the profit function $R(q)$.

Summarizing the above results obtained in a static environment, since the per extracted unit subsidy is $s^* = -p'(q_{so})q_{so} > 0$, the social optimum is achieved by the regulator by offering the scheme given by the linear subsidy function

$$S(q) = s^* q \tag{5}$$

From the firm's side, taking the subsidy as given by (5) the firm extracts the quantity q_{so}. As it becomes clear the minimum information required for the above linear subsidy is the marginal cost function and the demand function.

3 The Dynamic Case

3.1 *Achieving Environmental Optimality*

This section clarifies core concepts of environmental optimality and examines how a social planner, implementing a subsidy policy, can maximize ecological quality in a dynamic setting. To illustrate the method, we derive the standard optimal-control formulation for the problem of subsidizing the exploitation of natural resources. Under this framework the planner selects a time path of subsidies that trades off immediate gains from exploitation with the regeneration and depreciation of the environmental stock, thereby determining the socially optimal evolution of environmental quality over time.

At any instant t, the environment inherits a stock of natural resources from the past, which reflects its quality and the amenities it provides to citizens. We denote this stock by $E(t)$. Given this resource stock, the social planner is able to make policy decisions that may concern, for instance, the rate of resource extraction or the level of subsidies offered to exploiters. In what follows, we focus on the case where the planner's decision concerns the subsidy rate, denoted by $\sigma(t)$, with the objective of achieving ecological optimality. However, the framework can be readily applied to alternative cases, such as the direct control of extraction quantities. At each point in time, the endowed stock of resources $E(t)$, together with the subsidy decision $\sigma(t)$, generates a flow of social benefits enjoyed by citizens. We denote this benefit flow by $U(E(t), \sigma(t), t)$, where the utility function U captures the instantaneous rate at which welfare accrues as a result of both environmental quality and subsidy policies.

Now we focus on the initial time $t = 0$. The total benefits accrued over the planning horizon T are given by following integral (from now on we omit the time as an argument notation, unless it is necessary)

$$W(E_0, \sigma) = \int_0^T U(E, \sigma, t)dt$$

which aggregates the flow of instantaneous utilities over time. Here E_0 denotes the initial stock of environmental resources, while σ indicates the entire time path of subsidy decisions $\sigma(t)$ from the initial time up to the terminal time T. This formulation highlights that social welfare depends not only on the initial environmental conditions but also on the sequence of policy interventions undertaken throughout the planning period.

Remark 2 If the social planner begins the subsidy policy with an initial stock of environmental resources (viewed as natural capital) E_0 and subsequently follows the policy path σ then society obtains a total benefit W, given by the integral of instantaneous utilities over the planning horizon. These benefits accrue dynamically, depending on the time instant considered, the prevailing stock of environmental resources, and the subsidy decision undertaken at that instant.

The social planner is free, within certain limits, to choose the time path of the subsidy vector $\vec{\sigma}$; however, the planner cannot directly control the stock of natural resources at each instant. Instead, the evolution of the environmental capital is determined endogenously: the rate of change of the resource stock at any time depends on its current level, the date, and the subsidy decisions undertaken. This dynamic relationship can be expressed as

$$\dot{E} = \frac{dE}{dt} = f(E, \sigma, t) \tag{6}$$

where $\dot{E}$ denotes the instantaneous change in the resource stock.

The subsidy decisions made at any given instant have two effects: they influence the rate at which utility accrues at that instant, and they also affect the rate of change of the natural resource capital stock, determining the stock available to society in the future.

In this context, the problem reduces to selecting a time path of subsidies, denoted as $\vec{\sigma}$ such that the total value of social benefits W is maximized. This must account for the effect of subsidy choices on both the instantaneous utility and the evolution of the natural capital stock over time. A challenge arises because the decision concerns an entire time path of a variable, which is more complex than typical problems in basic calculus. Therefore, a different solution approach is needed. The strategy involves reducing the original problem to one where only a few numbers need to be determined—a scenario where techniques from basic calculus can be applied. In this study, we follow the approach that leads to the maximum principle, introduced by Pontryagin and his collaborators [5]. First, we define a function representing the value of benefits obtainable by society, starting at an arbitrary time with a given natural resource stock E and following an arbitrary subsidy policy σ until the terminal time T.

This is the following formula

$$W(E_t, \sigma, t) = \int_t^T U(E, \sigma, t)dt \tag{7}$$

which is a more general issue of the formula W presented earlier.

We proceed by breaking the expression W into two parts. We consider a short time interval of length Δ at time t. The meaning of the short Δ is that in this interval the social planner would not change the subsidy policy σ in his planning even it was possible to change. After that Δ introduction we can write the hole (bigger) problem as

$$W(E, \vec{\sigma}, t) = U(E, \sigma_t, t)\Delta + \int_{t+\Delta}^{T} U(E(t), \sigma, \tau)d\tau \qquad (8)$$

Expression (8) states that if the amount of natural resource capital available at time t is E and if the subsidy policy σ is followed thereafter, then the total value W can be viewed as the sum of two contributions.

The first term represents the contribution to social benefits over the short interval Δ beginning at time t during which the subsidy policy is held constant. It is calculated as the rate at which benefits accrue during the interval $U(E, \sigma_t, t)$ multiplied by the length of the interval Δ. This term depends on the current natural resource stock E, the time t and the current value of the subsidy policy σ_t.

The second term represents the contribution to social benefits over the remaining period, that is, the integral from $t + \Delta$ to the terminal time T. It is important to note that the starting natural resource stock for this integral is E$(t + \Delta)$ not E(t). This reflects the fact that the natural resource stock changes during the interval Δ under the influence of the subsidy policy σ_t.

To illustrate the solution technique, it is useful to rewrite the integral in (8) as the following equivalent sum:

$$W(E_t, \vec{\sigma}, t) = U(E, \sigma_t, t)\Delta + W(E_{t+\Delta}, \vec{\sigma}, t + \Delta)$$

We now introduce a new variable in the solution process: the value of social benefits assuming that the social planner knew the optimal subsidy policy $\vec{\sigma}$ from time t onward and adhered to it until the terminal time T. We denote this value by $V^*(E_t, t)$. The policy variable does not appear in the arguments because σ is already assumed to be chosen optimally, so V^* represents the maximum achievable value:

$$V^*(E_t, t) = \max W(E_t, \vec{\sigma}, t)$$

Following the previous technique of dividing the time interval, we consider the policy σ_t applied over the short interval from t to $t + \Delta$, and then the optimal policy thereafter. Using this approach, we can write:

$$V(E_t, \sigma_t, t) = U(E_t, \sigma_t, t)\Delta + V^*(E_{t+\Delta}, t + \Delta)$$

Remark 3 The formula above indicates that the value of following a given subsidy policy σ_t consists of two components: the benefits accrued during the initial period under σ_t and the maximum possible benefits that can be obtained from time $t + \Delta$ onward, starting with the natural resource stock $E(t + \Delta)$ that results from the policy applied during the initial period.

Clearly, if the social planner chooses the optimal subsidy policy at time t then the value $V(E_t, \sigma_t, t)$ coincides with the maximum value $V^*(E_t, t)$. We continue with the results.

3.2 Results

Proposition 2 *Along the optimal path of the subsidies policy $\sigma(t)$, any marginal increase of utility resulting from a change in the subsidy policy, $\frac{\partial U}{\partial \sigma_t}$, must be exactly offset by the marginal decrease in future utility that this change induces through its effect on the natural resource capital stock. Formally, this condition can be expressed as:*

$$\frac{\partial U}{\partial \sigma_t} - \frac{\partial V^*(E,t)}{\partial E}\frac{\partial f(E,\sigma_t,t)}{\partial \sigma_t} = 0$$

where $\dot{E} = f(E, \sigma, t)$ describes the evolution of the natural resource stock.

This condition captures the trade-off between the immediate benefit of adjusting the subsidy and the long-term effect on the resource stock, which in turn affects future utility.

Proof We proceed with the maximization taking the first derivative of the value function $V(E_t, \sigma_t, t)$ and equate this derivative to zero, under the assumption that the function is rather maximized. That is

$$\frac{\partial}{\partial \sigma_t}V(E_t,\sigma_t,t) = 0 \Rightarrow \Delta\frac{\partial}{\partial \sigma_t}U(E_t,\sigma_t,t) + \frac{\partial}{\partial \sigma_t}V^*(E_{t+\Delta}, t+\Delta) = 0 \tag{9}$$

Observing Eq. (9) we realize that we have a problem to solve, since the function V^* is still unknown, i.e., we are told to differentiate V^* with respect to σ_t, whereas it does not involve σ_t explicitly. We overcome this difficulty by writing the partial derivative of that function, i.e., the second term of (9) is written as:

$$\frac{\partial V^*}{\partial \sigma_t} = \frac{\partial V^*}{\partial E(t+\Delta)}\frac{\partial E(t+\Delta)}{\partial \sigma_t} \tag{10}$$

We try to analyze the latter right hand side expression starting from the last term. Since Δ is an infinitesimal time period the following approximation is useful, i.e., $E(t+\Delta) = E(t) + \dot{E}\Delta$, which in turn says that the amount of natural resources capital at the time instant $t+\Delta$ is equal to the amount of natural resources capital at time t plus the rate of change of that capital during the interval multiplied by the infinitesimal length of the time interval. Taking formula (6), at which $\frac{dE}{dt} = \dot{E}$ depends on σ_t, i.e., $\dot{E} = f(E, \sigma_t, t)$, then the following equation is valid

$$\frac{\partial E(t+\Delta)}{\partial \sigma_t} = \Delta \frac{\partial f}{\partial \sigma_t}$$

We inspect now the first term of the partial derivative (10), i.e. the term $\frac{\partial V^*}{\partial E}$. This derivative is the rate at which the maximum possible social benefits flow from time $t + \Delta$ on changes with respect to the amount of the capital of natural resources available at $t+\Delta$. Therefore, it is the marginal value of the capital of natural resources at time $t + \Delta$, or the amount by which a unit increment in the capital of natural resources occurring at that time would increase the maximum possible value of the social benefits W. Putting it differently, this value is the highest hypothetical unit price that the social planner would be willing to pay for an infinitesimal addition to the capital of the environmental resources. This marginal value of natural resources capital at time t is denoted by $\lambda(t)$ and is defined as

$$\lambda(t) = \frac{\partial}{\partial E} V^*(E, t)$$

We now insert the above formulas into the maximization condition (9) and

$$\Delta \frac{\partial U}{\partial \sigma_t} + \lambda(t+\Delta)\Delta \frac{\partial f}{\partial \sigma_t} = 0 \tag{11}$$

and furthermore, the constant Δ can be canceled out. Moreover, the marginal value of the capital of natural resources changes gradually over time and so, to a sufficiently good approximation

$$\lambda(t+\Delta) = \lambda(t) + \dot{\lambda}(t)\Delta$$

The interpretation of the above expression is that the marginal value of natural resources capital at time $t + \Delta$ is the marginal value at time t plus the rate at which it is changing during the interval multiplied by the length of the interval. Inserting the latter expression into Eq. (11) and cancelling the common factor Δ we have the following maximization condition

$$\frac{\partial U}{\partial \sigma_t} + \lambda(t)\frac{\partial f}{\partial \sigma_t} + \dot{\lambda}(t)\Delta \frac{\partial f}{\partial \sigma_t} = 0$$

We allow Δ to approach zero, then the term $\dot{\lambda}(t)\Delta \frac{\partial f}{\partial \sigma_t}$ approaches zero as well, therefore it must be neglected. The maximization condition now becomes at its final form as

$$\frac{\partial U}{\partial \sigma_t} + \lambda(t)\frac{\partial f}{\partial \sigma_t} = 0 \tag{12}$$

The proof of proposition 2 is completed, since now it is obvious, because the second term in (11) is the marginal effect of the current subsidy policy $\sigma(t)$ on the rate of growth of natural resources capital with the capital valued at its marginal worth, λ.

Remark 4 At each point in time, the social planner should choose the subsidy policy σ so that the marginal immediate benefit of the policy is exactly equal to the marginal long-term cost. The long-term cost is measured by the product of the marginal value of the natural resource capital and the effect of the subsidy on the accumulation of that capital.

The next result is recorded as:

Proposition 3 *Under the assumption of the optimal time path of natural resource capital accumulation, the depreciation of a unit of natural resource capital in a short time interval equals the sum of two factors:*

(a) *Its marginal contribution to the social benefits accrued during the interval.*
(b) *Its marginal contribution to the increase in the value of the natural resource stock at the end of the interval.*

Formally, the relationship of the depreciation with the sum of contributions can be expressed as:

$$-\dot{\lambda} = \frac{\partial U}{\partial E} + \lambda \frac{\partial f}{\partial E}$$

Proof We consider the subsidy policy σ_t, that is supposed is determined such that to satisfy the optimality conditions. Under the latter assumption, i.e., the subsidy policy σ_t is the optimal, then the value of the social benefits $V(E_t, \sigma_t, t)$ would be equal to its maximal possible value i.e., would be $V^*(E, t)$. Therefore,

$$V^*(E, t) = U(E, \sigma_t, t)\Delta + V^*(E(t + \Delta), t + \Delta)$$

The differentiation of the value $V^*(E, t)$ with respect to E, $\frac{\partial V^*}{\partial E}$, is by definition the marginal value of natural resources capital $\lambda(t)$. Therefore

$$\lambda(t) = \frac{\partial V^*}{\partial \mathrm{E}} = \Delta \frac{\partial U}{\partial E} + \frac{\partial V^*(E(t + \Delta), t + \Delta)}{\partial \mathrm{E}}$$

Using the chain rule for the second term:

$$\frac{\partial V^*(E(t + \Delta), t + \Delta)}{\partial \mathrm{E}} = \frac{\partial E(t + \Delta)}{\partial \mathrm{E}} \lambda(t + \Delta)$$

For a small interval Δ we approximate $E(t + \Delta) = E(t) + \dot{E}\Delta$ with $\dot{E} = f(E, \sigma_t, t)$ then

$$\frac{\partial E(t + \Delta)}{\partial \mathrm{E}} = 1 + \Delta \frac{\partial f}{\partial E}$$

Substituting and expanding $\lambda(t+\Delta) = \lambda(t) + \Delta\dot{\lambda}(t)$ we get

$$\lambda(t) = \Delta\frac{\partial U}{\partial E} + \left(1 + \Delta\frac{\partial f}{\partial E}\right)(\lambda + \dot{\lambda}\Delta) = \Delta\frac{\partial U}{\partial E} + \lambda + \Delta\lambda\frac{\partial f}{\partial E} + \Delta\dot{\lambda} + \dot{\lambda}\frac{\partial f}{\partial E}\Delta^2$$

The term Δ^2 is a very small quantity therefore can be neglected, and after some obvious simplifications we obtain

$$-\dot{\lambda} = \frac{\partial U}{\partial E} + \lambda\frac{\partial f}{\partial E} \tag{13}$$

This is the second result of our dynamic context, and in the model is the equation of motion of the shadow price of the natural resources capital.

Remark 5 From the mathematical view, $\dot{\lambda}(t)$ is the rate at which the value of one unit of the natural resources capital is changing. From the economic view, is the rate at which the natural capital appreciates, and $-\dot{\lambda}(t)$ is the rate at which the same unit of natural capital depreciates at time t.

The economic interpretation of Proposition 3 is that the depreciation (or loss in value) of natural resource capital at any moment equals the sum of:

(a) Its marginal contribution to current social benefits, and
(b) Its marginal contribution to the future value of the resource stock, weighted by the shadow price $\lambda(t)$

The following economic conclusion is apparent.

Corollary 2 *Each unit of natural resource capital gradually decreases in value at the same rate at which it is either extracted and used in production or accumulated for future use. Another interpretation of the natural resource capital depreciation* $-\dot{\lambda}(t)$ *is that it represents the loss in potential value if the acquisition of one unit of natural resource capital were postponed for a short interval.*

4 The Subsidy Addiction of the Firm and Subsidizing Policy

Becker and Murphy [3] argue that addiction extends beyond substances such as alcohol, narcotics, and tobacco to encompass a wide range of human activities, including work, consumption, and leisure. Within their framework of rational addiction, addictive behavior is not an anomaly but can be explained through standard rational choice theory. A key element of this framework is that current consumption is strongly influenced by past consumption, which is accumulated as a stock and subsequently shapes present preferences and utility.

Extending this reasoning to the firm, one may argue that enterprises are not only dependent on production, sales, or financial inputs from banks, but may also develop an addiction to government subsidies. From this perspective, subsidies can be conceptualized as a consumption good. This interpretation provides a theoretical foundation for explaining the persistence and cyclical nature of subsidy trajectories observed in practice.

In line with rational addiction theory [3, 4, 7–9], a firm that has consumed subsidies in the past is more likely to demand and rely on them in the present, since past subsidy inflows augment its current utility. More importantly, the cumulative effect of prior subsidies can be represented as a stock of subsidy consumption, which reinforces the firm's present and future dependence on continued support.

In the context of environmental policy, we consider subsidies directed at firms exploiting natural resources while ostensibly contributing to environmental protection. Here, a firm's past stock of subsidies not only conditions its current ability to secure new subsidies but also functions as a performance enhancer, lowering costs and supporting competitiveness. In this sense, subsidies exhibit the characteristics of an addictive good: once consumed, they create a self-reinforcing trajectory of dependence, with significant implications for subsidy policy design and long-term market efficiency.

In this section, we conceptualize environmental subsidies granted by the government as a form of value consumed by firms that exploit natural resources under the requirement of maintaining environmental quality. To capture this process, we introduce the firm's stock of past subsidy consumption, which we interpret as a performance booster influencing its current ability to benefit from and utilize subsidies. Consistent with the rational addiction framework, we assume that subsidies function as an addictive good, whereby past consumption accumulates into a stock that reinforces present performance.

Formally, we denote the performance booster of the representative firm at time t by $X(t)$. At the start of the period, the government defines the rules governing subsidy allocation [10], such that the initial stock of the performance booster is given by $X(0) = X_0$. In principle, a regulator could design subsidy rules to induce the firm to extract the socially optimal quantity of resources. However, this outcome requires the regulator to possess precise information about the demand schedule and thus the socially optimal extraction level q_{so}. In the absence of such information—a case we assume throughout this paper—the representative firm instead extracts some arbitrary quantity $\overline{q} \in (0, q_{so})$. Regardless of the quantity extracted, the firm accepts the subsidy rate offered by the government, as established in Corollary 1.

This subsidy inflow accumulates in the firm's performance index, reinforcing its dependence. As argued by Halkos et al. [11], subsidies consumed by economic agents may exhibit the characteristics of an addictive good, in line with the properties described by Becker and Murphy [3]. Let $\sigma(t)$ denote the subsidy received for the extracted quantity at time t. The evolution of the performance booster (state variable) is then expressed as:

$$\frac{dX}{dt} = \dot{X}(t) = \sigma(t) - \delta X(t) \tag{14}$$

$\sigma(t)$ is the current taken subsidy rate, $\delta > 0$ is the depreciation rate of the firm's accumulated performance booster.

Equation (14) indicates that the performance booster $X(t)$ increases with the flow of subsidies $\sigma(t)$ but depreciates over time at rate δ. Thus, while continuous subsidy inflows enhance the firm's performance capacity, the effect of past subsidies gradually diminishes unless reinforced by new allocations.

Moreover, the performance booster $X(t)$ can be interpreted as follows: past consumption of the environmental subsidy σ affects the firm's current utility through a learning-by-doing mechanism, and this effect is summarized in the booster index $X(t)$. Importantly, the current subsidy rate $\sigma(t)$ is not constant; it may vary both in magnitude and in sign. In particular, if $\sigma(t)$ becomes negative, the subsidy is transformed into a tax rate. At the beginning of the period under consideration ($t = 0$), the government announces the updating rule for the performance booster given in Eq. (14). After this announcement, the chosen parameters remain fixed for the duration of the period. Finally, note that the performance variable $X(t)$ can take either positive or negative values, depending on the current subsidy (or tax) rate. However, the initial value is constrained to be non-negative, i.e., $X_0 \geq 0$ for any finite t. This formulation thus provides a unified framework for analyzing both subsidy dependence and potential tax-induced disincentives, allowing us to capture the dual role of government policy as either a performance enhancer or a constraint on the firm's behavior.

The social planner, who is concerned with the state of the environment $E(t)$, incorporates the firm's performance booster $X(t)$ into the dynamics of environmental quality. In addition to the firm's impact, changes in environmental quality are also influenced by the environment's intrinsic growth function, denoted by $g(E(t))$. Accordingly, the equation of motion for environmental quality reflects two opposing forces: on the one hand, the natural growth of the environment, $g(E(t))$, which improves environmental quality; and on the other hand, the firm's performance effort, represented by $D(X(t))$, which degrades environmental quality through resource extraction. Formally, the dynamic evolution of environmental quality is expressed as:

$$\frac{dE}{dt} = \dot{E} = g(E(t)) - D(X(t)) \tag{15}$$

where $g(E(t))$ captures the regenerative capacity of the environment, and $D(X(t))$ represents the deterioration induced by the firm's activity.

Equation (15) illustrates that environmental quality evolves as a balance between natural growth and degradation caused by firm activity. While $g(E(t))$ enhances the environment over time, $D(X(t))$ captures the negative impact of the firm's performance booster, highlighting the trade-off between economic activity and environmental sustainability.

The damage (cost) function $D(X(t))$ (representing environmental degradation) is convex, meaning that:

$$D'(X) > 0, D''(X) \geq 0, D(0) = 0, D'(0) = 1$$

$D'(X) > 0$: Damage increases with X (more extraction/performance implies more environmental cost).

$D''(X) \geq 0$: The marginal damage is non-decreasing, i.e., each additional unit of performance causes at least as much damage as the previous.

$D(0) = 0$: no extraction/performance implies no damage.

$D'(0) = 1$: the initial rate of damage is normalized to 1.

With zero-depreciation, the evolution of the firm's extraction performance $X(t)$ is simplified to:

$$\frac{dX}{dt} = \dot{X}(t) = \sigma(t) \tag{16}$$

where $\sigma(t)$ is the control variable representing the firm's effort to increase performance.

The implications of the above Eqs. (15), (16) could be the following

1. Since there is no depreciation, the performance $X(t)$ is non-decreasing and purely driven by $\sigma(t)$.
2. If the firm wants to optimize something (e.g., profit minus environmental damage), the state equation is linear in $\sigma(t)$ but the damage function is convex. This creates the classic convex-concave trade-off in optimal control problems: linear control with convex cost.
3. The initial conditions are typically $X(0) = X_0$ and then $X(t) = X_0 + \int_0^t \sigma(s)ds$.

4.1 The Social Planner's Problem

The social planner, on the other hand, wants the best environmental approval $A(t)$. In other words, his objectives include all the appropriate actions that ensure better environmental quality, because good quality offers better amenities to the citizens and, therefore, better environmental approval. Environmental approval hinges on current environmental quality $E(t)$, as well as on the firm's performance index $X(t)$, and finally on the most recent subsidies' which offers, $\sigma(t)$. After all, the latter approval is rather a function of the mentioned variables, i.e., $A(E, X, \sigma)$.

Formal, the objective of the social planner seems like the following, discounting at the infinite time horizon integral

$$max_\sigma \int_0^\infty e^{-\rho t} \mathrm{A}(\mathrm{E}, \mathrm{X}, \sigma)dt \tag{17}$$

For the environmental approval the following usual assumptions are in effect, i.e.,

$$A_E > 0, A_{EE} < 0, A_X > 0, A_{XX} < 0, A_\sigma > 0, A_{\sigma\sigma} < 0$$

And these assumptions in turn means.

$A_E > 0, A_{EE} < 0$ more environmental quality implies more approval, but diminishing returns.

$A_X > 0, A_{XX} < 0$ higher firm performance means more approval, but diminishing returns.

$A_\sigma > 0, A_{\sigma\sigma} < 0$ more subsidies imply more approval, diminishing returns.

These above assumptions ensure strict concavity in each argument, which is important for uniqueness of the optimum.

After all, the social planner has to solve the following optimal control problem (17)

$$\int_0^\infty e^{-\rho t} A(E, X, \sigma) dt$$

Subject to the equations

$$\frac{dE}{dt} = \dot{E}(t) = g(E(t)) - D(X(t))$$

$$\frac{dX}{dt} = \dot{X}(t) = \sigma(t)$$

Here the control variable is the subsidy rate $\sigma(t)$ and is equivalent to the decision about what quantity of subsidy the regulator would offer taking into account the lack of information about the socially optimal quantities of environmental resources extracted, which lack of information is reflected in environmental quality.

The function $D(X(t))$ is the damage (cost) function associated with the performance of the extracting firm and it is supposed in convex form, as it is mentioned above, i.e., $D' > 0, D'' \geq 0, D(0) = 0, D'(0) = 1$. Moreover, it is worth noting that were the performance index of the extracting firm is small there is likely to be no deadweight costs (in Becker's sense, Becker, 1983), i.e., for $X \approx 0, D(X) \approx X$ and, due to the convex assumption, when $X > 0$ then $D(X) > X$. Note that the condition $D'' = 0$ is also compatible with cyclical subsidy strategies.

For the growth function $g(E(t))$ we accept that is in linear form $g(E) = rE, r \geq 0$, which is a good approximation of the population models, since the environmental amenities are in most cases renewable lively beings. Of course, the non-renewable resources case is not excluded for our model, since is the case at which, $r = 0$ Moreover, we assume that the growth rate is less than the discount rate, i.e., $r \leq \rho$ and the damage (cost) function is assumed in linear form, i.e., $D(X(t)) = DX(t)$.

4.2 The Solution Process

We proceed with the solution of the optimal control problem (17), forming the current value Hamiltonian function (omitting the arguments) as follows

$$H = A + \lambda(rE - D) + \mu\sigma \tag{18}$$

and sufficient conditions

$$\frac{\partial H}{\partial \sigma} = H_\sigma = \mathrm{A}_\sigma + \mu \tag{19}$$

$$\frac{d\lambda}{dt} = \dot{\lambda} = -\frac{\partial H}{\partial \mathrm{E}} + \rho\lambda = -\mathrm{A_E} + (\rho - r)\lambda \tag{20}$$

$$\frac{d\mu}{dt} = \dot{\mu} = -\frac{\partial H}{\partial \mathrm{X}} + \rho\mu = -\mathrm{A_X} + \rho\mu + \lambda D' \tag{21}$$

and transversality conditions

$$\lim_{t\to\infty} e^{-\rho t}\lambda(t)E(t) = 0 \tag{22}$$

$$\lim_{t\to\infty} e^{-\rho t}\mu(t)\mathrm{X}(t) = 0 \tag{23}$$

The stationary solution of the state variables E, X and the auxiliary variables λ, μ are as following below (the subscript ∞ indicates the stationary value).

$$E_\infty = D(X_\infty)/r \tag{24}$$

$$\sigma_\infty = 0 \tag{25}$$

$$\lambda_\infty = A_E/(\rho - r) \tag{26}$$

$$\mu_\infty = \left[(\rho - r)A_X - A_E D'\right]/[\rho - (\rho - r)] \tag{27}$$

$$\mathrm{A}_\sigma(\mathrm{E}_\infty, \mathrm{X}_\infty, 0) = -\mu_\infty \tag{28}$$

4.3 Stability Analysis

Now we proceed calculating the Jacobian matrix of the system of canonical Eqs. (24)–(28). As the next step we calculate the four eigenvalues of the above Jacobian matrix with the well-known Dockner's theorem [12]. The above two steps provide the necessary and sufficient conditions for stability and necessary conditions for the possibility of cyclical policies. The Jacobian matrix is, therefore,

$$J = \begin{bmatrix} r & -D' & 0\ 0 \\ -\frac{\mathrm{A}_{\sigma \mathrm{E}}}{\mathrm{A}_{\sigma\sigma}} & -\frac{\mathrm{A}_{\sigma \mathrm{X}}}{\mathrm{A}_{\sigma\sigma}} & 0\ -\frac{1}{\mathrm{A}_{\sigma\sigma}} \\ -\frac{\mathrm{A}_{\mathrm{EE}}\mathrm{A}_{\sigma\sigma}-\mathrm{A}_{\sigma \mathrm{E}}^2}{\mathrm{A}_{\sigma\sigma}} & -\frac{\mathrm{A}_{\mathrm{EX}}\mathrm{A}_{\sigma\sigma}-\mathrm{A}_{\sigma \mathrm{E}}\mathrm{A}_{\sigma \mathrm{X}}}{\mathrm{A}_{\sigma\sigma}} & \rho - r\ \frac{\mathrm{A}_{\sigma \mathrm{E}}}{\mathrm{A}_{\sigma\sigma}} \\ -\frac{\mathrm{A}_{\mathrm{EX}}\mathrm{A}_{\sigma\sigma}-\mathrm{A}_{\sigma \mathrm{X}}\mathrm{A}_{\sigma \mathrm{E}}}{\mathrm{A}_{\sigma\sigma}} & -\frac{\mathrm{A}_{\mathrm{XX}}\mathrm{A}_{\sigma\sigma}-\mathrm{A}_{\sigma \mathrm{X}}^2}{\mathrm{A}_{\sigma\sigma}} + \frac{\mathrm{A}_{\mathrm{E}}D''}{\rho - r} & D''\ \rho + \frac{\mathrm{A}_{\sigma \mathrm{X}}}{\mathrm{A}_{\sigma\sigma}} \end{bmatrix} \tag{29}$$

In order to apply Dockner's theorem it is necessary to compute the determinant of the Jacobian (29) and the following coefficient K

$$K = \begin{matrix} \frac{\partial \dot{E}}{\partial E} & \frac{\partial \dot{E}}{\partial \lambda} \\ \frac{\partial \dot{\lambda}}{\partial \mathrm{E}} & \frac{\partial \dot{\lambda}}{\partial \lambda} \end{matrix} + \begin{matrix} \frac{\partial \dot{\mathrm{X}}}{\partial \mathrm{X}} & \frac{\partial \dot{\mathrm{X}}}{\partial \mu} \\ \frac{\partial \dot{\mu}}{\partial \mathrm{X}} & \frac{\partial \dot{\mu}}{\partial \mu} \end{matrix} + 2 \begin{matrix} \frac{\partial \dot{\mathrm{E}}}{\partial \mathrm{X}} & \frac{\partial \dot{\mathrm{E}}}{\partial \mu} \\ \frac{\partial \dot{\lambda}}{\partial \mathrm{X}} & \frac{\partial \dot{\lambda}}{\partial \mu} \end{matrix} \tag{30}$$

Computing the determinant of the Jacobian (29) and the coefficient K of the above Dockner's theorem we record the following result.

Proposition 4 *The economic cyclical strategies for the proposed model (17) are feasible if the following conditions are met:*

(a) *The subsidy program produces low deadweight costs ($C' \approx 0$)*
(b) *A strong concavity of the rate of environmental quality with respect to the most recent subsidy concessions. This means that the marginal gain of environmental quality from granting an additional euro declines rapidly with respect to the concessions. Conversely any substantial deregulation carries with it large costs in terms of environmental quality*

Remark 6 (Some policy implications) Cyclical strategies (periodic increases and reductions in subsidies) are not always feasible. The model (17) implies that such cycles can occur under specific conditions:

Condition 1: Low Deadweight Costs $C' \approx 0$ (marginal deadweight loss is near zero), which implies that subsidies do not distort the market significantly. This condition has the consequence that the government/regulator can adjust subsidies up or down without imposing large efficiency losses. Moreover, this ensures that the system can oscillate without large economic penalties, which is consistent with linear or nearly-linear damage function $D(X(t)) = DX(t)$.

Condition 2: Strong Concavity of Environmental Response to Subsidies or equivalently $\mathrm{A}_{\sigma\sigma} < 0$, which economically has the economic meaning: Each additional euro of subsidy adds less and less environmental quality. Conversely, removing subsidies sharply reduces environmental quality, so deregulation is costly.

This condition 2 creates the feedback necessary for cycles:

(a) Subsidies increase implied marginal gain declines in turn implies incentive to reduce subsidies
(b) Reduction in subsidies implies environmental quality drops implying incentive to raise subsidies again

Mathematically, this shows up as complex eigenvalues in the Jacobian with non-zero imaginary parts, generating oscillatory behavior in the control variable $\sigma(t)$.

5 Concave Specifications for the Model

Let us specify the functional forms in the model as in the following.

We assume that the utility function that the social planner maximizes is in separable form for the state variables E, X but suffers from a quadratic cost arising from the control variable σ. We denote this utility function as

$$\mathrm{A}(\mathrm{E}, \mathrm{X}, \sigma) = aE + bX - \frac{1}{2}\gamma\sigma^2 \tag{31}$$

In other words, the social planner derives utility first from both sources, the stock of the natural resources' capital, and second from the performance augmentation of the firm that extracts natural resources but suffers from the costs by offering subsidies from the national budget.

As the environmental growth function we accept the logistic growth $g(E) = E(1-E)$ which is the more appropriate as it is recommended by biological models in the literature. The second state variable is the performance booster of the firm $X(t)$, i.e., the harvesting natural resources function of the representative firm. We assume moreover for the damage (cost) function that it is proportional to the performance booster, i.e., $D(X(t)) = X(t)$.

For the above specifications, the sufficient conditions for optimality are:

First, the Hamiltonian is (omitting the argument t).

$$H = aE + bX - \frac{1}{2}\gamma\sigma^2 + \lambda(E(1-E) - \mathrm{X}) + \mu\sigma \tag{32}$$

and the optimality conditions

$$H_\sigma = -\gamma\sigma + \mu = 0 \tag{33}$$

$$\dot{\lambda} = -\frac{\partial \mathrm{H}}{\partial \mathrm{E}} + \rho\lambda = -\alpha - \lambda + 2\lambda\mathrm{E} + \rho\lambda = \lambda(\rho - (1 - 2\mathrm{E})) - \alpha \tag{34}$$

$$\dot{\mu} = -\frac{\partial \mathrm{H}}{\partial \mathrm{X}} + \rho\mu = \lambda - b + \rho\mu \tag{35}$$

$$\lim_{t\to\infty} e^{-\rho t}\lambda(t)E(t) = 0 \tag{36}$$

$$\lim_{t\to\infty} e^{-\rho t}\mu(t)\mathrm{X}(t) = 0 \tag{37}$$

Solving (5.3), we get $\sigma = \frac{\mu}{\gamma}$ and therefore the canonical system in state and costate equations becomes

$$\dot{E} = E(1 - E) - X \tag{38}$$

$$\dot{X} = \frac{\mu}{\gamma} \tag{39}$$

$$\dot{\lambda} = \lambda(\rho - (1 - 2\mathrm{E})) - \alpha \tag{40}$$

$$\dot{\mu} = \dot{\rho}\mu + \lambda - b \tag{41}$$

5.1 Stability of the System

In order to apply the Hopf bifurcation theorem, which gives sufficient conditions for the existence of stable limit cycles for the system (38)–(41) the following three conditions are required to hold:

(i) the existence of a pair of purely imaginary roots of the characteristic equation for a particular parameter value;
(ii) the crossing of the imaginary axis at non-zero velocity when the bifurcation parameter is changed;
(iii) a negative coefficient of the so-called normal form, which guarantees stability of cycles.

Conditions (i) and (ii) ensure the existence of limit cycles. It remains to investigate the stability of the cycle through the condition (iii), which in turn is often impossible to check. Therefore, we restrict the current research to the necessary conditions for the application of Hopf theorem which detects the existence of limit cycles, and we let the sufficiency for further research.

Proposition 5 *The necessary condition for cyclical subsidy policies in the concave model is the rate of the environmental growth function $g(E)$ to be positive at equilibrium $g'(E_\infty) > 0$, but for its growing rate to be less than the discount rate of the* *social planner, loosely speaking the inequality condition* $\rho > g' = (1 - 2E_\infty) > 0$ *must hold.*

Proof The Jacobian matrix of the system (38)–(41) is the following

$$J = \begin{bmatrix} \frac{\partial \dot{E}}{\partial E} & \frac{\partial \dot{E}}{\partial X} & \frac{\partial \dot{E}}{\partial \lambda} & \frac{\partial \dot{E}}{\partial \mu} \\ \frac{\partial \dot{X}}{\partial E} & \frac{\partial \dot{X}}{\partial X} & \frac{\partial \dot{X}}{\partial \lambda} & \frac{\partial \dot{X}}{\partial \mu} \\ \frac{\partial \dot{\lambda}}{\partial E} & \frac{\partial \dot{\lambda}}{\partial X} & \frac{\partial \dot{\lambda}}{\partial \lambda} & \frac{\partial \dot{\lambda}}{\partial \mu} \\ \frac{\partial \dot{\mu}}{\partial E} & \frac{\partial \dot{\mu}}{\partial X} & \frac{\partial \dot{\mu}}{\partial \lambda} & \frac{\partial \dot{\mu}}{\partial \mu} \end{bmatrix} = \begin{bmatrix} 1 - 2\mathrm{E} & & -1\,0\,0 & \\ 0 & 0 & 0 & \frac{1}{\gamma} \\ 2\lambda & 0 & \rho - 1 + 2\mathrm{E} & 0 \\ 0 & 0 & 1 & \rho \end{bmatrix} \tag{42}$$

We apply Dockner's formula [12] to calculate the four eigenvalues of the Jacobian (42), $\xi_{1,\ldots,4}$

$$\xi_{1,\ldots 4} = \left(\frac{\rho}{2}\right) \pm \left[\left(\frac{\rho}{2}\right)^2 - \frac{\mathrm{K}}{2} \pm \frac{1}{2}(\mathrm{K}^2 - 4J)^{\frac{1}{2}}\right]^{\frac{1}{2}} \tag{43}$$

where the coefficient K is the sum of determinants,

$$\begin{aligned} K &= \left\| \begin{matrix} \frac{\partial \dot{E}}{\partial E} & \frac{\partial \dot{E}}{\partial \lambda} \\ \frac{\partial \dot{\lambda}}{\partial E} & \frac{\partial \dot{\lambda}}{\partial \lambda} \end{matrix} \right\| + \left\| \begin{matrix} \frac{\partial \dot{\mathrm{X}}}{\partial \mathrm{X}} & \frac{\partial \dot{\mathrm{X}}}{\partial \mu} \\ \frac{\partial \dot{\mu}}{\partial \mathrm{X}} & \frac{\partial \dot{\mu}}{\partial \mu} \end{matrix} \right\| + 2 \left\| \begin{matrix} \frac{\partial \dot{\mathrm{E}}}{\partial \mathrm{X}} & \frac{\partial \dot{\mathrm{E}}}{\partial \mu} \\ \frac{\partial \dot{\lambda}}{\partial \mathrm{X}} & \frac{\partial \dot{\lambda}}{\partial \mu} \end{matrix} \right\| = \\ &= \left\| \begin{matrix} 1 - 2E & 0 \\ 2\lambda & \rho - 1 + 2\mathrm{E} \end{matrix} \right\| + \left\| \begin{matrix} 0 & \frac{1}{\gamma} \\ 0 & \rho \end{matrix} \right\| + 2 \left\| \begin{matrix} -1 & 0 \\ 0 & 0 \end{matrix} \right\| = \\ &= (1 - 2E)(\rho - (1 - 2\mathrm{E})) \end{aligned}$$

Therefore,

$$K = g'(E)\big(\rho - g'(E)\big) \tag{44}$$

Condition (i) requires that the two of the four eigenvalues $\xi_{1,\ldots,4}$ to be purely imaginary, e.g. $\xi_{1,2} = wi$ Taking in mind (43) the latter is equivalent with the following conditions

$$K > 0, \quad \|J\| \geq 0$$

The determinant of (42) is:

$$\|J\| = \frac{2\lambda}{\gamma}$$

Since $g' = 1 - 2E$, condition $K > 0$ is satisfied only for positive growth $g' \geq 0$ and $\big(\rho - g'(E)\big) > 0$.

6 Conclusions

In this study, we conceptualize environmental resources as a fundamental form of national capital. We begin by analyzing a static monopolistic framework to examine the impact of resource exploitation on environmental quality. Recognizing that monopolistic extraction results in suboptimal outcomes, we propose a subsidy rule to correct this inefficiency.

We then extend the analysis to a dynamic setting using capital theory, treating environmental resources as a component of national capital and subsidies as a key policy instrument. To mitigate environmental degradation during resource exploitation, we derive several results based on the principle of optimality.

Proposition 2 shows that, along the optimal subsidy path, any marginal increase in current utility from the policy must be exactly offset by a marginal decrease in future utility due to changes in the natural resource capital stock.

Proposition 3 addresses the optimal accumulation and depreciation of natural resource capital. Specifically, the depreciation of one unit of resource capital over a short time interval equals the sum of two components: the marginal social benefits realized during the interval and the marginal increase in the future value of the resource capital.

Next, we introduce a novel perspective by modeling subsidies as an "addictive good" for resource extractors. Given the addictive characteristics of subsidies, we formulate the social planner's optimal control problem and analyze the conditions under which cyclical subsidy policies may arise.

Proposition 4 identifies the conditions for the emergence of a cyclical equilibrium. By specifying concave functional forms, Proposition 5 establishes the conditions under which such cyclical subsidy policies are stable.

Overall, our results offer a theoretical framework for designing subsidy policies that reconcile short-term utility gains with the long-term preservation of natural resource capital, thereby promoting sustainable resource management.

Disclosure of Interests The authors have no competing interests to declare that are relevant to the content of this article.

References

1. Hotelling, H.: The economics of exhaustible resources. J. Polit. Econ. **39**, 137–175 (1931)
2. Dorfman, R.: An economic interpretation of the optimal control theory. Am. Econ. Rev. **59**, 817–831 (1969)
3. Becker, G., Murphy, K.: A theory of rational addiction. J. Polit. Econ. **96**, 675–700 (1988)
4. Stigler, G.J., Becker, G.S.: De Gustibus Non Est Disputandum. Am. Econ. Rev. **67**, 79–90 (1977)
5. Pontryagin, L.S.: The Mathematical Theory of Optimal Processes (translated by Trirogoff, K.N.). Routledge, New York (1962). https://doi.org/10.1201/9780203749319

6. Chiang, A.C.: Elements of Dynamic Optimization. McGraw-Hill Inc (1992)
7. Becker, G.S.: A theory of competition among pressure groups for political influence. Q. J. Econ. **98** (1983)
8. Iannaccone, L.R.: Addiction and satiation. Econ. Lett. **21**, 95–99 (1986). https://doi.org/10.1016/0165-1765(86)90130-8
9. Leonard, D.: Market behavior of rational addicts. J. Econ. Psychol. **10**, 117–144 (1989). https://doi.org/10.1016/0167-4870(89)90060-3
10. Benchekroun, H., Long, N.V.: A class of performance-based subsidy rules. Japanese Econ. Rev. **59**, 381–400 (2008)
11. Halkos, G.E., Papageorgiou, G.J., Halkos, E.G., Papageorgiou, J.G.: Environmental regulation and economic cycles. Econ. Anal. Policy **64**, 172–177 (2019). https://doi.org/10.1016/j.eap.2019.07.005
12. Dockner, E.J.: Local stability analysis in optimal control problems with two state variables. In: Feichtinger, G. (ed.) Optimal Control Theory and Economic Analysis, pp. 89–113. North Holland, Amsterdam (1985)

The γ-Order Generalized Distributions for Environmental Economics Applications

Christos P. Kitsos, Constantinos-Symeon Nisiotis, and Ioannis S. Stamatiou

Abstract The target of this paper is to provide a compact review of the γ-order generalized distributions appropriate for a Probability Model-Oriented Data Analysis for Environmental Economics applications. The appropriate models are presented in compact line of thought, just to help researchers to apply the models and extract more information. Examples for the pollution factors in Athens and Thessaloniki, Greece, are provided and discussed.

Keywords γ-order generalized normal · Kolmolgorov–Smirnov test · Environmental economics data analysis

1 Introduction

In principle, Environmental Economics is one of the main sub-fields of Economics and is focused on environmental problems, see [7]. It has become a widely studied subject due to growing environmental concerns and high related to Data Analysis and therefore to Statistics. This is why in this paper, we propose the γ-order distributions to be adopted in the Data Analysis concerning the Environmental Statistical data, see for a Probability Model-Oriented Data Analysis in Sect. 4. In this section, we introduce very briefly the γ-order Generalized Normal distribution, which, due to

C. P. Kitsos (✉)
Department of Informatics, University of West Attica, Egaleo, Greece
e-mail: xkitsos@uniwa.gr

C.-S. Nisiotis
Department of Sociology, University of the Aegean, University Hill, Mytilene, Greece
e-mail: csnisiotis@aegean.gr

I. S. Stamatiou
Department of Surveying and Geoinformatics Engineering, University of West Attica, Aigaleo, Greece
e-mail: istamatiou@uniwa.gr

G. Halkos et al. (eds.), *Environmental Risk Analysis*, Springer Proceedings in Mathematics & Statistics 540, https://doi.org/10.1007/978-3-032-24155-9_5

the extra shape parameter defines a wide family of distributions, appropriate for the Environmental Analysis.

Both, Normal and Chi-square distributions, are very popular to applications: The Normal distribution is so essential to the Information Theory, [3], while the Chi-square tests, [22] for a theoretical approach, are widely used in Applied Statistics.

The multivariate Normal distribution, see [2] among others, attracts more interest in applications than the multivariate Chi-square [21]. One explanation might be that the multivariate Normal offers very useful results to many fields of interest, the Information Theory being one of them, [3] and the Central Limit Theorem, [22]. Even as an inverse Gaussian, as introduced in [23], Normal distribution is really very attractive, especially to the experimentalist. Let the probability density function (pdf)

$$\phi_\gamma(\boldsymbol{x}) = C \exp\left\{-\frac{\gamma-1}{\gamma}[Q(\boldsymbol{x})]^{\frac{\gamma}{2(\gamma-1)}}\right\}, \tag{1}$$

with

$$C = C_p(\boldsymbol{\mu}, \boldsymbol{\Sigma}; \gamma) = \frac{1}{\pi^{p/2}|\boldsymbol{\Sigma}|^{1/2}} \frac{\Gamma(\frac{p}{2}+1)}{\Gamma(p\frac{\gamma-1}{\gamma}+1)} \left(\frac{\gamma-1}{\gamma}\right)^{p\frac{\gamma-1}{\gamma}}, \gamma_0 = \frac{\gamma-1}{\gamma}, \gamma_0\gamma_1 = 1, \tag{2}$$

$$Q(\boldsymbol{x}) =< \boldsymbol{x} - \boldsymbol{\mu}, \boldsymbol{\Sigma}^{-1}(\boldsymbol{x} - \boldsymbol{\mu}) >, \ \boldsymbol{\mu} \in \mathbb{R}^p, \ \boldsymbol{\Sigma} \in \mathbb{R}^{p\times p}, \tag{3}$$

where $< \boldsymbol{x}, \boldsymbol{y} >:= \boldsymbol{x}^T\boldsymbol{y}$ is the square of the Mahalanobis distance of $\boldsymbol{x}$ from distribution $N_\gamma(\boldsymbol{\mu}, \boldsymbol{\Sigma})$, with $\boldsymbol{\Sigma}$ being positive definite. Referring to the random variable X following (1) we shall denote $\boldsymbol{X} \sim N_\gamma(\boldsymbol{\mu}, \boldsymbol{\Sigma})$ and we shall say that the random variable (rv) X is coming from the γ-order Generalized Normal (γ-GN) distribution. Figures 1 and 2 present the pdf, while Table 3 evaluates the cumulative distribution function (cdf) of different standardized γ-order Normals at traditional values. For an application of the defined in (1) distribution, on the real data concerning pollutants

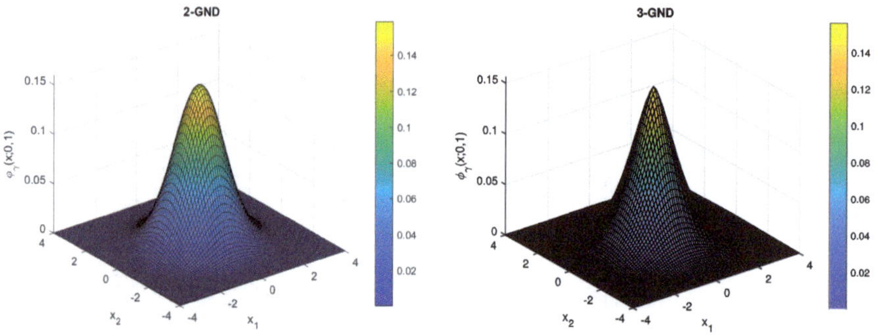

Fig. 1 The pdf of the standardized $\phi_\gamma(\boldsymbol{x})$ for $\gamma = 2$ (Normal), $\gamma = 3$ (fat-tailed) with $p = 2$

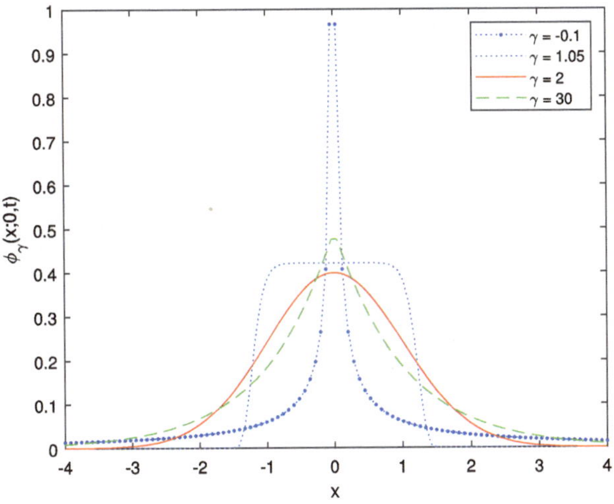

Fig. 2 The pdf of the standardized $\phi_\gamma(x)$ for $\gamma = 2$ (Normal), $\gamma = -0.1$ (near to Dirac), $\gamma = 1.05$ (near to Uniform), and $\gamma = 30$ (near to Laplace) with $p = 1$.

in Athens and Thessaloniki see Figs. 10, 11, 12, 13, 14, 15, 16, 17, 18, 19, 20, 21, 22, 23, 24, 25, 26, 27, 28, 29, 30, 31, 32, 33, 34, 35, 36, 37, 38, 39, and 40.

Therefore, with an extra shape parameter γ and the strong theoretical background of the Logarithmic Sobolev Inequalities, [12] proposes the γ-order Generalized Normal, $N_\gamma(\boldsymbol{\mu}, \boldsymbol{\Sigma})$, which appears to be a Kotz-type distribution. Notice that all the other definitions, discussed in Sect. 5, where they are applied and compared, in practice, with the γ-order Generalized Normal, are without a solid statistical background. Only the γ-GN distribution has a nice mathematical framework.

For a detailed discussion on the extremal function property due to the Logarithmic Sobolev Inequalities (LSI) for the introduced in (1) distribution, see [12].

For the probability function (1), see [13], and for different values of $\gamma \in \mathbb{R}$-[0,1] and $p \in \mathbb{N}$, the p-variable γ-GN distribution coincides with the Dirac, $D(\boldsymbol{\mu})$, the degenerate vanishing distribution $V(\boldsymbol{\mu})$, the Uniform, $U(\boldsymbol{\mu}, \boldsymbol{\Sigma})$, the Normal, $N(\boldsymbol{\mu}, \boldsymbol{\Sigma})$, and the Laplace $L(\boldsymbol{\mu}, \boldsymbol{\Sigma})$. Moreover, the cdf of ϕ_γ, say Φ_γ, for $z = \frac{x-\mu}{\sigma}$ is obtained and equals to

$$\Phi_\gamma(z) = \frac{1}{2} + \frac{\sqrt{\pi}\,\mathrm{sgn}(z)}{2\Gamma(\frac{\gamma-1}{\gamma})\Gamma(\frac{\gamma}{\gamma-1})} Erf_{\frac{\gamma}{\gamma-1}}\left\{\left(\frac{\gamma-1}{\gamma}\right)^{\frac{\gamma-1}{\gamma}} |z|\right\}, \tag{4}$$

with the generalized error function defined as, see [1],

$$Erf_a(x) = \frac{\Gamma(a+1)}{\sqrt{\pi}} \int_0^x e^{-t^a}\,dt. \tag{5}$$

Table 1 Values of the coefficient $\upsilon_{p,\gamma}$ in the evaluation of the covariance matrix of $N_\gamma(\boldsymbol{\mu}, \boldsymbol{\Sigma})$ in (6)

$p\backslash\gamma$	1.1 (Uniform)	1.8	2 (Normal)	2.2	3	30 (Laplace)
1	0.4868	0.9213	1	1.068	1.2680	1.9122
2	0.7453	1.7660	2	2.2130	2.8931	5.5795
3	0.9110	2.5682	3	3.4055	4.7703	10.9443

Furthermore, the covariance matrix of $N_\gamma(\boldsymbol{\mu}, \boldsymbol{\Sigma})$ can be evaluated as

$$\mathrm{Cov}(\boldsymbol{X};\gamma) = \frac{\Gamma((p+2)\frac{\gamma-1}{\gamma})}{\Gamma(p\frac{\gamma-1}{\gamma})}\left(\frac{\gamma}{\gamma-1}\right)^{2\frac{\gamma-1}{\gamma}}(\mathrm{rank}\boldsymbol{\Sigma})^{-1}\boldsymbol{\Sigma} = \upsilon_{p,\gamma}(\mathrm{rank}\boldsymbol{\Sigma})^{-1}\boldsymbol{\Sigma}. \tag{6}$$

In Table 1, values of the coefficient $\upsilon_{p,\gamma}$ in (6) have been evaluated. Note that for $\gamma = 2$ (Normal case) the coefficient $\upsilon_{p,\gamma}$ is proportional to p.

Example 1 Consider the covariance matrix $\boldsymbol{\Sigma} = \begin{bmatrix} 1 & 0.5 \\ 0.5 & 1 \end{bmatrix}$. Then obviously $p = 2$ and $\mathrm{rank}\boldsymbol{\Sigma} = 2$ thus taking into account Table 1 $\mathrm{Cov}(\boldsymbol{X};\gamma)$ takes the following form for values of γ equal to 1.8, 2, 2.2, and 3 accordingly:

$$\mathrm{Cov}(\boldsymbol{X};1.8) = \begin{bmatrix} 0.8830 & 0.4415 \\ 0.4415 & 0.8830 \end{bmatrix}, \mathrm{Cov}(\boldsymbol{X};2) = \begin{bmatrix} 1 & 0.5 \\ 0.5 & 1 \end{bmatrix},$$

$$\mathrm{Cov}(\boldsymbol{X};2.2) = \begin{bmatrix} 1.1065 & 0.5533 \\ 0.5533 & 1.1065 \end{bmatrix}, \mathrm{Cov}(\boldsymbol{X};3) = \begin{bmatrix} 1.4466 & 0.7233 \\ 0.7233 & 1.4466 \end{bmatrix},$$

while for values of γ equal to 1.1 and 30 it is:

$$\mathrm{Cov}(\boldsymbol{X};1.1) = \begin{bmatrix} 0.3726 & 0.1863 \\ 0.1863 & 0.3726 \end{bmatrix}, \mathrm{Cov}(\boldsymbol{X};30) = \begin{bmatrix} 2.7898 & 1.3949 \\ 1.3949 & 2.7898 \end{bmatrix}.$$

The eigenvalues of $\mathrm{Cov}(\boldsymbol{X};1.8)$ are $\lambda_1 = 0.4415, \lambda_2 = 1.3245$, for $\mathrm{Cov}(\boldsymbol{X};2)$ are $\lambda_1 = 0.5, \lambda_2 = 1.5$, for $\mathrm{Cov}(\boldsymbol{X};2.2)$ are $\lambda_1 = 0.5532, \lambda_2 = 1.6597$ and for $\mathrm{Cov}(\boldsymbol{X};3)$ are $\lambda_1 = 0.7233, \lambda_2 = 2.1698$. The eigenvalues of $\mathrm{Cov}(\boldsymbol{X};1.1)$ are $\lambda_1 = 0.1863, \lambda_2 = 0.5590$ and for $\mathrm{Cov}(\boldsymbol{X};30)$ are $\lambda_1 = 1.3949, \lambda_2 = 4.1847$. Note that by λ_1, we denote the minimum eigenvalue and by λ_2 the maximum eigenvalue of the matrix. The condition number for each matrix, which is the ratio λ_2/λ_1, is in every case equal to 3, that is $\mathrm{cond}[\mathrm{Cov}(\boldsymbol{X};\gamma)] = 3$ for $\gamma = 1.1, 1.8, 2, 2.2, 3,$ and 30. This result is really very useful and needs more investigation. Recall that the condition number is the main "measure of a matrix," not the determinant.

Consider the $N_\gamma(\mu, \sigma^2)$, $p = 1$, see [13], with position (mean) μ, positive scale parameter σ, extra shape parameter $\gamma \in \mathbb{R} - [0, 1]$ and pdf $\phi_\gamma(x; \mu, \sigma^2)$ coming from (1) to (3) and given by

$$\phi_\gamma(x; \mu, \sigma^2) = \frac{\lambda_\gamma}{\sigma\sqrt{\pi}} \exp\left\{-\gamma_0 \left(\frac{|x-\mu|}{\sigma}\right)^{\gamma_1}\right\}, \tag{7}$$

with

$$\lambda_\gamma = \frac{\Gamma(\frac{1}{2}+1)}{\Gamma(\frac{\gamma-1}{\gamma}+1)} \left(\frac{\gamma-1}{\gamma}\right)^{\frac{\gamma-1}{\gamma}} = \frac{\Gamma(\frac{1}{2}+1)}{\Gamma(\gamma_0+1)} \gamma_0^{\gamma_0}. \tag{8}$$

In Table 2, values of the coefficient λ_γ, see (8), in the evaluation of the pdf of $\phi_\gamma(x; \mu, \sigma^2)$ are given, see for details as well as relevant graphs and tables on [19].

In Table 3, the probability levels of the usual values (see a typical table for the Normal distribution ($\gamma = 2$)) for the γ-GN distributions have been calculated and presented. Notice that the almost Normal cases, $\gamma = 1.8, 2.2$, provide different (tail) probability levels. That is why we emphasize the fat-tail orientation of the γ-order Generalized Normal distribution.

It is easy to see that small violation from the Normal distribution, $\gamma = 2$, provides different tails $1 - \Phi_\gamma(z)$. These "fat tails" property of the γ-order GN can be proved helpful in Stock Market problems.

Table 2 Values of λ_γ for different γ.

γ	1.1 (Uniform)	1.8	2 (Normal)	2.2	3	30 (Laplace)
λ_γ	0.7462	0.6978	0.7071	0.7166	0.7492	0.8695

Table 3 Values of the cdf of the standard γ-order Generalized Normal, $\Phi_\gamma(z)$, for different values of γ, z around Normal distribution ($\gamma = 2$)

$\gamma \backslash z$	**−3**	**−2.75**	**−2.50**	**−2.25**	**−2**	**−1.75**	**−1.50**	**−1.25**	**−1**	**−0.75**	**−0.50**	**−0.25**
1.8	0.0005	0.0013	0.0034	0.0078	0.0167	0.0326	0.0589	0.0986	0.1538	0.2245	0.3087	0.4022
2	0.0013	0.0030	0.0062	0.0122	0.0228	0.0401	0.0668	0.1056	0.1587	0.2266	0.3085	0.4013
2.2	0.0025	0.0048	0.0090	0.0161	0.0277	0.0458	0.0726	0.1107	0.1622	0.2281	0.3083	0.4004

$\gamma \backslash z$	**0**	**0.25**	**0.5**	**0.75**	**1**	**1.25**	**1.5**	**2**	**2.25**	**2.5**	**2.75**	**3**
1.8	0.5	0.5978	0.6913	0.7755	0.8462	0.9014	0.9411	0.9833	0.9922	0.9966	0.9987	0.9995
2	0.5	0.5987	0.6915	0.7734	0.8413	0.8944	0.9332	0.9772	0.9878	0.9938	0.9970	0.9987
2.2	0.5	0.5996	0.6917	0.7719	0.8378	0.8893	0.9274	0.9723	0.9839	0.9910	0.9952	0.9975

Recall that the Laplace transform is related to the moment generating function in the continuous case. The Laplace transform of $N_\gamma(\mu, \sigma^2)$ has been evaluated in [15], where it has been proved that

$$\mathcal{L}\phi_\gamma(\xi) = \frac{e^{\xi\mu}}{\Gamma(\gamma_0)} \sum_{j=0}^{\infty} \frac{1}{(2j)!} (\xi\sigma(\gamma_1)^{\gamma_0})^{2j} \, \Gamma\left((2j+1)\gamma_0\right). \quad (9)$$

Therefore, the γ-GN is well-defined. Notice that the $N_\gamma(\mu, \sigma^2)$ can be adapted to generalize other known distributions such as the truncated Normal, the Lognormal, the Cauchy, the chi-square, the Rayleigh and the Maxwell–Boltzmann distributions, which are discussed in Sects. 2 and 3, see for details [19].

2 Other γ-Order Extensions Based on the Normal Distribution

The γ-order Normal distribution, as in (1), was the basis for developing more generalizations with main ingredient the Normal $N(\mu, \sigma^2)$ distribution. Following the Normal distribution extensions can be considered for the γ-GN $N_\gamma(\boldsymbol{\mu}, \boldsymbol{\Sigma})$ as in $N(\boldsymbol{\mu}, \boldsymbol{\Sigma})$ case. We provide a short description of these introduced distributions as we believe that the applied results in Sect. 5 can be also applied in Environmental Economics for the following mentioned distributions. Indeed:

2.1 The Truncated γ-Order Normal

A truncated distribution is usually considered to be a modified probability distribution with the domain of the random variable restricted to a specific interval. This is due to the fact that some data sets of certain observations are limited to certain ranges. In this paper, we shall refer only to the γ-order Generalized Normal truncation. Consider the univariate (with $p = 1$) rv $X \sim N_\gamma(\mu, \sigma^2)$. The truncated γ-order Normal to the right at $x = \tau$, $\tau \in [-3, 3,]$ see [11], extends the classical case and it is defined as

$$f_{\gamma,\tau^+}(x) = \begin{cases} 0 & \text{if } x > \tau \\ \frac{C}{\Phi_\gamma(\frac{\tau-\mu}{\sigma})} \exp\{-\frac{\gamma-1}{\gamma} |\frac{x-\mu}{\sigma}|^{\frac{\gamma}{\gamma-1}}\} & \text{if } x \le \tau \end{cases}$$

and the truncated γ-order Normal to the left at $x = \tau_0$

$$f_{\gamma,\tau_0^-}(x) = \begin{cases} 0 & \text{if } x < \tau_0 \\ \frac{C}{1-\Phi_\gamma(\frac{\tau_0-\mu}{\sigma})} \exp\{-\frac{\gamma-1}{\gamma} |\frac{x-\mu}{\sigma}|^{\frac{\gamma}{\gamma-1}}\} & \text{if } x \ge \tau_0 \end{cases}$$

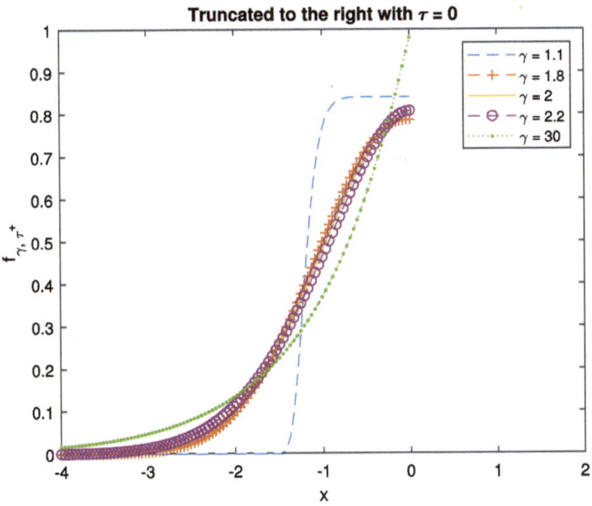

Fig. 3 Plots of the pdf of the truncated standardized γ-order Normal distribution to the right

with C as in (2) with $p = 1$. Plots for the pdf of the truncated standardized γ-order Normal distribution to the right, with $\tau = 0$ are illustrated in Fig. 3.

2.2 *The γ-Order Lognormal*

In principle, a Lognormal distribution is a continuous probability distribution of a random variable whose logarithm is normally distributed and is due to Sir Francis Galton (1822–1911). It can be applied in Environmental Economics where the collected data are very large (or very small) numbers. Consider the logarithm of a rv X that follows the γ-order Normal, that is, $\ln X \sim N_\gamma(\mu, \sigma^2)$. Then X is said to follow the γ-order Lognormal distribution, denoted by $LN_\gamma(\mu, \sigma)$, with pdf, [25],

$$f_{LN_\gamma}(x; \mu, \sigma) = \frac{1}{2x\sigma\Gamma(\frac{\gamma-1}{\gamma})\left(\frac{\gamma-1}{\gamma}\right)^{\frac{1}{\gamma}}} \exp\left\{-\frac{\gamma-1}{\gamma}\left(\frac{|\ln x - \mu|}{\sigma}\right)^{\frac{\gamma}{\gamma-1}}\right\}. \quad (10)$$

Notice that (10) and (1) keep the exponent as in classical situations, and this helps the researcher to identify the extension due to the shape parameter. The mean of $Z \sim LN_\gamma(\mu, \sigma)$ is given by, see [25, Theorem 20],

$$\mu = \mathbb{E}(Z) = \frac{e^\mu}{\Gamma(\gamma_0)} \sum_{j=0}^{\infty} \frac{\sigma^{2j}}{(2j)!} \gamma_0^{-2j\gamma_0} \Gamma((2j+1)\gamma_0) \quad (11)$$

Table 4 The mean μ and variance σ^2 of the standardized γ-order Generalized Lognormal distribution, $LN_\gamma(0, 1)$, for different values of γ

γ	1.1	1.8	2	2.2	3
μ	1.2626	1.5741	1.6487	1.7204	1.9866
σ^2	0.7236	3.2845	4.6708	6.5827	26.122

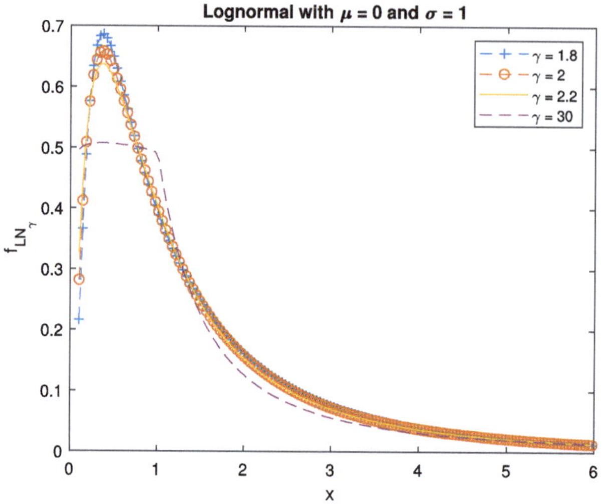

Fig. 4 Plots of the pdf of $LN_\gamma(0, 1)$ distribution for different values of γ

and the variance, $Var(Z)$, is given by

$$\sigma^2 = Var(Z) = \frac{e^{2\mu}}{\Gamma(\gamma_0)} \sum_{j=0}^{\infty} \frac{(2\sigma)^{2j}}{(2j)!} \gamma_0^{-2j\gamma_0} \Gamma((2j+1)\gamma_0) - \mathbb{E}^2(Z). \tag{12}$$

Table 4 provides the mean and variance of the standardized Lognormal distribution, $LN_\gamma(0, 1)$, for different values of γ. Despite their complicated form, the numerical calculations are not that difficult due to the MATLAB support. Moreover, plots for the pdf are illustrated in Fig. 4.

2.3 *The γ-Order Cauchy*

Among the rich in ideas and technicalities work, in different fields of Mathematics,Â ofÂ Augustin-Louis Cauchy (1789–1857) is the Cauchy distribution, which is also known in Physics as the Cauchy–Lorentz distribution,Â the Lorentz(ian) function, or as theÂ Breit–Wigner distribution, with the main parameters mean and variance undefined. The γ-order Cauchy distribution, to be defined below, can be applied in Environmental Economics when we have sharp growth and then sharp decay. In the data analysis presented in Sect. 5, there are cases where the γ-order Cauchy is the appropriate distribution to fit the data.

The γ-order Cauchy distribution is defined as the ratio of two γ-order Generalized Normals. Let $X \sim N_\gamma(0, 1)$, $Y \sim N_\gamma(0, 1)$. Then their ratio $Cau_\gamma(1) = \frac{X}{Y}$ follows the γ-order generalized Cauchy distribution with pdf, see [18],

$$f_{Cau_\gamma}(x) = \frac{K_\gamma}{(1 + |x|^{\gamma_1})^{2\gamma_0}}, \quad \gamma_0 = \frac{\gamma - 1}{\gamma}, \quad \gamma_1 = \frac{1}{\gamma_0}, \tag{13}$$

where

$$K_\gamma := \frac{\Gamma(2\gamma_0)}{2\Gamma(\gamma_0)\Gamma(\gamma_0 + 1)}, \quad \gamma_0 = \frac{\gamma - 1}{\gamma}. \tag{14}$$

The values of K_γ are presented in Table 5 and provide that $0.25 < K_\gamma < 0.50$ for $\gamma > 1$ while when approaching 0 from the left they are not bounded. Notice that for $\gamma < -1$ it is $0.5 < K_\gamma < 1.34$ and K_γ is, in principle, an increasing function of γ. In Fig. 5, the pdf of the standard γ-order Cauchy distribution is illustrated for different values of γ, recall (13).

When $\gamma = 2$ then $Cau_2(1)$ coincides with the known Cauchy distribution, that is,

$$f_{Cau_2}(x) = \frac{1}{\pi(1 + x^2)}. \tag{15}$$

Note that for γ close to 1 the uniform behavior is apparent, while for large values of γ the behavior of Laplace distribution is evident.

Consider $X_i, i = 1, 2, \ldots, n$ i.i.d random variables following the standardized 2-order generalized Cauchy distribution. Then Fisher's information is given by

$$I(\theta) = \frac{n}{2}. \tag{16}$$

So for 1 year collected data for a number of different pollutants the variance is $2/365 = 0.005$, when the Cauchy distribution is applied.

Table 5 Values of K_γ for different γ.

γ	**–40**	**-3.2**	**–2.1**	**–2**	**–1.9**	**–1.1**
K_γ	0.5127	0.6919	0.8268	0.8488	0.8739	1.3496
γ	**1.1**	**1.9**	**2**	**2.1**	**3.2**	**40**
K_γ	0.2530	0.3120	0.3183	0.3243	0.3720	0.4877

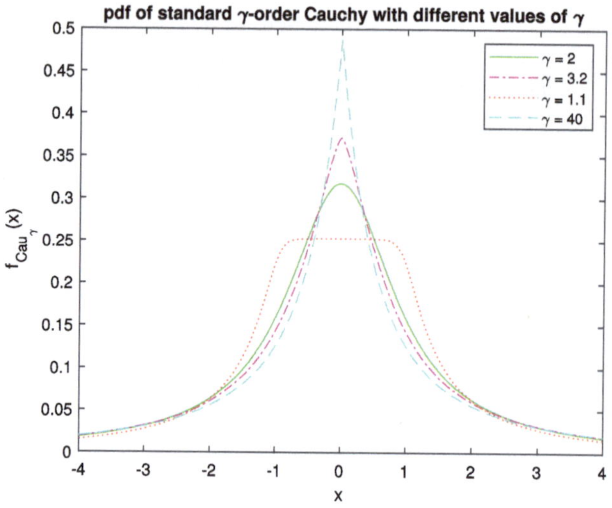

Fig. 5 Plots of the pdf of the standard γ-order Cauchy distribution

3 The γ-Order Extensions Based on the Chi-Square Distribution

In Statistical applications, the Chi-square and the Normal distribution are the most widely used continuous distributions. Both are also applicable in particular to Econometrics. Moreover, in Econometrics, there are cases where the Chi-square distribution is used even with one degree of freedom, see [9] for the Hausman test or Wald test. The γ-order Generalization has two steps. We briefly discuss them.

3.1 *The γ-Order Chi-Square*

Consider the sum of the independent identically distributed random variables $Z_i \sim N_\gamma(0, 1), i = 1, 2, \ldots, n$. The sum of the squares of Z_i, defines a new, but known distribution, the γ-order Chi-square.

$$Y_n = \sum_{i=1}^{n} Z_i^2. \tag{17}$$

Working with the convolution, the γ-order generalized Chi-square distribution was defined in [16], denoted by ${}_\gamma\mathcal{X}_n^2$, where for $\gamma = 2$, the Chi-square can be achieved. The pdf of the produced ${}_\gamma\mathcal{X}_n^2$ is equal to

Table 6 Values of the mean μ and variance σ^2 of γ-order Chi-square distribution, ${}_{\gamma}\mathcal{X}_n^2$, for different values of γ, n

	μ	σ^2	μ	σ^2	μ	σ^2	μ	σ^2	μ	σ^2
$\gamma \backslash n$	1		2		3		4		5	
1.8	0.9213	1.5173	1.7660	2.7832	2.5682	3.9207	3.3420	4.9767	4.0948	5.9747
2	1.0000	2.0000	2.0000	4.0000	3.0000	6.0000	4.0000	8.0000	5.0000	10.0000
2.2	1.0680	2.4965	2.2130	5.3544	3.4055	8.4492	4.6325	11.722	5.8866	15.1386

$$f_{{}_{\gamma}\mathcal{X}_n^2}(z) = \frac{\gamma_0^{n\gamma_0 - 1}}{2\Gamma(n\gamma_0)} z^{\frac{n}{2}-1} \exp\{-\gamma_0(\sqrt{z})^{\gamma_1}\}, \tag{18}$$

for any $z > 0$ (when $n = 1$) and any $z \geq 0$ (when $n \geq 1, n \in \mathbb{N}$).

The following theorem is stated and proved in [16] for the Laplace transform of ${}_{\gamma}\mathcal{X}_n^2$ with

$$\mathcal{L}f_{{}_{\gamma}\mathcal{X}_n^2}(\xi) = \frac{1}{\Gamma(n\gamma_0)} \sum_{j=0}^{\infty} \frac{1}{j!} \left(\xi(\gamma_1)^{2\gamma_0}\right)^j \Gamma((2j+n)\gamma_0). \tag{19}$$

The mean of $Z \sim {}_{\gamma}\mathcal{X}_n^2$ is given by

$$\mu = \mathbb{E}(Z) = \frac{(\gamma_1)^{2\gamma_0}\Gamma((2+n)\gamma_0)}{\Gamma(n\gamma_0)} \tag{20}$$

and the variance, $Var(Z) = \mathbb{E}(Z^2) - \mathbb{E}^2(Z)$, is given by

$$\sigma^2 = Var(Z) = \frac{(\gamma_1)^{4\gamma_0}}{\Gamma(n\gamma_0)} \left[\Gamma((4+n)\gamma_0) - \frac{\Gamma^2((2+n)\gamma_0)}{\Gamma(n\gamma_0)}\right]. \tag{21}$$

Table 6 provides the mean and variance of ${}_{\gamma}\mathcal{X}_n^2$ for different values of γ and n, while plots for the pdf are illustrated in Fig. 6. It is known that the classical Chi-square is important in Statistical Analysis, where variance is involved. It is clear that the ${}_{\gamma}\mathcal{X}_n^2$ can be applied in such general cases for the estimated γ-order variance. The target is to offer a theoretical background before such, computationally easy, applications.

3.2 *The γ-Order Generalized Chi-γ*

The generalization of the γ-order Chi-square is the γ-order Chi-γ. For independent variables $Z_i, i = 1, 2, \ldots, n$ from $N_\gamma(0, 1)$, i.e., $Z_i \sim N_\gamma(0, 1), i = 1, 2, \ldots, n$, consider the sum of their powers to γ_1, that is $Z_i^{\gamma_1}$,

Fig. 6 Plots of the pdf of ${}_{\gamma}\mathcal{X}_n^2$ distribution for different values of n and γ

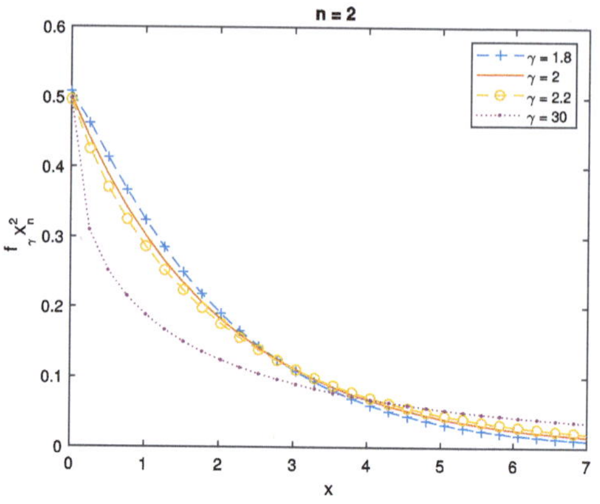

(a) With $n = 2$.

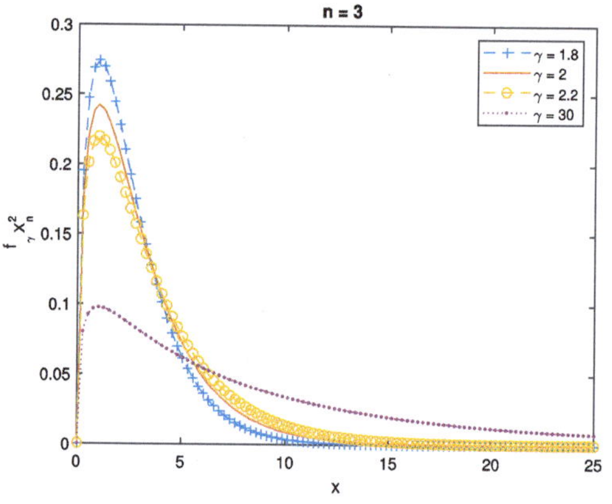

(b) With $n = 3$.

$$Y_n = \sum_{i=1}^{n} Z_i^{\gamma_1}. \tag{22}$$

Then the γ-order generalized Chi-γ, ${}_{\gamma}\mathcal{X}_n^{\gamma}$ is defined, see [17], while for $\gamma = 2$, the Chi-square can be achieved. The moments of ${}_{\gamma}\mathcal{X}_n^{\gamma}$ have been evaluated in [17] and are presented in Table 7, while plots for the pdf are illustrated in Fig. 7.

Finally, the γ-order Rayleigh and Maxwell–Boltzmann distributions have been defined in [16] and can be adapted for specific problems in Physics.

Table 7 Values of the mean μ and variance σ^2 of γ-order Chi-γ distribution, ${}_{\gamma}\mathcal{X}_n^{\gamma}$, for different values of γ, n

	μ	σ^2	μ	σ^2	μ	σ^2	μ	σ^2	μ	σ^2
$\gamma\backslash n$	1		2		3		4		5	
1.8	1	2.2500	1.9182	4.4728	2.6915	6.8862	3.2987	9.7355	3.7395	13.1275
2	1	2.0000	2.0000	4.0000	3.0000	6.0000	4.0000	8.0000	5.0000	10.0000
2.2	1	1.8333	2.0730	3.6492	3.2908	5.0762	4.7115	5.2855	6.3944	2.8066

Fig. 7 Plots of the pdf of ${}_{\gamma}\mathcal{X}_n^{\gamma}$ distribution for different values of n and γ

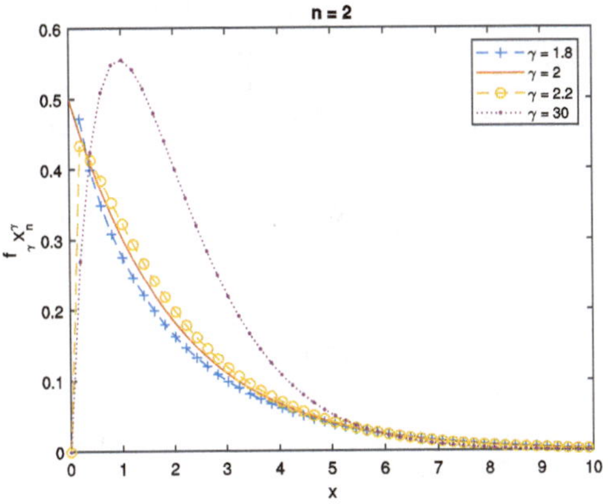

(a) With $n = 2$.

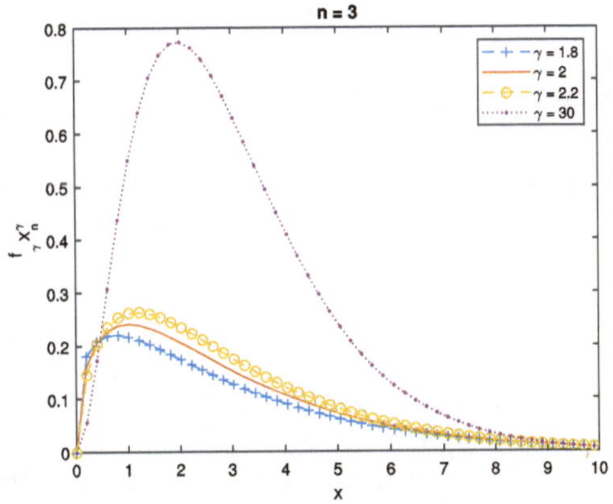

(b) With $n = 3$.

4 Information Measures

Among the merits of the γ-order generalized Normal, beyond the strong theoretical background due to the LSI, is the accordance it presents with the generalized Fisher's entropy type information, extending all the nice theoretical results, [3, 22]. What helps to the complicated proofs, which disappoint the experimentalist, and not only, is the international constant $(\frac{1}{\gamma-1})^{\frac{1}{\gamma-1}}$ which appears to a number of transformations in all the LSI development, [24].

Fisher's entropy type information $J(X)$ can be also generalized with an extra parameter α, [12], in the sense that we have used already γ. So, two extra parameters are involved and we provide the results, although we would like to comment that in Environmental Economics we can consider $\alpha = \gamma$.

$$J_\alpha(X) = \int_{\mathbb{R}^p} |\nabla \log f|^\alpha f dx = \int_{\mathbb{R}^p} |\nabla f|^\alpha f^{1-\alpha} dx, \quad \alpha \in \mathbb{R} - [0, 1] \tag{23}$$

and the Blachman–Stam inequality still holds for rv's X, Y from $N_\gamma(\boldsymbol{\mu}, \boldsymbol{\Sigma})$. Moreover, the defined $J_\alpha(X)$ offers an opportunity to extend the Cramér–Rao inequality, while for $\gamma = 2$ the classical one is obtained.

In the same line of thinking Shannon entropy, $H(X)$, and entropy power, $N(X)$, see [3] for details, have been extended, [12], in the sense that

$$N_\alpha(X) = \nu_\alpha e^{\frac{2}{p} H(X)}, \quad \nu_\alpha = \left(\frac{\alpha - 1}{\alpha e}\right)^{\alpha-1} \pi^{-\alpha/2} \left[\frac{\xi_{p,2}}{\xi_{p,\alpha}}\right]^{\alpha/p}, \tag{24}$$

with $\xi_{p,\alpha} = \Gamma(p\frac{\alpha-1}{\alpha} + 1)$. When $X \sim N_\gamma(\boldsymbol{\mu}, \boldsymbol{\Sigma})$ it holds, [12],

$$N_\alpha(X) = \left(\frac{\alpha - 1}{\alpha e}\right)^{\alpha-1} \left(\frac{\gamma}{\gamma - 1}\right)^{\alpha\frac{\gamma-1}{\gamma}} e^{p\frac{\gamma-1}{\gamma}} \left[\frac{\xi_{p,\gamma}}{\xi_{p,\alpha}}\right] |\boldsymbol{\Sigma}|^{\frac{\alpha}{2\gamma}}. \tag{25}$$

That is, in this paper, we present a number of extensions due to the γ-order Generalized Normal distribution emerged from the LSI. It is the common case in practice to consider $\alpha = \gamma$ which simplifies the above formulas.

The main characteristic of the $N_\gamma(\boldsymbol{\mu}, \boldsymbol{\Sigma})$ distribution, discussed is this paper, is that it offers a strong theoretical background, while others not, [20] for example. It also offers the grounds for generalizing a number of distributions—the truncated γ-order Normal, the γ-order Lognormal, the γ-order Cauchy, the γ-order Chi-square, γ-order generalized Chi-γ, which are already presented. In Environmental Economics, these Information Measures can be evaluated. It is easy for one to see that the calculations in Sect. 5 provide all the required quantities to support the evaluation of the Information Measures. These are not presented here to save some space and because we believe that there is an "Information Theory" line of thought to be followed in the analysis. This point is one of our future targets.

Therefore, the γ-order Generalized Normal distribution, as it has been emerged from the LSI, acts as a generator for new γ-order distributions, as the Normal and satisfies all the merits of multivariate Normal to Information Theory, for the introduced generalizations. Moreover, it covers the existing applications of the Normal distribution in other fields. We have reasons to believe that more extensions can be derived, thanks to the γ-order Generalized Normal distribution and we shall work on this direction, bearing applications in mind as in [8] and [14].

5 Applications and Discussion

Defining a Generalized Normal is discussed by [20] as an early, by trial and error, provided generalization. Moreover, [4] worked on the shape parameter on a generalized form of the Gaussian they provided, with no theoretical framework, as also [6] did, referring to "power exponential family," with nice figures and results, very similar as the one is a function of the other cf. (28) below, but without any justification why their introduced curve, close to [20], is defined, what gave rise to such a genesis. A hard attempt of trials and errors might be behind. In particular, the proposed generalization in [4] reads

$$gg(x;\mu,\sigma,p) = \frac{1}{2\Gamma(1+\frac{1}{p})A(p,\sigma)} \exp\left\{-\left[\frac{|x-\mu|}{A(p,\sigma)}\right]^p\right\}, \quad x \in \mathbb{R}, \tag{26}$$

where $\mu \in \mathbb{R}$ is the mean, $p > 0$ is a shape parameter and the constant term

$$A(p,\sigma) = \left(\frac{\sigma^2\Gamma(1/p)}{\Gamma(3/p)}\right)^{1/2},$$

acts as "a scaling factor that allows $Var(X) = \sigma^2$" where $\sigma > 0$, as it is declared in [4].

Moreover, the exponential power distribution, see [26] and [5] with the proper modifications in notations, was proposed as a generalization with

$$f_{EX}(x;\mu,\alpha,\beta) = \frac{\alpha}{2\beta\Gamma(1/\alpha)} \exp\left\{-\left[\frac{|x-\mu|}{\beta}\right]^\alpha\right\}, \quad x \in \mathbb{R}, \tag{27}$$

where $\mu \in \mathbb{R}$ is the mean, $\alpha > 0$ is the shape parameter, and β is the scale parameter with

$$\beta = \sigma\sqrt{\frac{\Gamma(1/\alpha)}{\Gamma(3/\alpha)}},$$

with σ being the standard deviation of the distribution above. The above (27) is a possible extension for the one variable Normal. Note that distribution (27) coincides with $N_\gamma(\mu, \sigma^2)$, see (7) below for the univariate case, with the particular value of parameters

$$\alpha = \gamma_1 \quad \text{and} \quad \beta = \frac{\sigma}{\gamma_0^{\gamma_0}}.$$

Moreover, note that distribution (26) coincides with $N_\gamma(\mu, \sigma^2)$, if

$$p = \gamma_1 \quad \text{and} \quad A(p, \sigma) = \frac{\sigma}{\gamma_0^{\gamma_0}},$$

which holds for all γ such that $\frac{\Gamma(\gamma_0)}{\Gamma(3\gamma_0)} = \gamma_0^{-2\gamma_0}$. Thus $\alpha = p$ and $\beta = A(p, \sigma)$.

These distributions, despite the nice technical improvements of the coefficients and the contribution in multivariate Normal, as in [6], offer nothing for their justification of existence. A proposed generalization for (26) in [6] provides a definition of their introduced model with no extra explanation, no justification. It is a nice result obtained perhaps by "trial and error." Note that the correspondence between the parameters in distributions (26) and (27) is such that

$$A(p, \sigma) = \beta \quad \text{and} \quad p = \alpha. \tag{28}$$

Therefore, due to (28), there is no real difference among the two proposed generalizations. That is why we believe that the γ-order Generalized Normal, related to the other proposed generalizations, offers a strong theoretical background, see [12].

There is a number of different data bases concerning different areas in the world. We are focused, in this study, to compare the air pollution between the most centralized areas in Greece, Athens, and Salonica. There are not for all the pollutants data (collected from the Greek Ministry of Environment and Energy), for the same period, between the cities, there is a uniform study on the provided data sets for specific areas within the city. We are eventually concentrated in one area of Athens (Patisia) and one in Salonica (Aristotelous). We briefly describe the data set, for detailed description and Stochastic Modeling Analysis; see [10]:

Both of the areas have data in each of the six time periods (1984–2000, 2001–2005, 2006–2010, 2011–2015, 2016–2020, and 2021). For the period 1984–2000, Aristotelous recorded measurements for four pollutants, and five for Patisia. For the next period, 2001–2005, five and six pollutants were recorded for Aristotelous and Patisia, respectively. For the remaining periods, 2 and 6 (2006–2010), 10 and 6 (2011–2015), 10 and 0 (2016–2020), as well as 5 and 0 (2021) were recorded for the areas (ordered as above), respectively. Common pollutants for the areas, by time period, are CO, NO, NO_2 and SO in 1984–2000, NO, NO_2 and SO_2 in 2001–2005, NO and NO_2 in 2006–2010, none in 2011–2015, NO and NO_2 in 2016–2020, and none for 2021. We continue our study with NO_2 pollutant.

In Tables 8 and 9, we summarized all the main position and dispersion measures for the similar to the family of the γ-oder Generalized Normal distribution, related to it, but with no theoretical background, and no members of this family of distribution. From these tables, there is enough evidence that only small differences in calculations

Table 8 Patisias's NO_2 summary statistics

	Classic	Normal	Exponential	Power		γ-order	Generalized	Normal	Generalized	Gaussian	
	μ_N	σ_N	μ_{EX}	α_{EX}	β_{EX}	μ_γ	σ_γ	γ	μ_{gg}	σ_{gg}	p_{gg}
Mean	91.797	29.158	91.601	42.891	2.327	91.597	30.303	1.966	91.598	29.172	2.328
Median	93.561	28.318	93.627	42.667	2.191	93.631	30.547	1.841	93.635	28.304	2.19
Trimmed Mean	91.988	29.083	91.739	42.846	2.314	91.737	30.27	1.883	91.738	29.097	2.314
1st Quartile	77.646	23.651	76.789	33.098	1.966	76.805	24.327	1.63	76.801	23.66	1.967
3rd Quartile	103.674	34.074	103.899	50.44	2.587	103.902	35.677	2.034	103.901	34.05	2.587
Min	56.379	14.52	57.29	17.876	1.228	57.29	13.311	1.347	57.29	14.539	1.226
Max	120.731	46.337	121.217	69.449	3.879	121.16	48.415	5.418	121.167	46.344	3.886
Range	64.352	31.818	63.927	51.574	2.651	63.87	35.104	4.071	63.876	31.805	2.659
IQR	26.029	10.422	27.11	17.343	0.621	27.097	11.35	0.404	27.1	10.39	0.62
Variance	306.966	58.376	309.305	147.276	0.355	309.14	68.062	0.475	309.158	58.354	0.356
Std Deviation	17.52	7.64	17.587	12.136	0.595	17.582	8.25	0.69	17.583	7.639	0.597
Skewness	−0.244	0.177	−0.186	0.129	0.686	−0.186	0.093	3.473	−0.186	0.18	0.693
Kurtosis	−0.926	−0.792	−0.979	−0.594	−0.018	−0.979	−0.611	14.612	−0.979	−0.79	0

Table 9 Aristotelous' NO_2 summary statistics

	Classic	Normal	Exponential	Power		γ-order	Generalized	Normal	Generalized	Gaussian	
	μ_N	σ_N	μ_{EX}	α_{EX}	β_{EX}	μ_γ	σ_γ	γ	μ_{gg}	σ_{gg}	p_{gg}
Mean	72.905	23.257	72.642	31.286	1.958	72.645	22.644	2.374	72.645	23.258	1.957
Median	69.164	20.431	68.971	29.312	1.95	68.975	20.546	2.054	68.977	20.436	1.949
Trimmed Mean	72.383	22.318	72.135	30.287	1.942	72.138	21.865	2.322	72.138	22.315	1.941
1st Quartile	60.735	17.167	60.313	21.699	1.53	60.309	16.333	1.837	60.311	17.224	1.529
3rd Quartile	84.397	23.769	84.707	34.599	2.194	84.697	24.997	2.902	84.709	23.771	2.196
Min	41.451	11.309	41.649	15.07	1.286	41.647	10.868	1.497	41.643	11.308	1.287
Max	116.898	57.727	115.814	71.491	3.01	115.814	53.102	4.495	115.823	57.838	3.013
Range	75.447	46.419	74.166	56.421	1.724	74.167	42.234	2.997	74.18	46.53	1.726
IQR	23.662	6.603	24.394	12.9	0.665	24.388	8.664	1.065	24.398	6.548	0.667
Variance	412.048	110.474	391.274	197.202	0.218	391.369	100.835	0.661	391.44	110.716	0.218
Std Deviation	20.299	10.511	19.781	14.043	0.467	19.783	10.042	0.813	19.785	10.522	0.467
Skewness	0.555	1.828	0.51	1.629	0.451	0.51	1.721	1.174	0.51	1.835	0.452
Kurtosis	–0.599	2.971	–0.651	2.44	–0.715	–0.652	2.63	0.466	–0.652	3.001	–0.707

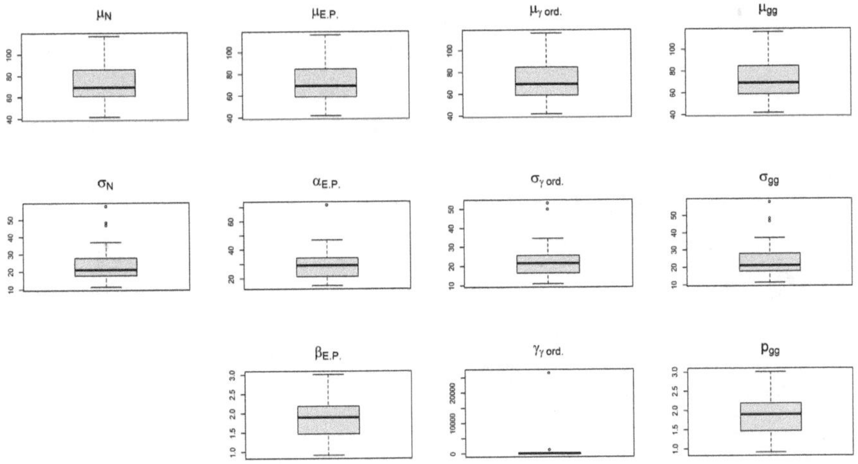

Fig. 8 Aristotelous' NO_2 parameter estimates boxplot for years 1994–2021

exist, despite the merit of the LSI background of the γ-order Generalized Normal, which is also a Kotz-type distribution. The γ-order Generalized Normal acts as the "mother distribution" to all the distributions described and introduced above.

Example 2 (Aristotelous) Parameters for each distribution adapted, were estimated, for every available year. In total, data set for 36 years (1984–2021) was used for the boxplots. First row boxplots are essential the same plot, small deviations observed are due to optimization procedures, see Fig. 8. γ parameter of γ-order GN for years 2000 and 2010, are estimated extremely big (1232.3 and 26587.6 correspondingly) and are excluded from summary statistics, see Table 9 in the Appendix.

For years 1995, 1996, 2000, 2002, 2004, 2008, 2009, 2010, 2014, and 2021, a more parametrically complex distribution required to describe NO_2 data, based on AIC value comparisons. Years 2000 and 2010 are removed, since optimization procedures results in extreme parameter values. There was a strong influence to the whole analysis.

Example 3 (Patisia) Parameters for each distribution adapted, were estimated, for every available year. In total, data set for 36 years (1984–2021) were used for the boxplots. First row boxplots are essential the same plot, small deviations observed are due to estimation accuracy, see Fig. 9. γ parameter of γ-order GN for year 1995 and β parameter of Exponential Power for year 1987, are estimated extremely big (7558.1 and 780286.5 correspondingly).

For years 1985, 1989, 2003, 2004, 2007, 2011, 2012, 2015, and 2017, a more parametrically complex distribution required to describe NO_2 data, based on AIC value comparisons. Years 1985 and 1987 are removed, since optimization procedures results in extreme parameter values.

In almost all cases, the three generalized distributions have some common values. It seems that the target of the γ-GN is quite different from the others: the target was

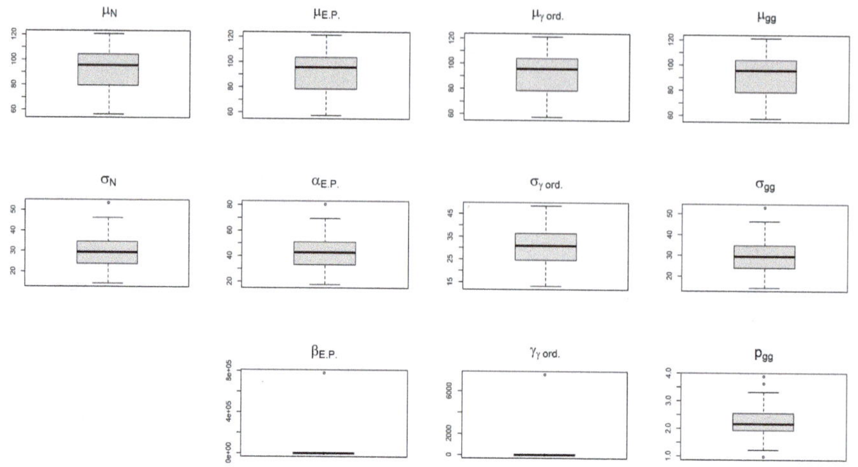

Fig. 9 Patisia's NO_2 parameter estimates boxplot for years 1994–2021

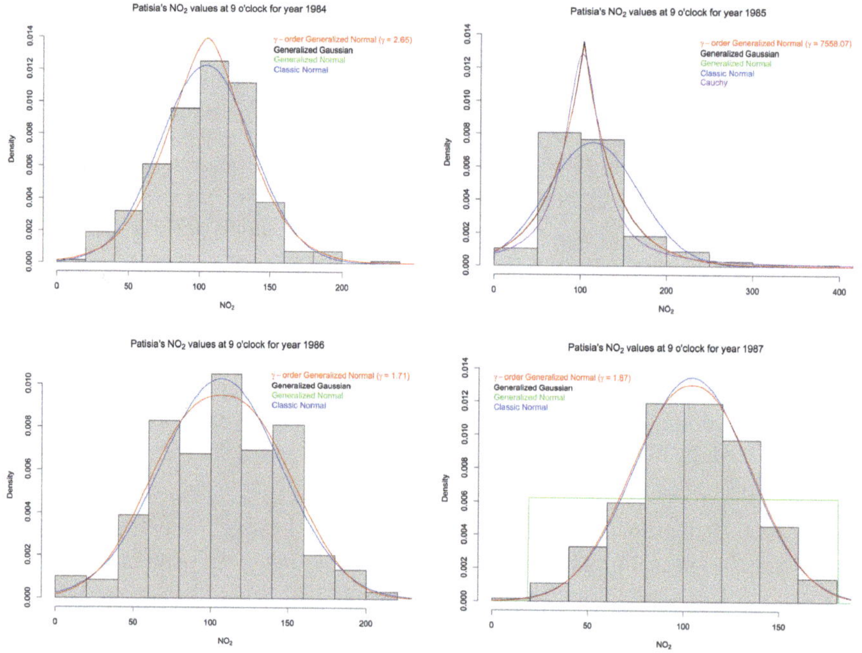

Fig. 10 Patisia's NO_2 values at 9 o'clock for years 1984–1987

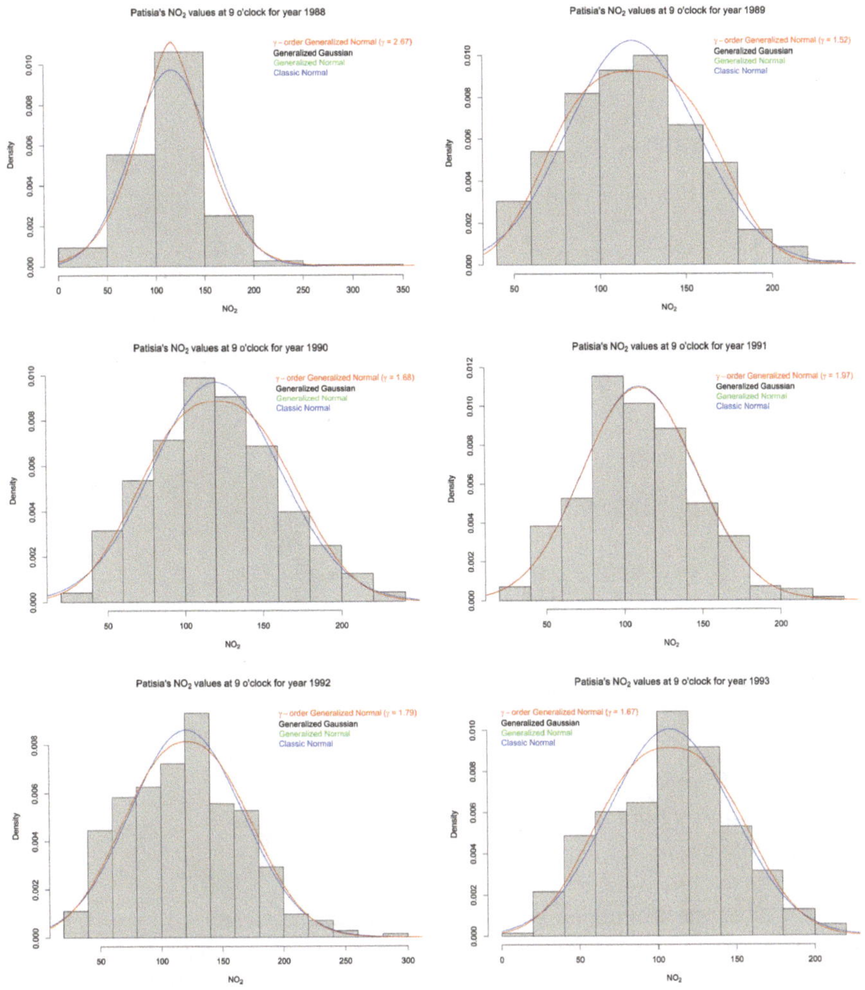

Fig. 11 Patisia's NO_2 values at 9 o'clock for years 1988–1993

to fit the data better, while the other attempts, it seems, have a target to be close to the classical Normal, which could be not the appropriate one for the data analysis in Environmental Economics and to any fat-tail oriented economical data. Classic Normal distribution differs from the others. For year 1987, however, it becomes clear the weakness of Generalized Normal in optimization procedures. Although, initial values are the same for all the distributions, Generalized Normal parameters fail to fit observed NO_2 data for this year.

Due to data for years 1985, 1995, 1996, and 2000, we also tried the γ-order Cauchy distribution see Fig. 12.

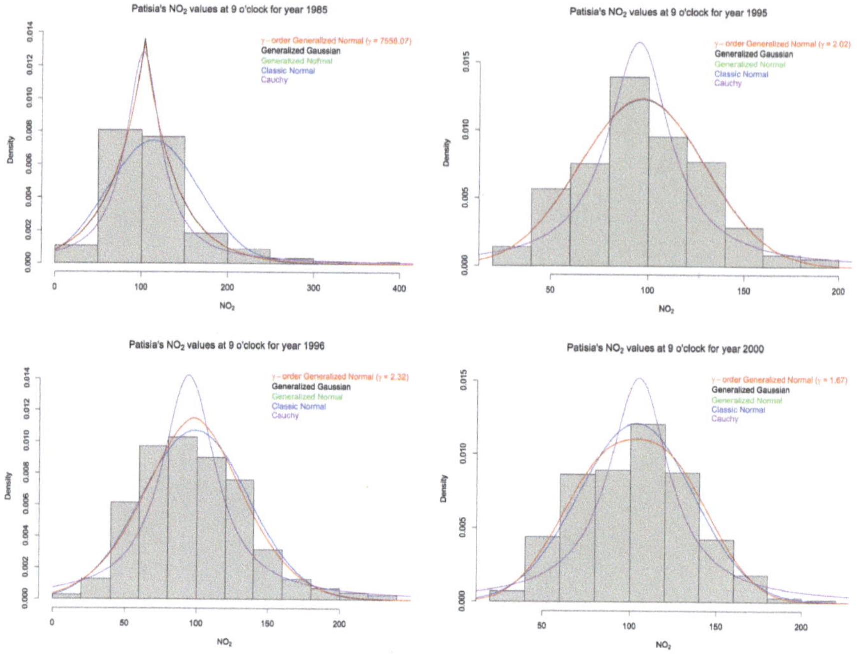

Fig. 12 Patisia's NO_2 values at 9 o'clock for years 1985, 1995, 1996, and 2000

Example 4 (Comparison) Consider the null hypothesis

H_0 : "the true cdf of Patisia's NO_2 is equal to Aristotelous' cdf"
versus the alternative one

H_1 : "the two true cdfs are not equal."

The values of the Kolmogorov–Smirnov test statistic are (in most cases) small enough, as they are evaluated below, to reject the corresponding H_0 at the denoted figure. All the provided figures below are placed for a better study in the Appendix. These are only a portion of the produced figures below, which provide a particular information for the study of pollution, between Athens and Thessaloniki:

1. $ks = 0.125$ (p-value$= 0.008$), see Fig. 13.
2. $ks = 0.271$ (p-value< 0.001), see Fig. 14.
3. is not small enough, $ks = 0.091$ (p-value$= 0.095$), to reject H_0, see Fig. 15.
4. $ks = 0.229$ (p-value< 0.001), see Fig. 16.
5. $ks = 0.223$ (p-value$= 0.001$), see Fig. 17.
6. $ks = 0.273$ (p-value< 0.001), see Fig. 18.
7. $ks = 0.184$ (p-value$= 0.036$), see Fig. 19.
8. $ks = 0.416$ (p-value< 0.001), see Fig. 20.
9. $ks = 0.476$ (p-value< 0.001), see Fig. 21.
10. $ks = 0.325$ (p-value< 0.001), see Fig. 22.
11. $ks = 0.484$ (p-value< 0.001), see Fig. 23.

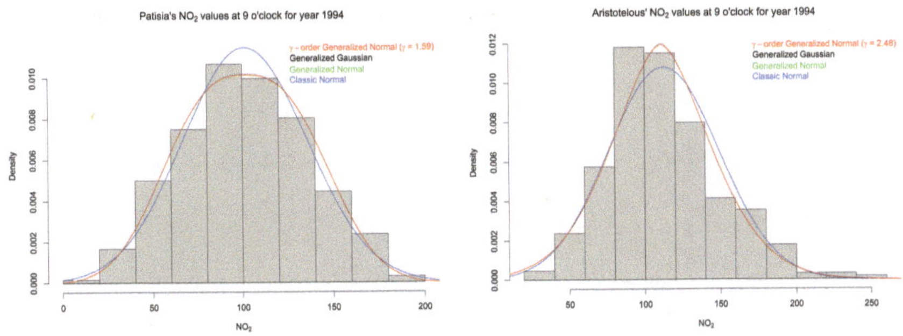

Fig. 13 Patisia's and Aristotelous' NO_2 values at 9 o'clock for year 1994

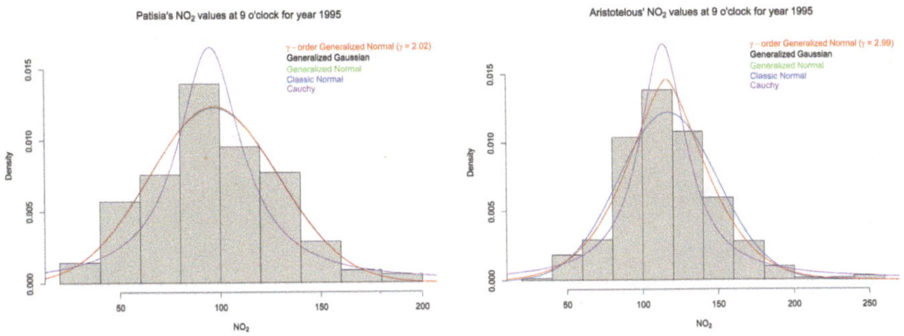

Fig. 14 Patisia's and Aristotelous' NO_2 values at 9 o'clock for year 1995

12. $ks = 0.169$ (p-value< 0.001), see Fig. 24.
13. $ks = 0.182$ (p-value< 0.001), see Fig. 25.
14. $ks = 0.430$ (p-value< 0.001), see Fig. 26.
15. $ks = 0.630$ (p-value< 0.001), see Fig. 27.
16. $ks = 0.752$ (p-value< 0.001), see Fig. 28.
17. $ks = 0.531$ (p-value< 0.001), see Fig. 29.
18. is not small enough, $ks = 0.772$ (p-value< 0.001), to reject H_0, see Fig. 30.
19. is not small enough, $ks = 0.121$ (p-value $= 0.011$), to reject H_0, see Fig. 31.
20. $ks = 0.156$ (p-value< 0.001), see Fig. 32.
21. $ks = 0.188$ (p-value< 0.001), see Fig. 33.
22. $ks = 0.103$ (p-value $= 0.032$), see Fig. 34.
23. $ks = 0.417$ (p-value< 0.001), see Fig. 35.
24. $ks = 0.353$ (p-value< 0.001), see Fig. 36.
25. $ks = 0.228$ (p-value< 0.001), see Fig. 37.
26. $ks = 0.471$ (p-value< 0.001), see Fig. 38.
27. $ks = 0.573$ (p-value< 0.001), see Fig. 39.
28. $ks = 0.270$ (p-value< 0.001), see Fig. 40.

In the present, paper we investigated the possibility of applying the γ-order generalized Normal distribution in Environmental Data Analysis. A very big number

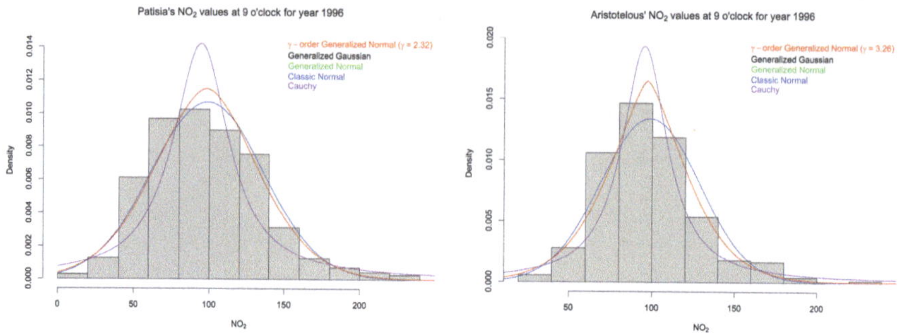

Fig. 15 Patisia's and Aristotelous' NO_2 values at 9 o'clock for year 1996

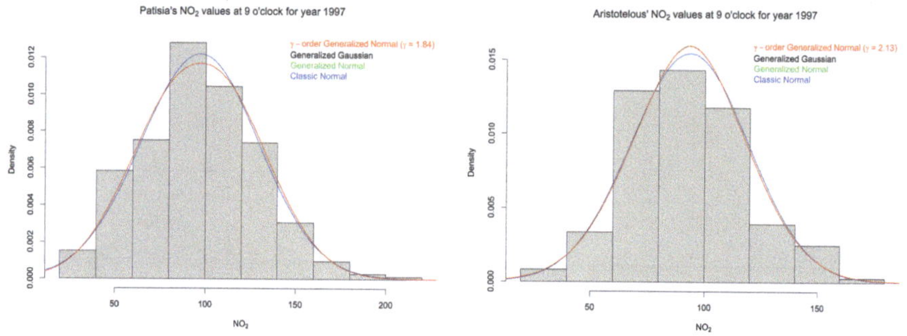

Fig. 16 Patisia's and Aristotelous' NO_2 values at 9 o'clock for year 1997

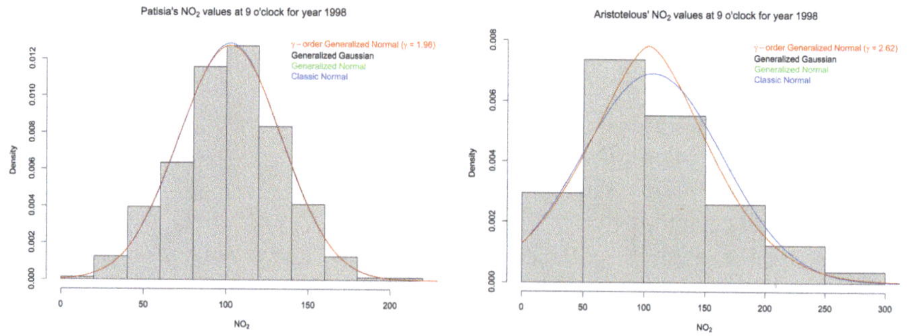

Fig. 17 Patisia's and Aristotelous' NO_2 values at 9 o'clock for year 1998

of tables have been produced and we presented the main ones as above. We believe that we extract more information and better models with this family of distributions which provides fat tails and it has emerged from the LSI. We hope that we will be able to provide more results on data analysis soon. Those who like to have computational support with MATLAB can contact Dr. Ioannis Stamatiou.

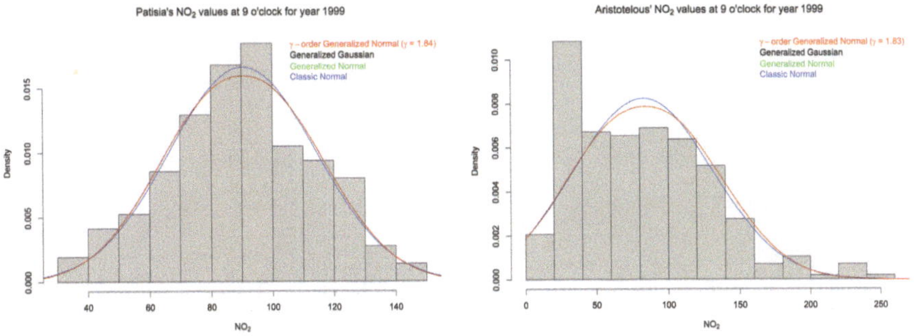

Fig. 18 Patisia's and Aristotelous' NO_2 values at 9 o'clock for year 1999

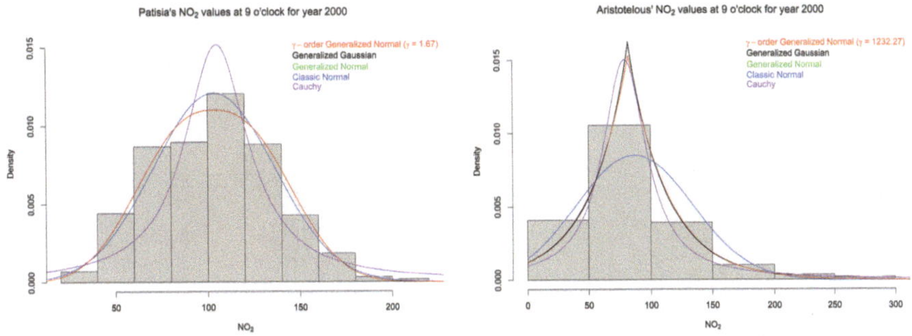

Fig. 19 Patisia's and Aristotelous' NO_2 values at 9 o'clock for year 2000

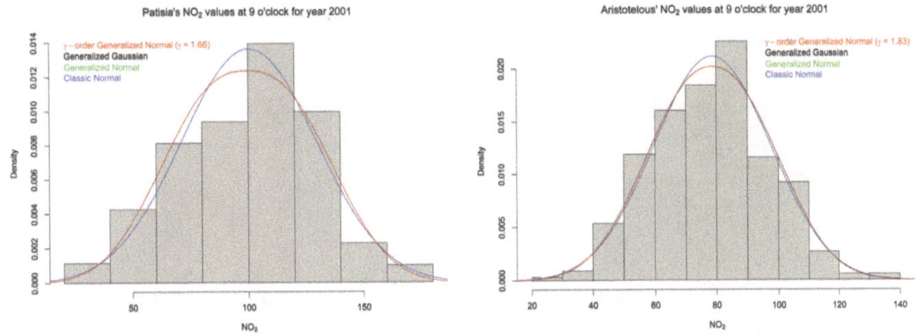

Fig. 20 Patisia's and Aristotelous' NO_2 values at 9 o'clock for year 2001

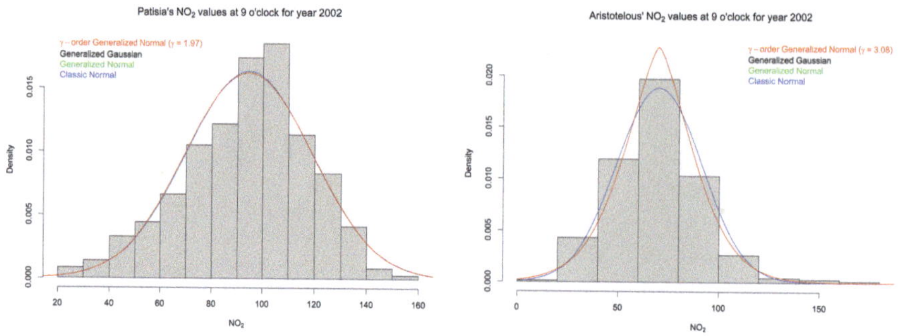

Fig. 21 Patisia's and Aristotelous' NO_2 values at 9 o'clock for year 2002

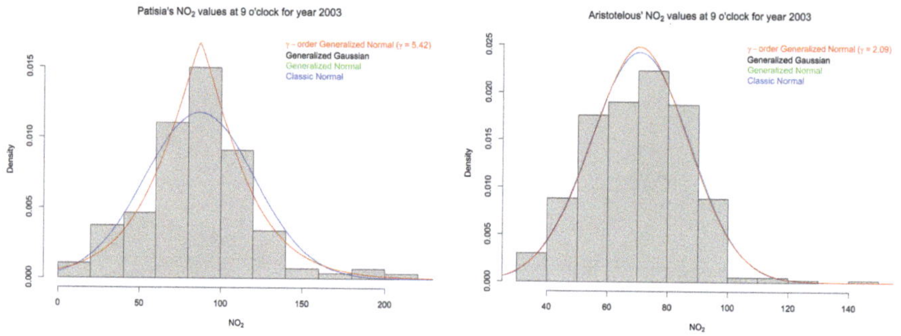

Fig. 22 Patisia's and Aristotelous' NO_2 values at 9 o'clock for year 2003

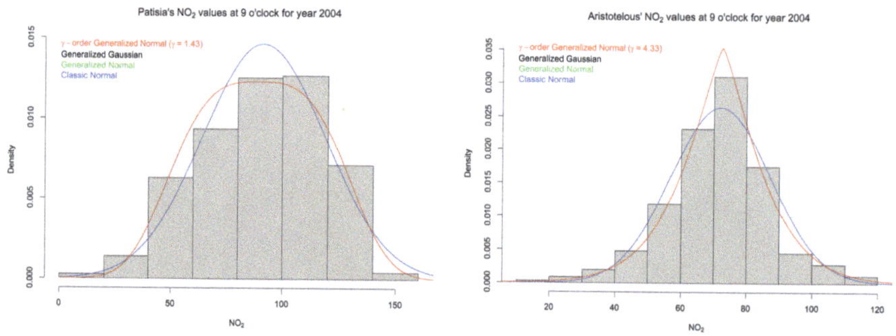

Fig. 23 Patisia's and Aristotelous' NO_2 values at 9 o'clock for year 2004

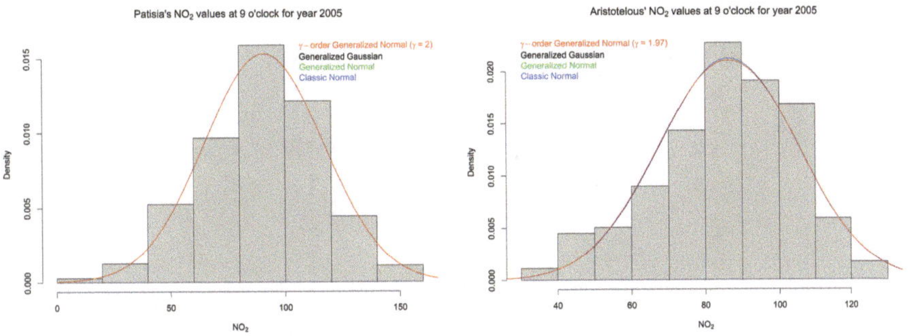

Fig. 24 Patisia's and Aristotelous' NO_2 values at 9 o'clock for year 2005

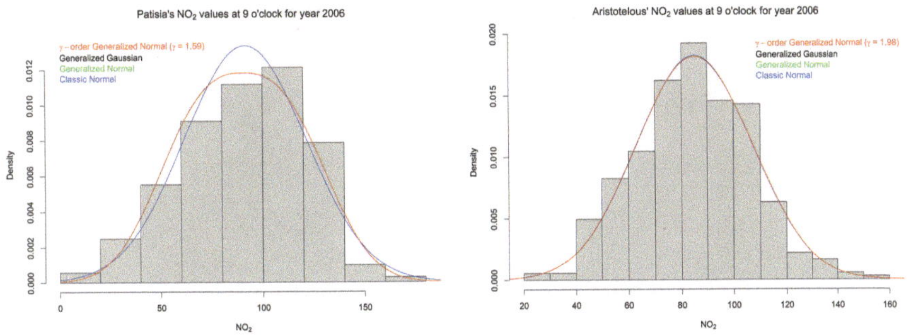

Fig. 25 Patisia's and Aristotelous' NO_2 values at 9 o'clock for year 2006

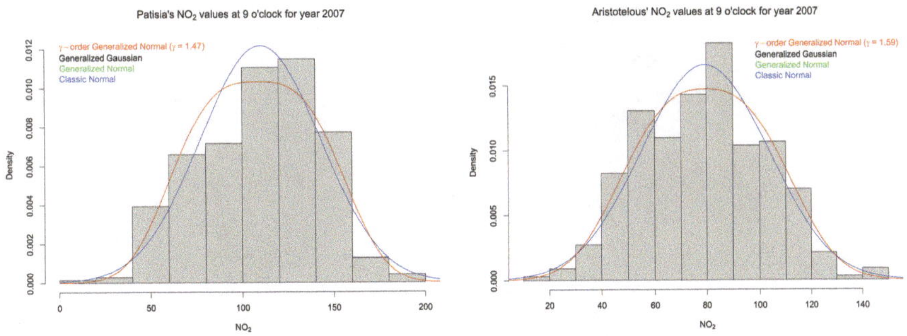

Fig. 26 Patisia's and Aristotelous' NO_2 values at 9 o'clock for year 2007

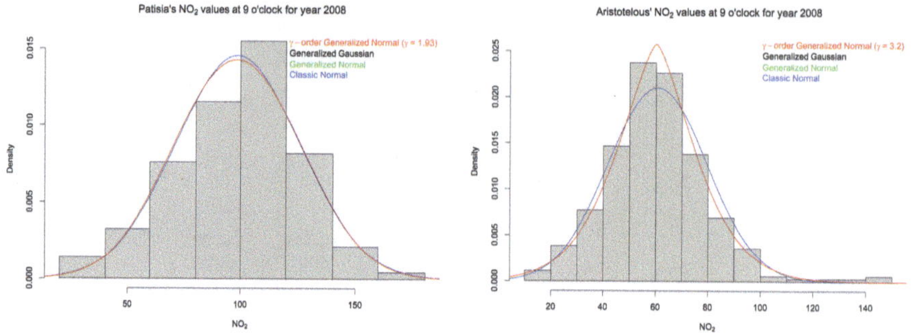

Fig. 27 Patisia's and Aristotelous' NO_2 values at 9 o'clock for year 2008

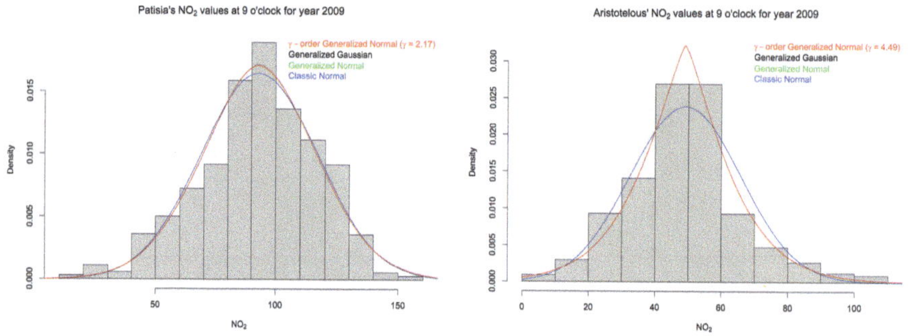

Fig. 28 Patisia's and Aristotelous' NO_2 values at 9 o'clock for year 2009

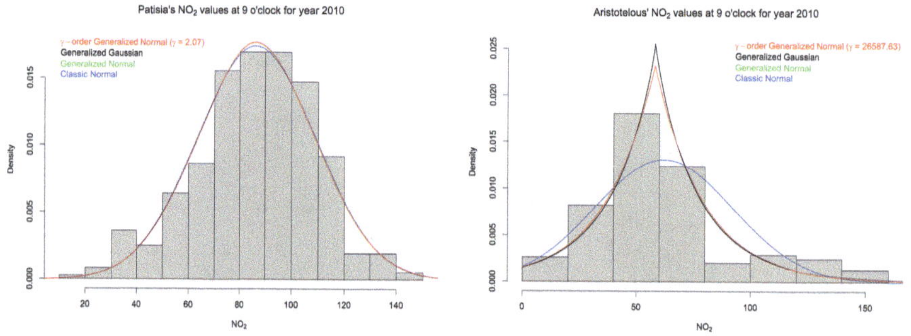

Fig. 29 Patisia's and Aristotelous' NO_2 values at 9 o'clock for year 2010

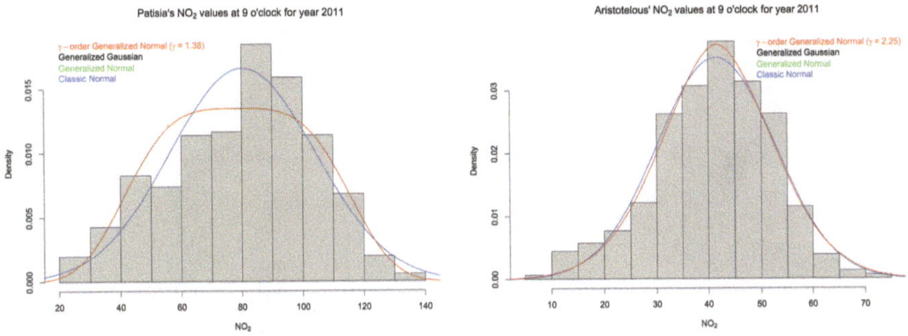

Fig. 30 Patisia's and Aristotelous' NO_2 values at 9 o'clock for year 2011

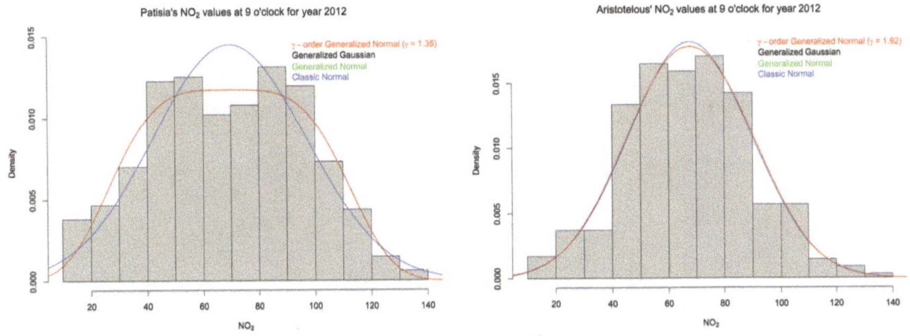

Fig. 31 Patisia's and Aristotelous' NO_2 values at 9 o'clock for year 2012

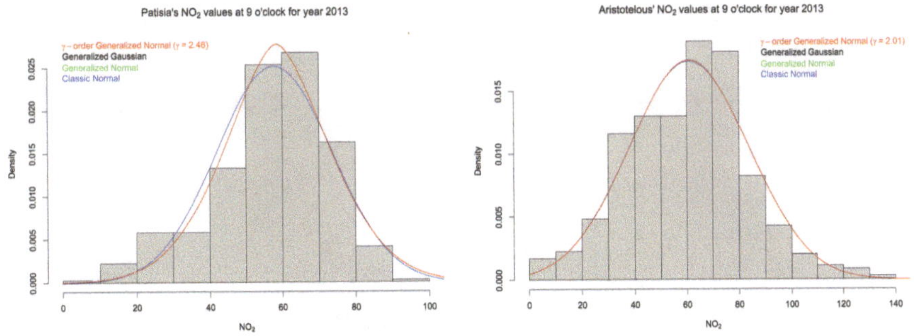

Fig. 32 Patisia's and Aristotelous' NO_2 values at 9 o'clock for year 2013

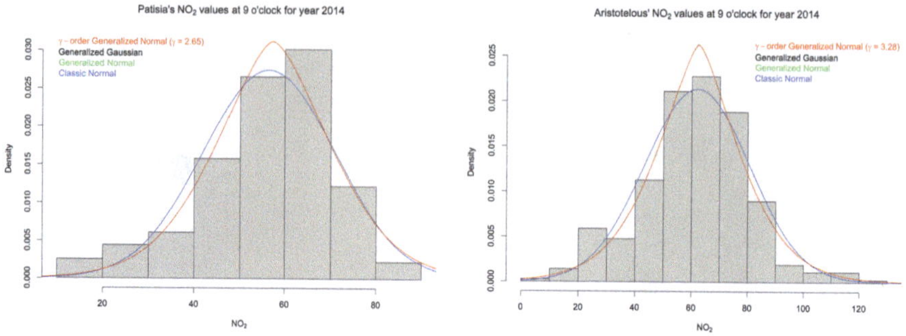

Fig. 33 Patisia's and Aristotelous' NO_2 values at 9 o'clock for year 2014

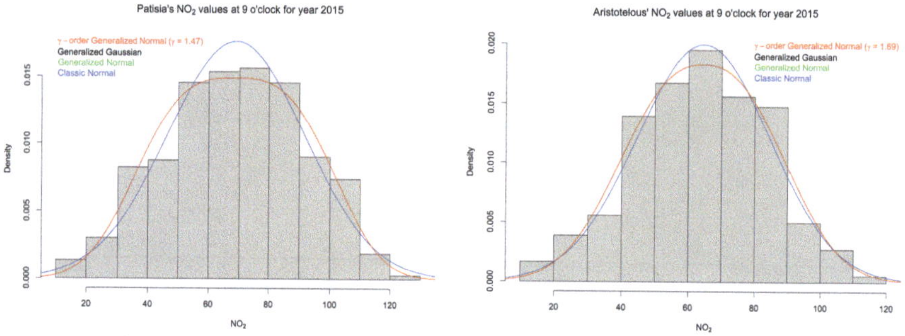

Fig. 34 Patisia's and Aristotelous' NO_2 values at 9 o'clock for year 2015

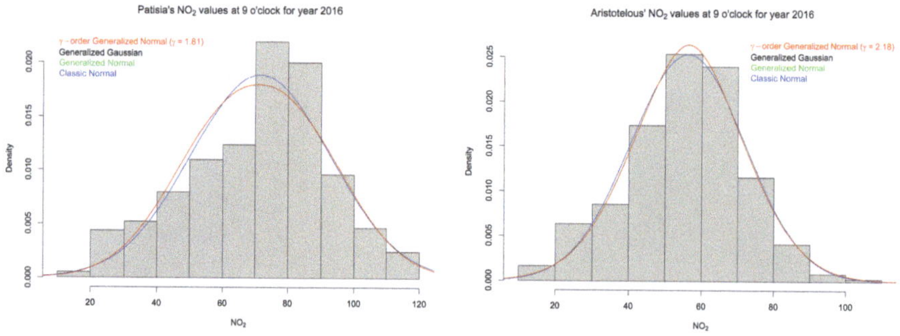

Fig. 35 Patisia's and Aristotelous' NO_2 values at 9 o'clock for year 2016

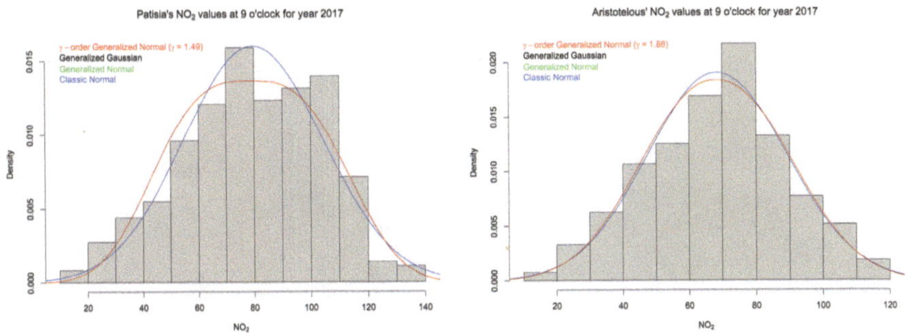

Fig. 36 Patisia's and Aristotelous' NO_2 values at 9 o'clock for year 2017

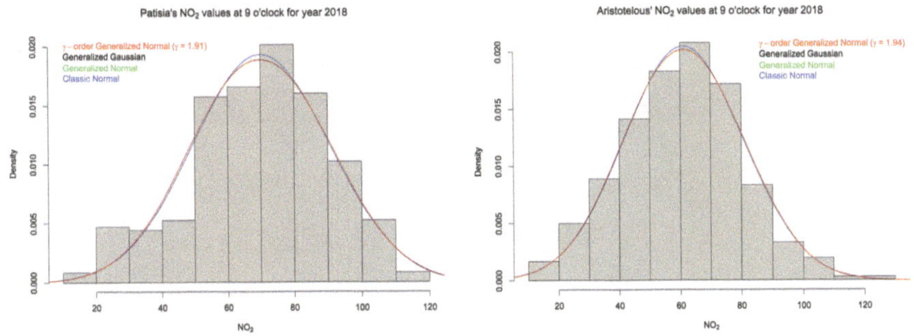

Fig. 37 Patisia's and Aristotelous' NO_2 values at 9 o'clock for year 2018

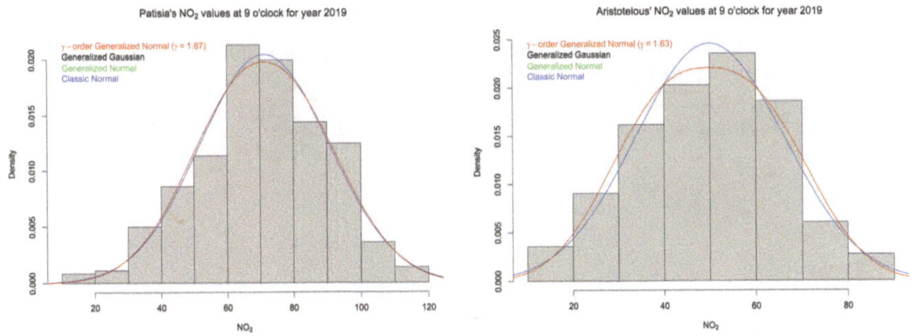

Fig. 38 Patisia's and Aristotelous' NO_2 values at 9 o'clock for year 2019

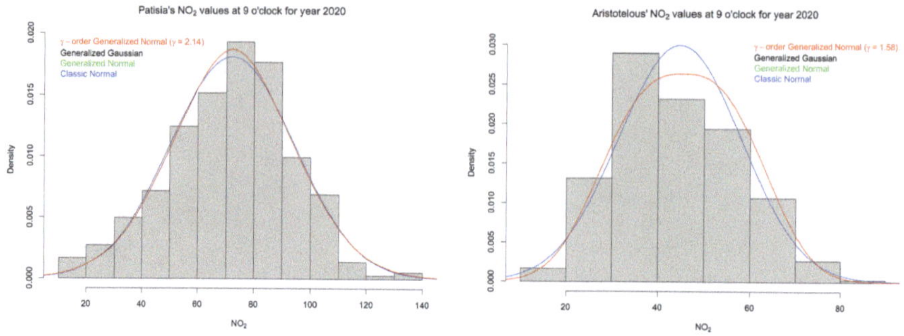

Fig. 39 Patisia's and Aristotelous' NO_2 values at 9 o'clock for year 2020

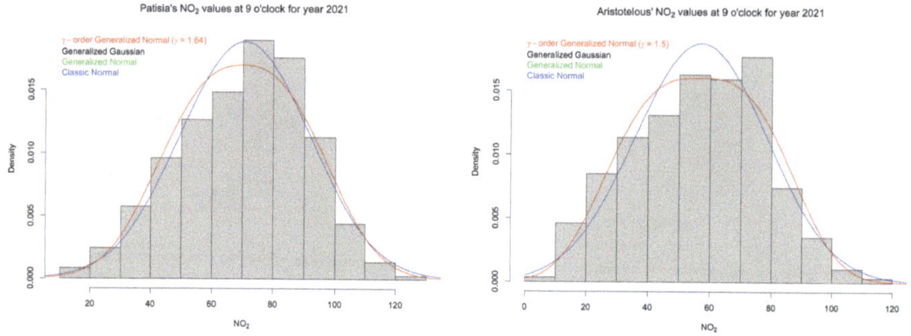

Fig. 40 Patisia's and Aristotelous' NO_2 values at 9 o'clock for year 2021

Acknowledgements We would like to thank all the referees for their helpful comments which lead us to the final version of the paper.

Disclosure of Interests. There are no competing interests of the authors to be declared.

6 Appendix

Appendix: Probability Model Comparison of NO_2 in different areas, different times

Note that we have more results for other pollutants, but here we present only for NO_2.

Note also that for years 1985, 1989, 2003, 2004, 2007, 2011, 2012, 2015, and 2017, the three parameter distributions (γ-order Generalized Normal and Generalized Gaussian) fit the data better (based on AIC values) than any other. Table below presents γ-order Generalized Normal parameters' estimates of Athens and Salonica data for each year available.

	Patisia			Aristotelous		
year	γ	μ	σ	γ	μ	σ
1984	2.655	105.273	29.824			
1985	7558.073	103	37.346			
1986	1.713	106.705	41.347			
1987	1.872	103.873	30.359			
1988	2.669	115.832	37.479			
1989	1.517	119.7	42.158			
1990	1.678	120.72	44.183			
1991	1.974	109.9	36.399			
1992	1.788	121.16	48.415			
1993	1.675	107.923	42.773			
1994	1.59	101.553	38.258	2.478	11 0.9	34.381
1995	2.024	96.971	32.258	2.995	115.814	28.901
1996	2.323	97.829	35.436	3.264	96.912	25.905
1997	1.839	97.001	33.757	2.135	93.161	25.241
1998	1.961	102.249	31.271	2.625	103	53.102
1999	1.843	90.339	24.676	1.83	83.772	49.975
2000	1.674	103.446	35.366	1232.27	82	32.602
2001	1.658	99.421	31.557	1.826	78.636	19.484
2002	1.966	94.273	24.574	3.082	69.75	18.613
2003	5.418	86.898	27.111	2.094	70.471	16.182
2004	1.432	89.147	31.629	4.332	72.45	12.502
2005	2.003	90.66	26.089	1.966	86.273	18.951
2006	1.592	90.228	32.957	1.98	85.005	21.961
2007	1.47	108.089	37.658	1.59	79.439	26.502
2008	1.927	98.323	27.736	3.199	60.089	16.508
2009	2.165	92.989	23.587	4.495	49	13.763
2010	2.065	86.267	22.625	26587.627	58	21.505
2011	1.382	78.429	28.715	2.247	41.647	10.868
2012	1.347	69.382	33.096	1.92	67.854	22.268
2013	2.462	58.582	14.791	2.015	60.969	22.877
2014	2.652	57.29	13.311	3.283	62.204	16.275
2015	1.466	68.559	26.152	1.69	64.277	21.412
2016	1.814	70.927	21.971	2.181	56.273	15.292
2017	1.495	78.316	28.33	1.858	68.2	21.466
2018	1.908	69.826	20.941	1.937	61.088	19.681
2019	1.868	71.14	19.958	1.629	49.702	17.633
2020	2.138	72.274	21.558	1.577	45.334	14.728
2021	1.643	69.884	22.973	1.497	56.555	24.264

There is a number of figures as the comparison tries to cover, different pollutants, different years, two different industrial cities. Those who need more figures and information may contact Ass. Prof. C-S. Nisiotis.

References

1. Abramowitz, M., Stegun, I.A.: Handbook of Mathematical Functions with Formulas, Graphs, and Mathematical Tables, 10th edn. National Bureau of Standards, Wiley, Washington DC (1972)
2. Anderson, T.W.: An Introduction to Multivariate Statistical Analysis, 3rd edn. Wiley Series in Probability and Statistics, Wiley-Interscience (2013)
3. Cover, T.M., Thomas, J.A.: Elements of Information Theory, 2nd edn. Wiley, New York (2006)
4. Domínguez-Molina, J.A., González-Farías, G., Rodríguez-Dagnino, R.M.: A practical procedure to estimate the shape parameter in the generalized Gaussian distribution. Universidad de Guanajuato, ITESM Campus Monterrey, Guanajuato **27**(3), 1–27 (2003)
5. Dytso, A., Bustin, R., Poor, H.V., Shamai, S.: Analytical properties of generalized Gaussian distributions. J. Stat. Distrib. Appl. **5**(1), 1–40 (2018)
6. Gómez, E., Gomez-Viilegas, M.A., Marín, J.M.: A multivariate generalization of the power exponential family of distributions. Communications in Statistics-Theory and Methods **27**(3), 589–600 (1998)
7. Halkos, G.E.: Economics and the Environment: Methods of valuation and management. Liberal Books (2013)
8. Halkos, G.E., Kitsou, D.C., Marín, J.M.: Weighted location differential tax in environmental problems. Environ. Econ. Policy Stud. **20**, 1–15 (2018)
9. Halkos, G.E.: Econometrics. Disigma Pub. (2019)
10. Kitsos, C.P., Nisiotis, C-S.: A time series analysis comparing the main air pollutants of Athens and Salonica in Greece submitted (2025)
11. Kitsos, C.P., Vassiliadis, V.G., Toulias, T.L.: MLE for the γ-order Generalized Normal Distribution. Discussiones Mathematicae-Probability and Statistics **34**(1–2), 143–158 (2014)
12. Kitsos, C.P., Tavoularis, N.K.: Logarithmic Sobolev Inequalities for Information measures. IEEE Trans. Inform. Theory **55**(6), 2554–2561 (2009)
13. Kitsos, C.P., Toulias, T.L.: On the family of the γ-ordered Normal Distributions. Far East J. Theoret. Stat. **35**(2), 95–114 (2011)
14. Kitsos, C.P., Nyamsi, U.E.: Applying the Generalized Normal distribution for Modelling Asset Returns Data, 5th International Congress of Applied Statistics, Istanbul (2024)
15. Kitsos, C.P., Stamatiou, I.S.: Laplace transformation for the γ-order generalized Normal $N_\gamma(\mu, \sigma^2)$. Far East Journal of Theoretical Statistics **68**(1), 1–21 (2024a)
16. Kitsos, C.P., Stamatiou, I.S.: The γ-order generalized Chi-square distribution ${}_\gamma X_n^2$. Research in Statistics **2**(1), 1–9 (2024b)
17. Kitsos, C.P., Stamatiou, I.S.: The γ-order generalized Chi-γ distribution ${}_\gamma X_n^\gamma$, REVSTAT–Statistical Journal, Accepted - June (2024). Retrieved from https://revstat.ine.pt/index.php/REVSTAT/article/view/737
18. Kitsos, C.P., Stamatiou, I.S.: The γ-order Generalized Normal and Differential Equations, to appear in Non-Linear. Geometry, and Differential Equations by Springer, Analysis (2025a)
19. Kitsos, C.P., Stamatiou, I.S.: On the emerged from LSI γ-order generalized Normal distribution. J. Indian Soc. Probab, Stat (2025b)
20. Nadarajah, S.: A Generalized Normal Distribution. J. Appl. Stat. **32**(7), 685–694 (2005)
21. Royen, T.: Expansions of the multivariate Chi-square distribution. J. Multivar. Anal. **38**(2), 213–232 (1991)
22. Schervish, M.J.: Theory of Statistics, Springer, New York, 1995 corrected second printing (1997)
23. Schrödinger, E.: Zur Theorie Der Fall-und Steigversuche an Teilchen Mit Brownsher Bewegung. Physikalische Zeitschrift **16**(220), 289–295 (1915)
24. Takashi, S., Takasi, S.: Applied analysis: Mathematical Methods in Natural Science, Imperial College Press (2001)
25. Toulias, T.L., Kitsos, C.P.: On the generalized lognormal distribution. J. Probab. Stat. **2013**, 15 (1/432642) (2013)
26. Yu, S., Zhang, A., Li, H.: A Review of Estimating the Shape Parameter of Generalized Gaussian Distribution. J. Comput. Inf. Syst. **8**(21), 9055–9064 (2012)

Measuring Uncertainty, Transfer Entropy, and G-Causality in Environmental Economics

George Halkos and Christos Kitsos

Abstract Human activities have created environmental degradation with the internalization of the resulting externalities having been the main concern for policy-makers worldwide. Uncertainty and unconvincing scientific evidence of various biophysical processes are present in many planned environmental policies. An important source of model uncertainty is accounted by entropy with the typical normal distribution being inadequate in such analyses challenging for more sensible approximations. The problems are, from one side the fat tails characteristic in this area and on the other side what probability density function (pdf) to be chosen. The choice of the appropriate probability model describing the phenomenon that is the pdf is a main priority in any decision-making planning. Here we pay attention to the entropy and even more on the transfer entropy in Environmental Economics and the existing underlying uncertainty based on the probability theory. We show that the γ-order Generalized Normal distribution covers both requests, due the "International constant" $(\gamma/(\gamma-1)) * * (\gamma/(\gamma-1))$, leading to a number of pdf, and the Logarithm Sobolev Inequalities (LSI), which provide a solid background.

Keywords Uncertainty · Environmental economics · Transfer entropy · G-causality

1 Introduction

Uncertainty and entropy are two terms first adopted in the development of Thermodynamics with the pioneering work of Schrodinger [32]. Afterward they were adopted in Physics, then in Mathematics and Statistics and now are applied in various fields, covering Thermodynamics, with the pioneering work of Landau and Lifshitz

G. Halkos (✉)
Laboratory of Operations Research, Department of Economics, University of Thessaly, Volos, Greece
e-mail: halkos@uth.gr

C. Kitsos
University of West Attica, Egaleo, Greece

G. Halkos et al. (eds.), *Environmental Risk Analysis*, Springer Proceedings in Mathematics & Statistics 540, https://doi.org/10.1007/978-3-032-24155-9_6

[30] followed by Mandl [31] among others. An uncertain situation means not being known beyond doubt, not having complete knowledge. So it is clear that uncertainty is not the error, but it is somehow related to it. In principle, uncertainty is the lack of certainty and as far as the Environmental Economics is concerned, Halkos and Kitsos [17] tackled the problem investigating a number of possible causes of uncertainty providing possible measurements of it.

At the early stages of developing, such a theoretical approach was not all satisfied. A typical example is Caratheodory's [4] axiomatic formulation of the second law of Thermodynamics, which is considered as one of the standard forms of formulation of the law. However, it was oriented more as advanced Mathematics than Physics. The formulation was strongly criticized by Max Planck. We avoid considering such problems from other research fields, but still each problem needs its own Mathematical foundation, especially Physics [24]. Skeptics often remain reluctant to expand established beliefs, even when the emerging framework is theoretically solid, and closely reflect the new developments.

Entropy describes the disorder of a system, the "change within a closed system." It is based on the Greek prefix "en" implying within and the "trope" related to the root, here meaning the "change." The "closed system" is a request from the physical point of view of entropy. We understand entropy when the temperature of everything in the room evens out. Entropy is central to the second law of thermodynamics, known as Caratheodory's theorem (among others, Zachmanoglou [36]).

In principle and in mathematical terms, uncertainty links a real-valued function of events in a probability space, which depends on the probability law of the events under consideration. Moreover, events with probability one have zero uncertainty, as we are "certain" that it will take place. Besides as the probability of the event drops, the uncertainty of the event increases. It is thus required that, for the independent events, the uncertainty of their joint occurrence should equal the sum of their individual uncertainties. Looking for a measurable function, reflecting these imposed important requirements, we can conclude that the information of event E, $I(E)$, linked with its probability, $P(E)$, obeys

$$I(E) = -clog \cdot [P(E)] \tag{1}$$

with $c > 0$ a given constant.

In this paper we try to apply these ideas, providing more emphasis on the entropy and particularly to *transfer entropy* (TE), working on a family of distributions, to be used in various environmental problems. We have already worked with γ-order generalized normal $N_\gamma(\mu, \Sigma)$ [17], introduced by Kitsos and Tavoularis [26], and it has been already adopted for an extension to TE under the N_γ (μ, Σ) [20]. We shall try to extend such a work to Environmental Economics working with N_γ (μ, Σ) and the extensions on Transfer Entropy (TE).

The structure of the paper is as follows. In Sect. 2 we formalize the existent background, while in Sect. 3 we discuss the appropriate points for N_γ (μ, Σ). Section 4 is devoted to the extension that N_γ (μ, Σ) can offer to entropy and uncertainty. In

Sect. 5, we introduced the transfer entropy especially to Environmental Economics (TEEE). The last section concludes the paper.

2 Background

Uncertainty and entropy, as well as the fundamental approaches in Quantum Statistical Mechanics, share a solid background from Physics with Statistics. In 1865, Rudolf Clausius (1822–1888) introduced the term entropy, using Greek terminology as the energy has still the same orientation as etymologists believe. Around 1875, Ludwig Boltzmann (1844–1906) put entropy into the probabilistic framework and developed the statistical explanation of the Second Law of Thermodynamics. But the fundamental push to probability was given by Shannon [34], working on formation theory [5] as we will refer to it in Sect. 3.

Environmental concerns have been considered as a crucial and substantial part of assessing energy conversion systems. The thermodynamic approach implies a profound evaluation of the notion of sustainability as the second law of Thermodynamics counteracts the ability of any open and developing system to preserve itself sustainably without rewarding itself of an uninterrupted supply of low entropy that is high specific exergy input. Exergy in the form of available energy is a crucial notion in thermodynamics. If a society is considered as an open system, its ability to develop in a sustainable manner is a function not only on how it uses the conventional energy sources but also on the rate at which it exploits renewable energy sources [33].

In economic theory, among other fields, multiple sources of uncertainty, including model selection, data-related issues, and the choice of appropriate mathematical specifications, contribute to the overall uncertainty, as extensively discussed in Halkos and Kitsos [10, 17]. The three different research areas Risk Analysis, Uncertainty, and Entropy are supporting different lines of thought, through Statistics and are not identical. Having different initial points, Risk Analysis has arisen from cancer and now is involved in many areas, such as Environmental Risk; while the other two approaches originated in Physics, were subsequently improved by Statistics, and are demonstrated here to be important in Environmental Economics, adopting their Statistical evolution.

Heisenberg's uncertainty principle, in the first introduction in [19] as a part of the mathematical framework of quantum physics, stated that: It is not, in general, possible to predict the value of a quantity with arbitrary certainty, even if all the initial conditions are well imposed and specified. Under uncertainty the more precisely the position Δx of some particle is determined, the less precisely its momentum Δp can be predicted and concluded that $\Delta_X \Delta_P \approx h$ with h being the full Planck constant. Later, Weyl [35] proved that

$$\sigma_X \sigma_P \geq \frac{1}{2} \frac{h}{2\Pi} \tag{2}$$

with the reduced Planck's constant. See also Kitsos and Stamatiou [23].

There are cases where the information needs to be evaluated under the line of thought as in (1). In such a case for a measurable partition, τ, the space of the information function, $I(\tau)$, can be considered as

$$I(\tau) = (-c)\sum_{\mathrm{E}\in\tau} \log[\mathrm{P(E)}]\mathrm{X_E}$$

with X_{E} being the characteristic function and c chosen to yield logarithm to base 2, while the expected value of $I(\tau)$ provides the entropy of the partition.

It was Shannon [34] who thought to provide the information in units of bits. Therefore, he adopted the base-2 measure of entropy, using the logarithm of base 2, in such a way that for a given random variable (rv) X, with probability density function $p(x)$ for the discrete case the entropy of X, $H(X)$, is

$$H(X) = -\sum_{\chi\mathrm{E}\mathbb{R}} p(x)\log p(x) = E\big[-\log p(x)\big] \tag{3}$$

with E[.] denoting, as usually, the expected value.

Under the same line of thought, the conditional entropy of the rv Y given the rv X is defined as

$$H(Y|X) = -\sum_{\chi,y\in\mathrm{R}} p(x,y)\log\big[1/p(y|x)\big] \tag{4}$$

while the joined entropy can be, briefly, defined as

$$H(X,Y) = H(X|Y) + H(X) \tag{5}$$

When the comparison of two distributions is requested, the Kullback–Leibler divergence, D_{KL}, is adopted, provided that $p(x)$ is the "true" distribution probability and the $q(x)$ probability distribution is used to evaluate "how far" it deviates from $p(x)$¸ then:

$$\begin{aligned} &D(p,q) := D_{KL}(p(x)\|q(x)) = \\ &= \sum_{x\in X} p(x)\log\tfrac{p(x)}{q(x)} = \sum_{x\in X} p(x)\log p(x) - \sum_{x\in X} p(x)\log q(x) = \\ &= -\sum_{x\in X}\big[-p(x)\log p(x)\big] - \sum_{x\in X} p(x)\log q(x) = \\ &= E\big[-\log(q(x))\big] - E\big[-\log(p(x))\big] = H_q(X) - H_p(X) \end{aligned} \tag{6}$$

The first line is the notation, the second line is the definition, and the third line is a trivial new presentation, which with the last line we can see that the definition is the subtraction of the entropy of X due to "valid" $p(x)$, $H_p(X)$ from the entropy due to "examined" $q(x)$, $H_q(X)$.

In continuous case, the summation is replaced by the integration and is usually referred to as "differential entropy." Notice that the Kullback–Leibler divergence, $D = D_{KL}$ as in (6), with a number of notations, with the compact presentation $D = D(p, q)$, will be proved useful in the sequence; it is not symmetric. Actually, the distance measure in Statistics is not exactly what we mean as "distance" in Linear Algebra.

But the lack of symmetry does not create any problem in Environmental Studies. We usually observe the source point of pollution, say at the origin $O(0, 0, 0)$, and the pollution is moved to point $P_1(X_1, Y_1, Z_1)$ in (Euclidean) distance $OP_1 = L_1$. We are working on assuming continuous distribution functions (df) to approach the pollution at every point of investigation. At point O the distribution function of the initial pollution, we say that it is f_0, while at point P_1 it is f_1; therefore the K-L divergence is

$$D_{01} = D_{01}(f_0, f_1)$$

and can be evaluated. Following the same direction, a sequence of points $P_i(X_i, Y_i, Z_i)$, $i = 1, 2, \ldots, n$ is selected.

Therefore, we can evaluate the K-L divergence between the selected points, and each one from the origin, i.e.,

$$\begin{aligned} D_{oi} &= D_{oi}(f_0, f_i), i = 1, 2, \ldots, n-1 \\ D_{ij} &= D_{ij}\left(f_i, f_j\right), j \neq i \end{aligned}$$

We can say that the influence of the source of the pollution vanishes when

$$f_i \approx f_{i+1} f_\iota(x) = f_{i+1}(x)$$

The two successive points $P_i(X_i, Y_i, Z_i)$ and $P_{i+1}(X_{i+1}, Y_{i+1}, Z_{i+1})$ provide the same **df** and thus the K-L divergence is zero, practically very small as

$$D_{oi}(f_o, f_i) = D_{oi+1}(f_i, f_{i+1})$$

Equivalently

$$\begin{aligned} &|D_{oi} - D_{oi+1}| < \varepsilon, i = 1, 2, .., n - q\varepsilon > 0. \\ &\text{and } \left|D_{ij} - D_{ij+1}\right| < \varepsilon', j = i + 1, \varepsilon' > 0. \end{aligned} \tag{7}$$

The evaluation of the K-L distance at stage n is crucial, while we can have an information data set of "how far," in terms of Euclidean distance, we are from the origin. It is clear that the distance from the origin depends on the pollution measurements the investigator collects. If we keep the measurements every d, say, Euclidean distance, the total distance, from the origin, at the end is $d_T = (n - 1)d$.

It would be desirable to work with a specific "large family" of distributions to keep the evaluation of distribution easier. We decided to work with the family of the

γ-order Generalized Normal distributions, with shape parameter γ

$$G_\gamma = \{\mathbb{N}_\gamma(\mu, \Sigma), \gamma \in \mathbb{R} - [0, 1]\} \tag{8}$$

We introduce the definition and probability characteristics of this family in the next section, while Sect. 4 is devoted to the extension of uncertainty and entropy measures that can be obtained for the family G_γ.

3 γ-Ordered Generalized Normal Nγ (μ, Σ)

The multivariate normal distribution (p variables) is well known, with a large number of applications, in various fields of interest [1] with Econometrics being one (Halkos, [11] among others). There are some attempts to create a generalized form, by trial and error, working on the coefficients. The only one emerged with an extra parameter, γ, from a Logarithm Sobolev Inequality (LSI) by Kitsos and Tavoularis [26], provided an extra shape parameter, and a strong theoretical background for its existence. These were a number of technicalities to generalize the well-known $N(\mu, I\sigma^2)$, the p-variate normal with the vector $\mu \in \mathbb{R}^{\mathrm{P}}$ and covariance matrix $\Sigma \in \mathbb{R}^{p \times p}$.

The γ-order generalized normal with an extra shape parameter

$$\mathbb{R} - [0, 1] \rightarrow \gamma \in \mathbb{R} - [0, 1]$$

will be denoted by $N_\gamma(\mu, \Sigma)$ and density function $f_\gamma(x)$ equals

$$f_\gamma(x) = C_P[\det(\Sigma)]^{-\frac{1}{2}} \exp\left\{-\frac{\gamma - 1}{\gamma}[Q(x)]^{\frac{\gamma}{2(\gamma-1)}}\right\} \tag{9}$$

with $x, \mu \in \mathbb{R}^P$ and

$$Q(x) = (x - \mu)\Sigma^{-1}(x - \mu)^{\mathrm{T}}$$

with the transpose $a^T \in \mathbb{R}^{p \times 1}$, when $a \in \mathbb{R}^{1 \times p}$

$$C_P = \pi^{-\frac{p}{2}} \frac{\Gamma(\frac{\mathrm{P}}{2} + 1)}{\Gamma\left(p\frac{\gamma}{\gamma-1} + 1\right)} \left(\frac{\gamma - 1}{\gamma}\right)^{\mathrm{P}\frac{\gamma-1}{\gamma}} \tag{10}$$

when $\gamma = 2$ it is obvious that the γ-order generalized normal is reduced to the usual p-variable normal. Moreover $N_\gamma(\mu, \Sigma)$ is a Kotz-type distribution in the sense that

$$N_\gamma(\mu, \Sigma) = Kotz\left(1, \frac{\gamma - 1}{\gamma}, \frac{\gamma}{2(\gamma - 1)}; \mu, \Sigma\right)$$

with Kotz(.) being the appropriate distribution type, and its parameters.

Notice that the density function $f_\gamma(x)$, as in (9), the γ-order generalized normal distribution emerged as an extreme function to Euclidean Logarithm Sobolev Inequality (LSI), and it is not an artificial achievement. This is clear as the term $\gamma_0 = \frac{\gamma-1}{\gamma}$ and its inverse γ_1, as well as the term $(\gamma_1)^{\gamma_1}$ known as "international constant" are complex to be "constructed." Therefore, there is a solid background for the $N_\gamma(\mu, \Sigma)$ [26]. The family of distributions has been tackled by Kitsos and Toulias [29].

Estimation of the parameters would be a problem, and the MLE, as was considered by Fisher [6], for the γ-order generalized Normal, has been considered $f_\gamma(x)$; the γ-order generalized normal distribution emerged as an extreme function to Euclidean Logarithm Sobolev Inequality (LSI), and it is not an artificial achievement.

The distribution function (9), with normalizing factor (10), forms the family of Generalized Normal distributions as in (8). Therefore, different values of γ provide different curves, mainly "fat tailed" distributions.[1] In particular as γ approaches 1 and infinity, the $N_\gamma(\mu, \sigma^2 \mathrm{I_P})$ approach the uniform distribution with $p = 1$ and the p-variate uniform equals

$$U(x) = \frac{\Gamma\left(\frac{p}{2}+1\right)}{\pi^{p/2}\sigma^p}\ , \quad |\mathrm{X} - \mu| \le \sigma \tag{11}$$

Moreover as $(\gamma \to \infty)$, the p-variate Laplace is obtained with

$$L(x) = \frac{\Gamma\left(\frac{p}{2}+1\right)}{\pi^{p/2}\sigma^\rho}\frac{1}{p!}\exp\left\{-\frac{|x-\mu|}{\sigma}\right\}\ , \quad |\mathrm{X} - \mu| \le \sigma \tag{12}$$

Furthermore, a number of well-known distributions are included in family (8) of the γ-order generalized Normal distributions, defined by (9) and (10), the Dirac distribution being one of them (for details see Kitsos and Toulias [28]). For these two important distributions in Environmental Economics studies, the Kullback–Leibler divergence, for given values of γ, can been evaluated. Indeed, the following holds.

Theorem 1 *The K-L information between two uniforms with $\sigma_1 \ge \sigma_2$ and two Laplace as in (11) and (12) is*

$$D(U_1, U_2) = p \ln \frac{\sigma_1}{\sigma_2} \tag{13}$$

$$D(L_1, L_2) = p\left[\ln \frac{\sigma_1}{\sigma_2} - 1 + \frac{\sigma_1}{\sigma_2}\right] \tag{14}$$

In this section the Kullback–Leibler divergence, for the given introduced γ-order generalized Normal, for different values of γ has been evaluated. Next step is the

---

[1] To work with "fat-tailed" or "heavy-tailed" distributions, see Furioli et al. [7]. Inequalities (LSI) of the distributions discussed are rather complicated to be applied in practical problems, such as the Cauchy distribution.

introduction of the appropriate Entropy measures for the distribution function (9), generator of the family of distributions as in (8).

4 Entropy and Uncertainty for $N_\gamma(\mu, \Sigma)$

The Shannon Entropy, H(.), of the random variable $X \sim N_\gamma(\mu, \Sigma)$, [26] is

$$H_\gamma(\mathrm{X}) = p\frac{\gamma - 1}{\gamma} + \log \frac{(\det(\Sigma))^{\frac{1}{2}}}{Cp} C_P \textit{ as in } (10) \quad (15)$$

Depending on γ every member of the family G_γ obtains the Entropy measure, which in case of $\gamma = z$ coincides with the p-variate normal $N(\mu, \Sigma)$:

$$H_2(\mathrm{X}) = \frac{1}{2}\log\left\{(2\pi e)^p|\det(\Sigma)|\right\} \quad (16)$$

With $\gamma = 1$, the Laplace case of the entropy for the multivariate Laplace is

$$H_L(\mathrm{X}) = p + \log \frac{p!\pi^{\frac{p}{2}}}{\Gamma\left(\frac{p}{2}+1\right)}[\det(\Sigma)]^{\frac{1}{2}} \quad (17)$$

When $p = 1$ for the uniform distribution $U(\mu - \sigma, \mu + \sigma)$, for the normal, $N(\mu, \sigma^2)$ and for the Laplace $L(\mu, \sigma)$ it can be evaluated respectively as:

$$\begin{aligned} h_U(\mathrm{X}) &= \log 2\sigma \\ h_{\mathrm{N}}(\mathrm{X}) &= \log{(2\pi e\sigma)^{\frac{1}{2}}} \\ h_L(\mathrm{X}) &= 1 + \log 2\sigma \end{aligned} \quad (18)$$

Proposition 1 *For the differential entropy of Uniform, Normal, and Laplace distributions, it holds*

$$h_U^{(x)} h_U(\mathrm{X}) < h_N^{(x)} < h_{\mathrm{N}}(\mathrm{X}) < h_L(\mathrm{X}) h_L^{(x)} \quad (19)$$

Proof Indeed, with $\sigma > 0$ it is easy to see that it holds
$2\sigma < 2\pi\sigma < 2\pi e\sigma$thus: $h_U(X) = \log(2\sigma) < \log(2\pi e\sigma) < \frac{1}{2}\log(2\pi e\sigma) = h_N$
and: $h_U(X) = \log(2\sigma) < 1 + \log(2\sigma) = h_L$.

From (16) it is easy to be verified that the more the involving parameters p, the larger the (differential) entropy is, i.e., there is an order of the entropy depending on the number of the involving parameters. This implies that in Environmental Economics problems we are working with the accurate number of parameters, although this is another problem:

$$H_2^P(\mathrm{X}) \leq H_2^{p+1}(\mathrm{X}) \tag{20}$$

The following theorem is due to Kitsos and Toulias [27], and provides the evaluation of the K-L divergence between generalized Normal, from the defined family as in (8).

Theorem 2 *Let two γ-order generalized normal distributions* $N_\gamma\left(\mu_1, \sigma_1^2 \mathrm{I_P}\right)$ *and* $N_\gamma\left(\mu_2, \sigma_2^2 \mathrm{I_P}\right)$, *with density functions* f_1, f_2 *respectively, be from the family* G_γ. *Their K-L information is*

$$D_\gamma(f_1, f_2) = C_p \frac{1}{\sigma_1^P}\left\{\frac{p}{2}\left[\log\left(\frac{\sigma_2^2}{\sigma_1^2}\right)\mathrm{I}_1 - \mathrm{I}_2 + \mathrm{I}_3\right]\right\} \tag{21}$$

with the integrals I_i, $i = 1, 2, 3$ *defined as*

$$\mathrm{I}_1 = \int_{\mathbb{R}^P} \exp\left[-q_1(x)\right] d_x, \text{ with } q_i(x) = \frac{\gamma - 1}{\gamma}\left[\frac{1}{\sigma_1}\|x - \mu_i\|\right]^{\frac{1}{\gamma-1}}, \quad i = 1, 2$$

and $I_2 = \int_{\mathbb{R}^P} \exp\left[-q_1(x)\right] q_1(x) d_x$

$$I_3 = \int_{\mathbb{R}^P} \exp\left[-q_1(x)\right] q_2(x) d_x$$

Notice that ‖.‖ *represents the appropriate norm. It is clear that the entropy is a function only of the standard deviation (the variance in principle).*

5 Transfer Entropy and G-Causality

Uncertainty is related to entropy, which is associated with entropy type Fisher's information, describing different information measures. When we refer to observed data and possible causalities between them, we refer to G-causality (G-C) in honors of Granger [8] pioneering work and his Nobel Prize Lecture Granger [9].

There is obviously uncertainty to be measured: what is caused from a process as the new result. Therefore, we are looking for a "causality measure" and possible bounds for it. Since the time of Heisenberg [19], the bounds on the uncertainty are investigated.

As such a measure of uncertainty in G-C, we chose the entropy type of information so popular in Cryptography. It is necessary to assign greater weight to the tails, for this purpose, the $\mathrm{N}_\gamma(\mu, \mathrm{I}\sigma^2)$, introduced in Sect, is adopted. 3. The γ-order normal, $N_\gamma(\mu, I\sigma^2)$ as above, offers fat tails and an accepted interpretation for the application, as it is a normal distribution with an extra "shape parameter γ." This is why it was adopted working on the equivalence of G-C and Transfer Entropy (*TE*) [20]. The fact that Granger Causality and Transfer Entropy are equivalent for Normal variables was a main result for the developed background [3].

In principle the transfer entropy of Y to X given W is defined due to the "history" of the event [20, 21] as

$$TE\{Y \to X \mid W\} = H\left(X \mid X^{-}W^{-}\right) - H\left(X \mid X^{-} \oplus Y^{-} \oplus W^{-}\right) \tag{22}$$

where the "–" is also referred to as "the history" or to "past." Please, recall the definition of entropy H(.) and the conditional entropy, H(.|), see (4), as well as the definition ⊕, which Barnet et al. [2] proved that, in a compact presentation, the *TE* for the *G-C* is reduced to Granger-causality Transfer Entropy (*GTE*) and holds:

$$GTE\{Y \to X \mid W\} = \ln\left[\Theta\left(X \mid X^{-} \oplus W^{-}\right)\right] - \ln\left[\Theta\left(X \mid X^{-} \oplus Y^{-} \oplus W^{-}\right)\right] \tag{23}$$

With the nxn covariance matrix, and the n × m matrix of cross covariance defined as

$$\mathrm{Gov}\left[X_i, X_j\right] = \Theta(X) \in \mathbb{R}^{n\times n}$$

$$\mathrm{Gov}\left[X_i, Y_j\right] = \Theta(X, \mathrm{Y}) \in \mathbb{R}^{n\times m}$$

Respectively, for the given jointly distributed multivariate vectors

$$X = (X_1, \ldots, X_n) \in \mathbb{R}^n, (Y_1, \ldots, Y_m) \in \mathbb{R}^m,$$

$$X \oplus Y = (X_1, \ldots, X_n, Y_1, \ldots, Y_n) \in \mathbb{R}^{(n+m)}$$

Now, given Θ(X) the covariance matrix of random variable $X \in \mathbb{R}^n$ and Θ(X, Y)) $\in \mathbb{R}^{nxm}$, the cross covariance of the random vectors X and Y with $m \neq n$, we define

$$\Theta(X|Y) = \Theta(\mathrm{X}) - \Theta(\mathrm{X, Y})[\Theta(\mathrm{Y})]^{-1}\Theta(\mathrm{X, Y})^{\mathrm{T}} \tag{24}$$

under the assumption that the inverse of Θ(ϒ) exists in $\mathbb{R}^{mxm}$. Notice that $\Theta(X|Y) \epsilon R\ nxn$ as both parts on the right-hand side of (24) as in *nxn*. It is easy to see, with the formulation we offer, the similarity of (22) and (23).

According to Granger [9], a consequence of this attempt is that the casual variables can help to forecast the effect variable after other data has been first used. Although this statement is too optimistic, still we have more information for the "future stage" being at the "current stage."

Therefore, the extension to Transfer Entropy problem for the γ-order Normal distribution $N_\gamma(\mu; \Sigma)$ was necessary and Hlavackova–Schlinder et al. [21] provided a number of results for it. The important result, for our development, is provided in the Appendix A Theorem, which evaluates, although in a complicated way, the fact that

$$TE_{\gamma}\{W \to U \mid V\} \tag{25}$$

exists and it is evaluated.

The crucial result from this Theorem, in Appendix A, is that it generates the results for the classical Normal distribution, which are provided for the case of $\gamma = 2$: the transfer entropy with normally distributed and spherically contoured joint distribution vanishes.

Proposition 2 *In the case of the normal distribution* $N(\mu, \sigma^2 \mathrm{I}_q)$:

$$TE_2\{W \to U \mid V\} = 0 \tag{26}$$

Recall that the conditional mutual information $I(U,W|V)$ *vanishes, if and only if (iff) the conditional random variables* $U|V$ *and* $W|V$ *are independent. Therefore, transfer entropy can be expressed also by means of conditional mutual information:*

$$TE_2\{W \to U \mid V\} = \mathrm{I}\,(\mathrm{U}, \mathrm{W}|\mathrm{V}) \tag{27}$$

This is a crucial result as the transfer entropy with normally distributed variable with a spherically contoured joint distribution is zero. So we need more general forms, generalized multivariate normal $N_\gamma(\mu, \sigma^2)$, $\gamma \neq 2$, *elliptically contoured normal applied in Theorem of Appendix A.*

Proposition 3 *The Transfer Entropy.*
$TE\gamma\{W \to U \mid V\}$ with $U \oplus V \oplus W \sim$ Laplace $L^q(\mu, \sigma^2, \mathrm{I}_2)$,

$$i.e.\ \text{when}\ \gamma \to \pm\infty,\ \text{with}\ q = 1 + m + n$$

exists and can be evaluated.

Proposition 4 *Under certain conditions for given pdf p, q, namely p is the joint pdf of X and Y and p exists for stationary random processes Y and X, depending on the estimated value of x, up to time t = i. Given the "history" of x and y, i.e., the pdf*

$$p = p\big(x_{est}|x^-, y^-\big)\big)\ \text{ with }\ p \sim N_\gamma\big(\mu_0, \sigma^2 \mathrm{I}_r\big)$$

and the pdf q is

$$q = q(y) = p\big(x_{est}, y^-\big)p\big(x_{est},\ x^-\big)/p\big(x^-\big)\ \text{ with }\ q \sim N_\gamma\big(\mu_0, \sigma_1^2 \mathrm{I}_r\big)$$

Then:

$$TE\{Y \to X\} = D_{KL}(p\|q) = r\left\{\log\left(\frac{\sigma_0}{\sigma_1}\right) - \frac{\gamma-1}{\gamma}\left[1-\left(\frac{\sigma_1}{\sigma_0}\right)^{\frac{\gamma}{\gamma-1}}\right]\right\} \tag{28}$$

For details in the case of $\sigma_0 \neq \sigma_1$, *see Kitsos and Toulias* [27].

Following Jeffreys' [22] recipe for Symmetric Transfer Entropy (STE), we can define it as $STE\{Y \to X\} = 1/2\{ D_{KL}(p \parallel q) + D_{KL}(q \parallel p)\}$.

Due to this essential point, (28) can be reduced to a symmetric one as

$$STE\{Y \to X\} = r/2\left\{\left[2-\left(\frac{\sigma_1}{\sigma_0}\right)^{\frac{\gamma}{\gamma-1}} - \left(\frac{\sigma_1}{\sigma_0}\right)^{\frac{-2\gamma}{\gamma-1}}\right]\right\} \tag{29}$$

This is an easy relation to be calculated, to obtain symmetric transfer entropy in cases where symmetry is essential in Environmental Economics.

From the above-mentioned useful theoretical results, the following points must be considered with a special attention:

i. Keeping γ stable, then

$$D_{KL}^{r} < D_{KL}^{r+1}.\pi = 1, 2, 3\ldots \tag{30}$$

Due to (30) the Laplace distribution provides the lower bound, when the shape parameter tends to infinity:

ii. For given r = m + n + 1, $\mu_1 = \mu_0$ and for $\gamma_1 < \gamma_2$ then

$$D_{KL,\gamma 1} > D_{KL,\gamma_2} \tag{31}$$

iii. Notice that the value of the shape parameter participates with a special weight to the analysis we would like to proceed with. For value $\gamma = 2$ it vanishes! Therefore, the classical Normal distribution is inadequate for such applications. A rough explanation can be justified as "there is any disorder between Normals," that is we need a more realistic approach—the fat-tailed family of distributions as the family (8) with members as in (9).
iv. The exact calculations through given states of the process are difficult. To obtain $TE\{X \to Y\}$ is simple through the existent results, especially (28), quite easy to be calculated in the way of the involved transfer entropy.

6 Discussion

This paper is based on our beliefs that quantified methods and optimization procedures need the Mathematical background, and a common understanding and link with the area of interest. Our main concern here is in Environmental Economics in which there is no doubt that an underlying uncertainty exists based on the probability theory.

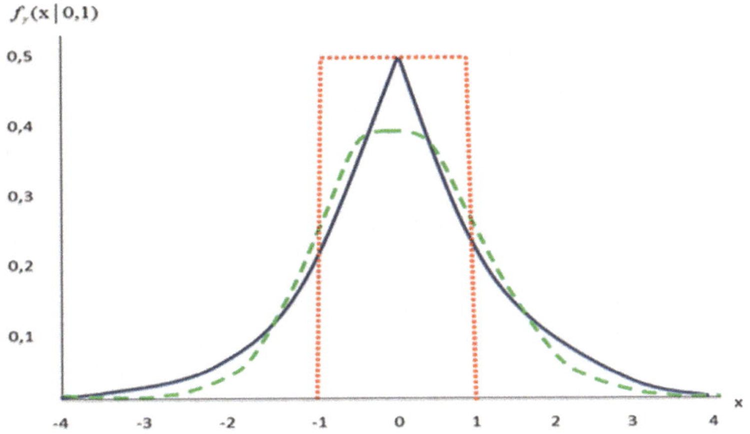

Fig. 1 Relationship among uniform, normal, and laplace graphically

An essential source of model uncertainty measured by entropy comes into view from the imposed assumptions and the assumed distributions to be considered. We are referring to entropy as it is strongly related to a system's disorder and uncertainty and is related to variance. As it is applied to a lot of engineering problems, we believe that this idea can be adopted in Economics in addition to a measure of existing uncertainty.[2]

It is clear that the entropy depends only on the variance–covariance matrix Σ or σ in case $p = 1$. In practice this means that the uncertainty is irrelevant to mean value μ (of the pollution center to an industry, say) but depends on the standard deviation (the experimental error), that is how much it is expanding from the center. The Uniform distribution can be adopted if it is assumed that pollution levels are (almost) the same around the area [a, b], whereas the Laplace when a "sharp explosion" around the center is assumed and much lower far from it. Estimates of (9) can be obtained in practical situations. Figure 1 is a graphical representation of the relationship among Uniform, Normal, and Laplace in the univariate case.

In simple words, this implies that the choice of the appropriate probability model describing the phenomenon that is the probability density function (pdf) is a main priority in any decision-making planning. The typical normal distribution is inadequate in such cases demanding more realistic approximations with the fat-tailed family of distributions as presented here to facilitate in these directions.

That is to say, a problem associated with the normal distribution is the "fat tails" and there are situations where the assumed normal distribution in tails "contains more probability" than the usual 0.05. This is spot on in a number of economic

[2] The researcher, in principle, may not be a Mathematician, so no need to know the existing development, but the solid results and the computational applications [25]. Exploring the economy–environment relationship in the case of air pollution [12] but also the effects of government expenditure on the environment [18], and coping with energy poverty [16] and exergy [33] will be helped among others by the proposed analysis.

applications and surely in various environmental problems where pollution has an effect on the "tails" more than 0.05. This implies an urgent need for a distribution that can accomodate "fat tails". Consequently, we seek a broader class of distributions, in which the normal distribution (representing the "ideal situation") is included as a special number, this requiring a generalization of normal. In these lines, the γ-order Generalized Normal Distribution (γ-GND) emerged from a completely mathematical problem with Logarithm Sobolev Inequality (LSI) providing a solid background for it.

Our target is not to clarify a number of fundamental ideas in other fields, but to improve their understanding, and applicability in Environmental Economics, where a very useful tool is in hands: the γ-order Generalized Normal distribution. To the three main pillars, uncertainly–G-causality–Transfer Entropy, we add Environmental Economics, to obtain new results, to create new theoretical developments, being helpful to the researcher to understand the investigating phenomenon much better. This will assist in the construction of adequate confidence intervals that will smooth the progress of the planning of effective environmental policies.[3]

7 Conclusions

In this paper we emphasize on the entropy and more specifically on transfer entropy. We have already shown that assuming the γ-order generalized normal may help extending to transfer entropy. One problem is the fat tails characteristic, in this area, and another problem is what pdf to be chosen. So the family of distributions, the γ-order Generalized Normal distribution, covers both requests, due the "International constant" $(\gamma/(\gamma - 1)) * * (\gamma/(\gamma - 1))$, which leads to a number of pdf, and the Logarithm Sobolev Inequalities (LSI), which provide a solid background.

Therefore uncertainty was extensively discussed through its origins, for the Environmental Economics problem. Moreover G-causality is essential to Economic problems and related to Transfer Entropy definition, highly related with the Information Theory. That is why we tried to clarify and simplify the Physical ideas: to be adopted in Environmental Economics, as they are not that complicated any more, and the appropriate packages can offer easy calculations. At the same time, the concepts are essential to the field of our concern: Environmental Economics.

[3] For instance in the case of transfrontier pollution and assuming linearity, the total annual depositions (AD_i) of air pollution (like sulphur) in country i will be given as [13–15] $AD_i = \Sigma_j d_{ij}(1 - \alpha_j)E_j + B_i \ \forall \ i, j \ i, j = 1, \ldots, N$, where α_i is the abatement efficiency coefficient in country i, d_{ij} is the transfer coefficient from country j to i, indicating what proportions of emissions from any source country is ultimately deposited in any receiving country, B_i is the level of the so-called background deposition attributable to natural sources (such as volcanoes, forest fires, and biological decay) in receptor-country i, or to pollution remaining too long in the atmosphere to be tracked by the model, i.e., is probably attributable not only to natural sources but also to emissions whose origin cannot be determined.

Disclosure of Interests It is now necessary to declare any competing interests or to specifically state that the authors have no competing interests.

Appendix A

Transfer Entropy Evaluation for the γ-Order Generalized Normal Distribution

Theorem: The transfer entropy for $U \oplus V \oplus W \sim N_\gamma(\mu, \sigma^2 \mathrm{I}_q)$ is

$$TE_\gamma\{W \to U \mid V\} = \log A - B(q,\gamma)S_1(q,\gamma)S_2(q)$$

With.

$$A = A(q,m,n,\gamma) = \frac{\Gamma\left(\frac{q}{2}\right)\Gamma\left(\frac{m}{2}\right)\Gamma\left(\frac{1+m}{\gamma_1}\right)\Gamma\left(\frac{m+n}{\gamma_1}\right)}{\Gamma\left(\frac{q}{\gamma_1}\right)\Gamma\left(\frac{m}{\gamma_1}\right)\Gamma\left(\frac{1+m}{2}\right)\Gamma\left(\frac{m+n}{2}\right)} \quad and \quad \gamma_1 = \frac{\gamma}{\gamma-1}$$

$$B = B(q,\gamma) = \frac{\Gamma\left(\frac{q}{2}\right)}{2^{\frac{q}{2}}\Gamma\left(\frac{q}{\gamma_1}\right)\gamma_1^{\frac{q+\gamma_1}{\gamma_1}}}$$

$$S_1(q,\gamma) = \sum_{k=0}^{\infty}\frac{\left(\frac{\gamma_1}{2}\right)_k}{k!}\sum_{\lambda=0}^{k}\binom{k}{\lambda}(-1)^\lambda \gamma_1^{\frac{2(k-\lambda)+q}{\gamma_1}}\Gamma\left(\frac{2(k-\lambda)+q}{\gamma_1}\right)$$

$$S_2(q) = \sum_{I_q^{k-\lambda}}^{\infty}\frac{(k-\lambda)!}{i_1 i_2 \ldots i_q}P_q(m,n)$$

with the usual notation valid $(r)_s = r(r-1)\ldots(r-s+1)$ and $I_q^{k-\lambda}$ being a set of indices such that

$$I_q^{k-\lambda} = \left\{in(i_1, i_2, \ldots, i_q) \in N^2, \sum_{j=1}^{q} i_j = k-1, i_1 \in (k,\lambda) \;\; or \; t_{m+2} + \cdots + t_q \in (k,\lambda)\right\}$$

The P_q (m, n) are well-defined polynomials depending on m, n $\in N^*$ being or not being both odd or even numbers; see Hlavackova–Schlinder et al. (2016), and their Theorem 6.2 for details.

References

1. Anderson, T.W.: An Introduction to Multivariate Statistical Analysis. Wiley India (2004)
2. Barnett, L., et al.: Granger causality and Gaussian variables. Phys. Rev. Lett. **103**(23) (2009)
3. Barnett, L., et al.: Granger causality and transfer entropy are equivalent for Gaussian variables. Phys. Rev. Lett. **103**(23), 238701 (2009). https://doi.org/10.1103/PhysRevLett.103.238701

4. Caratheodory C: The Second Law of Thermodynamics, Benchmark Papers on Energy 5, Edited by Joseph Kestin, DH&R Inc., Stroudsburg, This article was translated expressly for the Benchmark Volume by J Kestin, Brown Univ. from "untersuchungenuber die Grundlagen der Thermodyna. Math. Ann. 67, 355–386 (1909)
5. Cover, T.M., Thomas, J.A.: Elements of Information Theory. John Wiley & Sons, Inc., New York, Chichester, Brisbane, Toronto, Singapore (1991)
6. Fisher, R.A.: Theory of statistical estuation. Proc. Camb. Philol. Soc. **22**, 700–725 (1925)
7. Furioli, G., et al.: Fokker-Planck equations and one-dimensional functional inequalities for heavy tailed distribution. Milan J. Math. **90**, 177–208 (2022)
8. Granger, C.W.J.: Investigating causal relations by econometric and cross-spectal methods. Econometrics. **37**, 424–430 (1969)
9. Granger CWJ: Time series analysis cointegration and applications. Nobe Lecture, Dec. 8, 2003 in: Les Prix Nobel. The Nobel Prizes 2003. In: Frangsmyn T (ed.) Nobel Foundation, Stockholm, 2004. pp. 360–366 (2003)
10. Halkos, G., Kitsos, C.: Uncertainty in environmental economics: the problem of entropy and model choice. Econ. Anal. Policy. **60**, 127–140 (2018). https://doi.org/10.1016/j.eap.2018.09.004
11. Halkos, G.E.: Econometrics. Theory and Applications (in Greek). Disigma Publications, Athens (2019)
12. Halkos, G.E.: Exploring the economy—environment relationship in the case of Sulphur emissions. J. Environ. Plan. Manag. **56**(2), 159–177 (2013). https://doi.org/10.1080/09640568.2012.657756
13. Halkos, G.E.: Incomplete information in the acid rain game. Empirica. **23**(2), 129–148 (1996). https://doi.org/10.1007/BF00925336
14. Halkos, G.E.: Optimal abatement of Sulphur emissions in Europe. Environ. Resour. Econ. **4**(2), 127–150 (1994). https://doi.org/10.1007/BF00692200
15. Halkos, G.E.: Sulphur abatement policy. Energy Policy. **21**(10), 1035–1043 (1993). https://doi.org/10.1016/S0301-4215(06)80006-6
16. Halkos, G.E., Gkampoura, E.-C.: Coping with energy poverty: measurements, drivers, impacts, and solutions. Energies. **14**(10), 2807 (2021). https://doi.org/10.3390/en14102807
17. Halkos GE, Kitsos CP: Mathematics vs statistics in tackling environmental economics uncertainty. Munich Pers. RePEc Arch. No.: 85280, (2018)
18. Halkos, G.E., Paizanos, E.A.: The effects of fiscal policy on CO2 emissions: evidence from the U.S.A. Energy Policy. **88**, 317–328 (2016). https://doi.org/10.1016/j.enpol.2015.10.035
19. Heisenberg, W.: Über den anschaulichen Inhalt der quantentheoretischen Kinematik und Mechanik (in German). Z. Phys. **43**(3), 172–198 (1927)
20. K, H.–.S.: Equivalence of Granger causality and transfer entropy: a generalization. Appl. Math. Sci. **5**(73), 3637–3648 (2011)
21. K, H.–.S., et al.: Evaluating transfer entropy for normal and γ-order normal distribution. Br. I. Math. Comp. Sci. **17**(5), 1–20 (2016)
22. Jeffreys, H.: An invariant form for the prior probability in estimation problems. Proc. Roy. Soc A. **186**, 453–461 (1946)
23. Kitsos, C., Stamatiou, I.: On the emerged from LSI γ-order generalized normal distribution. J. Indian Soc. Probab. Stat. (2025). https://doi.org/10.1007/s41096-025-00241-z
24. Kitsos CP: Technological Mathematics and Statistics, vol. 2 (in Greek). New Tech Publications, Athens (2015)
25. Kitsos, C.P., Nyamsi, U.E.: Applying the generalized normal distribution for modelling asset returns data. In: The Fifth International Congress of Applied Statistics (UYIK-2024) Istanbul 21–23 May 2024. Turkey (2024)
26. Kitsos, C.P., Tavoularis, N.: Logarithm Sobolev inequalities for information measures. IEEE Trans. Inf. Theory. **55**(6), 2554–2561 (2009)
27. Kitsos, C.P., Toulias, T.: New information measures for the generalized normal distribution. Information. **1**, 13–27 (2010)

28. Kitsos, C.P., Toulias, T.L.: Bounds for the generalized entropy type information measure. J. Commun. Comput. **9**(1), 56–64 (2012)
29. Kitsos, C.P., Toulias, T.L.: On the family of the γ-ordered normal distributions. Far East J. Theor. Stat. **35**(2), 95–114 (2011)
30. Landau, L.D., Lifshitz, E.M.: Statistical Physics. Pergamon, London (1959)
31. Mandl, F.: Statistical Physics, 2nd edn. Wiley (1988)
32. Schrodinger, E.: Statistical Thermodynamics. Cambridge University Press (1946)
33. Sciubba, E.: A thermodynamic measure of sustainability. Front. Sustain. **2** (2021). https://doi.org/10.3389/frsus.2021.739395
34. Shannon, C.: A mathematical theory of communication. Bell System Tech J. **27**, 379–423 (1948)
35. Weyl, H.: Gruppentheorie und Quantenmechanik. Hirzel, Leipzig (1928)
36. Zachmanoglou, E.C.: Caratheodory's theorem on the second law of thermodynamics. SLAM. J. Appl. Math. **25**(4), 592–596 (1973)

Environmental Risk Footprint on Cryptocurrencies and Money Evolution in Inflationary Crises

Nikolaos A. Kyriazis

Abstract This paper investigates the dynamic interplay between environmental risk indices—as expressed by the physical risk index and the transition risk index—and major cryptocurrencies during the Covid-19 crisis and the Russia-Ukraine conflict when inflation soared. Using daily data and the innovative Quantile-VAR methodology, this study addresses the gap in literature about inflationary pressures strengthening (or not) the impact of environmental degradation on promising digital successors of traditional fiat money. Estimations reveal that Bitcoin, Bitcoin Cash and Bitcoin SV along with Litecoin are the strongest sources of systemic risk during the pandemic while Ripple takes the lead in bull markets and Zcash gains prominence in war times. Transition risk has stronger effects than physical risk in the Covid-19 and this is more obvious in bear markets but reverses during conflicts. Costs arising from physical destruction could bring about new leaders in the cryptocurrency markets and the modern financial landscape overall in inflationary crises. Our findings could serve as a compass for policymakers, regulators and investors in environmentally rapidly evolving times.

Keywords Environment · Physical risk index · Transition risk index · Major cryptocurrencies · Dynamic connectedness · Inflationary crises

1 Introduction

Environmental degradation has become one of the problems of major concern in recent years and has triggered increasing interest for policymakers and investors. Furthermore, environmental issues have brought about and boosted a remarkable bulk of research by academics in the fields of economics and finance due to the large footprint that climate deterioration exhibits on investments and the perspectives of money evolution [22, 56, 17]. The green label of modern financial instruments is

N. A. Kyriazis (✉)
Department of Accounting and Finance, University of Thessaly, Larissa, Greece
e-mail: knikolaos@uth.gr

G. Halkos et al. (eds.), *Environmental Risk Analysis*, Springer Proceedings in Mathematics & Statistics 540, https://doi.org/10.1007/978-3-032-24155-9_7

increasingly considered to constitute a desirable characteristic that raises the attractiveness of investments and carries the potential of generating bull market phenomena and causing euphoria in financial markets, expressed in the form of investor optimism [2, 4]. In the advent of consecutive recent crises, environmentally friendly investments are examined through the lens of their capacities to serve as shields for protecting investors from the bear markets of traditional investments during crises. Moreover, green investments are considered to be capable of functioning as profit-makers useful for counteracting reductions in the purchasing capacity generated by inflationary pressures rooted in health and geopolitical crises.

Special emphasis should be given on whether environmental risk could be a significant determinant of asset performance and how this could evolve in alternative economic and market conditions. Interest on environmental impacts by a macro-monetary perspective increases due to the limited ability of fiscal and monetary authorities to find and apply an efficient policy mix for fighting impoverishment as the consecutive stages of expansionary action taking have left very low eligibility to further fight crises [14]. The aggressive monetary responses to these crises have infused large levels of liquidity in major as well as developing economies and have set into question the capacity of major international currencies to serve as stores of value and their eligibility to remain the predominant money forms in international transactions and global reserves. This has brought to the epicentre the necessity of setting the spotlight on digital forms of money—which are characterized by lower transaction costs and substantially higher flexibility in transactions—that at the same time enable investors to reap potential benefits from their high volatility [5, 23, 43].

This study looks into the evolving relation between two major environmental indices and seven cryptocurrencies of very large popularity and capitalization. To be more precise, this paper investigates whether these popular cryptocurrencies function as trustworthy shields from inflation and environmental risks and the level by which climate degradation could prove to be a major driver of their market performance. This investigation aims to provide insights into whether environmental evolution could constitute a crucial determinant of the evolution of the current monetary outlook into a remarkably more modern and sophisticated monetary and financial landscape. In such an outlook, digital money forms would prevail and gain prominence along with or at the detriment of existing currencies. This potential is set under scrutiny by employing the innovative Quantile-VAR framework that enables investigation of dynamic connectedness by (i) pairwise and (ii) network perspectives. This provides a clearer view on whether environmental evolution also promotes monetary evolution and financial transformation.

The dynamic interplay between environmental risk and the cryptocurrency market evolution is based on the perception that crises—in the form of financial, monetary, environmental or other type of uncertainty—generate higher risk appetite to investors and make them seek for riskier investments. This theory is founded on [26]. These authors support that crises bring about large changes in the state of the economy and lead to radical changes (volatility) in the levels of consumption and GDP. A significant part of this volatility reflects the perception that people have the opportunity to achieve substantially higher levels of wealth. This is rooted in the great opportunity that risky financial innovations—like cryptocurrencies—offer to achieve great performance in a risky investment portfolio. Differently said, crises fuel the over-optimistic sentiment of investors and make larger the 'pseudo-wealth' component of expected wealth. This pseudo-wealth can only be achieved with assets prone to present bubbles, which means very large deviations of their nominal price from their fundamental value. Thereby, cryptocurrencies can safely serve for widely-adopted money and investments when crises occur.

This paper is connected with previous academic research that focuses on environmental issues and their impact on the real economy [31, 28], the connection of environmental degradation and performance in financial markets [48], cryptocurrency markets [35, 36, 47, 52], the evolution of money forms [7, 54, 16, 18, 17, 20, 21] and the effects of expansionary monetary policy and inflationary phenomena on financial markets [6]. The contribution of this study is three-fold. Firstly, it investigates the impacts of the environmental crisis on the latest aspect of the financial markets. Secondly, it employs two advanced specifications of the innovative Quantile-VAR methodology to conduct estimations. Thirdly, it makes separate estimations about the Covid-19 health crisis and the Russia-Ukraine geopolitical crisis that both caused price increases. Therefore, it provides insights into how the environmental crisis along with these two inflationary crises could pave the way towards the establishment and evolution of cryptocurrencies, which highly probably constitute the 'next big thing' in monetary economics and could have a crucial role on the transformation of the existing financial landscape. The 'pseudo-wealth' phenomenon gains prominence as crises evolves and casting light on its drivers and instruments could be of major interest for understanding the new status quo in financial economics.

Empirical results indicate that four major cryptocurrencies—Bitcoin, Bitcoin Cash and Bitcoin SV as well as Litecoin—serve as leaders of cryptocurrency markets and main sources of systemic risk during the pandemic. Notably, another leader, which is Ripple becomes the driving force of cryptocurrencies in bull markets. Moreover, Zcash emerges as the driving force of cryptocurrency markets during the Russia-Ukraine conflict. It is also noteworthy that the transition risk—expressed in the form of higher costs for fighting pollution—illustrates a larger footprint on the cryptocurrency markets in the pandemic than physical risks that represents costs due to natural disasters. This reverses during the Russia-Ukraine conflict. Both environmental indices are more influential in bear cryptocurrency market conditions, illustrating that poor performance makes cryptocurrencies more vulnerable to environmental factors. Overall, environmental risk factors along with differently sourced inflation could serve as drivers of major cryptocurrency markets and determine their leading forces.

The remainder of this study is structured as follows. Section 2 provides the main literature on environmental impacts on the real economy and financial markets as well as high-quality studies on existing and innovative money forms. Furthermore, Sect. 3 lays out the data and the methodology adopted. Section 4 presents and analyses the empirical outcomes and provides the economic implications of estimations. Finally, Sect. 5 concludes and suggests avenues for further research.

2 Literature Review

Environmental effects on the real economy and financial markets attract an increasing bulk of academic research and bring to the surface the necessity for including climate risks in the group of crucial factors that drive the evolution of financial economics. Early and credible climate policies are necessary for preventing disorderly transition policies that would deteriorate economic performance and generate financial instability. Apart from this, difficulties in energy access could have detrimental effects on people of very low income while the adoption of renewables and sustainable development goals could mitigate such phenomena [28]. Moreover, the adoption of renewable energy sources leads to greater productivity performance in the European Union by almost three times [29]. Industrialized economies like Switzerland, Denmark, Germany, the Czech Republic and Austria achieve better 'environmentally friendly' performance than other economies [27, 30]. Similarly, it is argued that the environmental performance index displays a U-shaped connection with economic growth in the Globe and Middle East and African countries while this relation is N-shaped in the case of Europe [31]. Interestingly, technological environmental risks illustrate a negative footprint on developed economies while the effects on the least developed countries are more severe [32].

An innovative aspect of environmental economics is behavioural environmental economics that focus also on psychological factors for explaining economic issues. This field is believed to be crucial for maximizing and reaping environmental benefits and solving the problem of limited resources [33]. The willingness to pay for environmental goods and services relies on dominant psychological factors like environmental attitudes, values, beliefs, concern and norms [25]. Moreover, normative and hedonic motives have a significant negative impact on both psychological and structural barriers to environmental behaviour, motives and environmental knowledge whereas gain motives increase perceived constraints [24].

Environmental risk—and especially transition risk—constitutes the axis for a proliferating bulk of academic research in the field of economics and finance. According to [56], investors' portfolios benefit from the transition risk as it more accurately forecasts changes in energy markets than the actual climate risk. The anticipation component of the transition risk also fluctuates about its centre value, decreasing as fears of physical harm rise. When transition costs decrease as a result of stricter environmental protection laws, there is less uncertainty, while it is still significant [55]. Firms with significant transition risk have lower market values when climate challenges receive more attention [48]. Besides this, higher transition risk despite being considered a menace for economies, is found to result into modest losses during crises and larger losses in good economic conditions as concerns the banking sector [1]. When it comes to other green financial assets, major economies like Germany, Sweden, Luxembourg, the Netherlands, France, the US and the UK serve as leaders of green bond markets [27, 30].

There is a group of studies that reveal positive impacts of the 'environmentally friendly' character on the popularity and market prices of investment assets. Green stocks are rewarded by the market when higher climate risks emerge but this impact weakens in the case of emerging economies. Furthermore, stocks with low exposure to transition risk perform better financially. On the contrary, it is higher transition risk that leads to elevated stock price crash risk. This effect strengthens in the case of firms greatly exposed to natural disasters while weakens when firms display high ESG performance. Moreover, the green label decreases stock prices and increases risk premia but the green character could prove to be beneficial in the future [17]. By a different viewpoint, it is argued that higher carbon emissions result into better stock performance in all sectors and the majority of countries and carbon premia are considered to be larger in less developed economies [8].

There is an extant literature focusing on the idiosyncratic character of cryptocurrencies [10, 46, 37, 11, 19, 53] and the impacts of uncertainties from a wide array of sources on cryptocurrencies [38, 40, 41, 42 43, 44, 50, 52]. It is supported that including Bitcoin in an economy along with money decreases welfare [35] and that Bitcoin exhibits limited capacity to fight macroeconomic shocks [49]. The two largest-cap cryptocurrencies—Bitcoin and Ethereum—are considered to be clearly bubbly in some periods [13] and optimistic investor sentiment about them feeds these bubbles [51]. Moreover, it is influential on traditional money forms and safe havens [34] and is especially influential (negatively) on the US dollar and (positively) on major US stock indices [3]. While they are inferior hedges than gold for confronting market panics [57] they benefit from the end of strict monetary policies [36] for acquiring larger shares in investors' preferences.

3 Data and Methodology

This paper examines the dynamic connectedness and the prospect for alterations in the leading forces of the major cryptocurrency markets through the prism of environmental impacts and inflationary pressures. This is the reason why examination takes place between each of the two major environmental indices—the physical risk index (PRI) and the transition risk index (TRI)—and seven of the largest-cap cryptocurrencies. To conduct these estimations, daily closing prices are used that include both the latest inflationary crises. More specifically, data about the Covid-19 health crisis (22 January 2020 to 23 February 2022) and the Russia-Ukraine conflict (24 February 2022 to 29 December 2023) are adopted. Econometric procedures take place for each of the crises separately in order to enable comparisons. Data limitations apply as December 29, 2023 is the latest date with available values of the PRI and the TRI indices. Market values of cryptocurrencies are expressed in US dollars. Data about the environmental risk indices are downloaded from the www.policyuncertainty.com database and are based on [9]. The closing prices of cryptocurrencies are extracted from the Coinmarketcap.com database. All values are transformed into returns for the purposes of estimations.

Figures 1a and 1b illustrate the evolution the environmental indices and major cryptocurrencies, respectively, expressed in returns. It can be seen that both the

Physical Risk Index (PRI) and the Transition Risk Index (TRI) are very volatile, exhibiting numerous ups and downs as time evolves. A large number of sign alterations is obvious. As concerns major cryptocurrencies, they experience strong bull markets at the end of 2021 and prices keep being elevated in early 2022. Afterwards, bear markets prevail while weak upwards tendencies appear in late 2024 due to increased optimism about cryptocurrency rallies in 2024 and 2025.

Tables 1 and 2 display the descriptive statistics of returns of the environmental indices and the major cryptocurrencies during the Covid-19 and the Russia-Ukraine war, respectively. Ripple (0.535) is the most profitable asset during the pandemic and Zcash (0.403), Bitcoin (0.386) and Monero (0.359) follow. Ripple (72.423) is also the most volatile asset and Zcash (55.608) follows while Bitcoin (22.625) is by far the safest investment. When it comes to the war period, cryptocurrency performance is significantly lower. Bitcoin SV (0.192), Ripple (0.112) and Monero (0.111) are the top performers while Bitcoin SV (64.667), Ripple (30.646) and Bitcoin Cash (27.457) are the riskiest cryptocurrencies. All variables display non-normality in distributions as revealed by the skewness and excess kurtosis tests and the Jarque–Bera statistic. Furthermore, the ERS test demonstrates stationarity in returns. The

Fig. 1 **a** Daily values evolution of the Physical Risk index (PRI) and the Transition Risk Index (TRI) **b** Daily closing prices evolution of major cryptocurrencies (Bitcoin, Litecoin, Ripple, Bitcoin Cash, Bitcoin SV, Zcash and Monero)

Fisher-Gallagher weighted portmanteau test rejects the hypothesis of an adequately fitted ARCH process.

For the purposes of conducting estimations, the innovative Quantile Vector Autoregression (QVAR) methodology is employed which has been previously used in [15] and [45]. This is outlined as follows:

$$y_t = \mu(t) + \sum_{j=1}^{p} \Phi_j(\tau) y_{t-j} + u_t(\tau) \tag{1}$$

where y_t and y_{t-j} constitute $k \times 1$ dimensional endogenous variable vectors, τ takes values between zero and unity and stands for the quantile examined, p displays the lag length of the Quantile-VAR model, $\mu(\tau)$ is a conditional mean vector of $k \times 1$ dimensions, $\Phi_j(\tau)$ is the Quantile-VAR coefficient matrix of $k \times k$ dimensions and $u_t(\tau)$ represents the $k \times 1$ dimensional error vector featuring by a $k \times k$ dimensional variance–covariance matrix, which takes the symbol $\Sigma(\tau)$.

Besides this, the impact generated by a shock in variable i on variable j is laid out as:

$$\psi_{ij}^{g}(H) = \frac{\Sigma(\tau)_{ii}^{-1} \sum_{h=0}^{H-1} \left(e_i\prime \Psi_h(\tau)\Sigma(\tau)e_j\right)^2}{\sum_{h=0}^{H-1} \left(e_i\prime \Psi_h(\tau)\Sigma(\tau)\Psi_h(\tau)\prime e_j\right)} \tag{2a}$$

$$\widetilde{\psi_{ij}^{g}}(H) = \frac{\psi_{ij}^{g}(H)}{\sum_{j=1}^{k} \varphi_{ij}^{g}(H)} \tag{2b}$$

where e_i stands for a zero-vector illustrating the value of 1 on the ith position. The result of this normalization is:

$\sum_{j=1}^{k} \tilde{\psi}_{ij}^{g}(H) = 1$ and $\sum_{\iota,j=1}^{k} \tilde{\psi}_{ij}^{g}(H) = k$.

Notably, the joint spillover effect generated from a specific variable i to the system of the remaining variables j represents the joint connectedness TOWARDS the group of variables (Eq. 3):

$$C_{i\to j}^{g}(H) = \sum_{j=1, i\neq j}^{k} \tilde{\psi}_{ji}^{g}(H) \tag{3}$$

By a reverse perspective, causal spillovers sourced from the group of variables and directed towards the specific variable are the joint connectedness FROM others (Eq. 4):

$$C_{i\leftarrow j}^{g}(H) = \sum_{j=1, i\neq j}^{k} \tilde{\psi}_{ij}^{g}(H) \tag{4}$$

Table 1 Descriptive statistics in the Covid-19 (in returns)

	Bitcoin	Litecoin	Ripple	Bitcoin cash	Bitcoin SV	Zcash	Monero
Mean	0.386	0.33	0.535	0.203	–0.005	0.403	0.359
Variance	22.625	42.391	72.423	46.479	44.741	55.608	39.204
Skewness	–0.872***	–0.617***	2.308***	0.169	–0.727***	–0.388***	–0.759***
Ex.Kurtosis	9.715***	5.331***	23.980***	12.191***	9.018***	4.028***	10.151***
JB	2208.2***	678.7***	13,516.6***	3371.2***	1891.2***	381.4***	2387.8***
ERS	5.463***	–4.198***	5.536***	–4.191***	–1.982**	–5.828***	–5.695***
Q(10)	16.663***	11.792**	8.784	14.199***	23.040***	9.318*	25.388***
Q2(10)	12.824**	20.904***	6.348	18.126***	35.936***	20.829***	55.190***

Notes ***, ** and * display significance at 1%, 5% and 10%, respectively; Skewness: D'Agostino (1970) test; Kurtosis: Anscombe and Glynn (1983) test; JB: Jarque and Bera (1980) normality test; ERS: Stock et al. (1996) unit-root test with constant; Q(20) and $Q^2(20)$: Fisher and Gallagher (2012) weighted portmanteau test

Table 2 Descriptive statistics in the Russia-Ukraine conflict (in returns)

	Bitcoin	Litecoin	Ripple	Bitcoin Cash	Bitcoin SV	Zcash	Monero
Mean	0.081	0.041	0.112	0.099	0.192	–0.134	0.111
Variance	12.107	22.702	30.646	27.457	34.667	25.953	16.841
Skewness	–0.377***	0.032	4.861***	1.077***	2.803***	–0.197*	–0.072
Ex.Kurtosis	7.694***	5.181***	63.617***	8.342***	20.842***	3.859***	9.011***
JB	1197.7***	538***	83,006.3***	1487.6***	9335.8***	301.6***	1627.9***
ERS	–5.615***	–3.859***	–2.043**	–6.024***	–4.711***	–4.995***	–3.410***
Q(10)	1.847	9.930*	11.340**	7.095	4.142	5.970	5.946
Q2(10)	4.174	17.890***	0.344	46.154***	5.428	22.074***	26.413

Notes ***, ** and * display significance at 1%, 5% and 10%, respectively; Skewness: D'Agostino (1970) test; Kurtosis: Anscombe and Glynn (1983) test; JB: Jarque and Bera (1980) normality test; ERS: Stock et al. (1996) unit-root test with constant; Q(20) and $Q^2(20)$: Fisher and Gallagher (2012) weighted portmanteau test

Thereby, the net joint spillover impact (net joint connectedness) generated by variable i on the system is calculated by subtracting the joint connectedness FROM others from the joint connectedness TO others (Eq. 5):

$$C_i^g(H) = C_{i \to j}^g(H) - C_{i \leftarrow j}^g(H) \tag{5}$$

It is important to underline that the Total Connectedness Index (TCI) of [12] takes values between [0, 1] and reflects systemic risk. Namely, higher values of the TCI index reflect higher levels of network interconnectedness (Eq. 6):

$$TCI(H) = \frac{\sum_{i,j=1,i \neq j}^{k} \tilde{\psi}_{ij}^g(H)}{k - 1} \tag{6}$$

The advantage of quantile regression is that it considers the entire conditional distribution of the variables and makes estimations at various quantile levels $\tau \in (0, 1)$. This enables to find asymmetries in distributions, tail phenomena and heterogeneity in different parts of the distribution. Characteristics like these are frequent in financial economics and reflect non-linearity, special forms of fluctuations and out of the ordinary phenomena. This is what makes the Quantile-VAR framework appropriate for examination of relations where mean estimates offer low levels of insights. Selection of lower, middle and upper quantiles for estimations makes feasible to examine the response of variables at specific quantiles that stand for bear, normal or bull market conditions. This allows to produce fruitful outcomes about the behaviour of variables in alternative extreme conditions where systemic risk tends to be higher.

This study benefits from the adoption of all three specifications of the Quantile-VAR framework as this allows to better estimate the footprint of alternative climate risk types on a group of important cryptocurrencies. These digital currencies could function as complements or substitutes in pairs or vis-à-vis the group of the remaining cryptocurrencies. Disentangling the spillover relationship in alternative environmental risk levels and cryptocurrency market conditions provides clearer comparisons regarding the potential for currency leadership. Moreover, it provides insights into the extent by which environmental risks affect this leadership and strengthen the potential that inflationary eras lead to radical reformation of the financial landscape. This large array of different conditions and scenarios offers a spectrum of findings and helps the construction of a compass for policymakers and investors.

4 Econometric Outcomes

Results concerning the net pairwise directional connectedness and network connectedness demonstrate a range of important findings about the interplay between (i) the Physical Risk Index and (ii) the Transition Risk Index the major cryptocurrencies considered. All estimations are conducted using a 60-day rolling window and a

10-day forecast horizon. Econometric procedures are applied at the lower (5th) quantile of return distributions to represent bear markets (poor performance), the middle (50th) quantile to count for normal market conditions and the upper (95th) quantile to stand for bull markets (good performance). Such estimations are conducted separately for the Covid-19 and the Russia-Ukraine conflict for the purposes of comparisons.

4.1 Outcomes About the Covid-19 Period

Figure 2a lays out the findings from the estimations by net pairwise directional connectedness between environmental risk indices and major cryptocurrencies at the lower (5th) quantile while Fig. 2b and c reveal results about the middle (50th) and the upper (95th) quantiles, respectively. In general, the Transition Risk Index is found to exhibit tighter linkages than the Physical Risk Index with each of the cryptocurrencies examined. Notably, both environmental indices show more persistent connectedness with major cryptocurrencies at the later phases of the Covid-19 crisis. This indicates that environmental risk has a larger footprint on the markets of digital currencies in conditions of accumulated inflation pressures. Thereby, it is more reasonable to expect that environmental uncertainty is a significant driver of financial evolution only when devaluation of conventional currencies is so large that there is high motivation to seek for alternative money forms. Strong inflationary phenomena combined with environmental risk have a considerable impact on cryptocurrency markets even when the latter are in bear conditions.

The PRI is most influential in the spring of 2021 while the TRI demonstrates its largest and most durable footprint in the summer of 2021. This indicates that the stages covering the effort for recovery of economies from the pandemic make environmental risks more influential on cryptocurrency markets. This has been probably promoted by the stagflation phenomena fuelled by the Covid-19 crisis which fed the exigency for fresh liquidity forms to finance the costs for the green transition. Remarkably, net pairwise connectedness levels do not exceed the limits of -40 and 80 in the majority of cases.

Outcomes about the middle quantile (Fig. 2b) reveal substantially lower levels of interaction between environmental risk and cryptocurrencies in normal risk levels and market conditions. The differences in the size of the PRI and the effects are more pronounced as transition risk is obviously more influential on major cryptocurrencies than in bear markets. Notably though, the spillover impact of environmental risk on digital currencies is larger in the initial phase of the Covid-19 when market pessimism is the norm. These findings display that the risk of having to bear costs for adapting firms to environmental changes (transition risk) is normally more linked with money evolution than the cost of making up for climate-sourced disasters. This strengthens the viewpoint that the transition to the environmentally friendly economy via greening firms acts as a stronger catalyst for monetary evolution than the need for recovery from random natural disasters. Interestingly, the dynamic connection

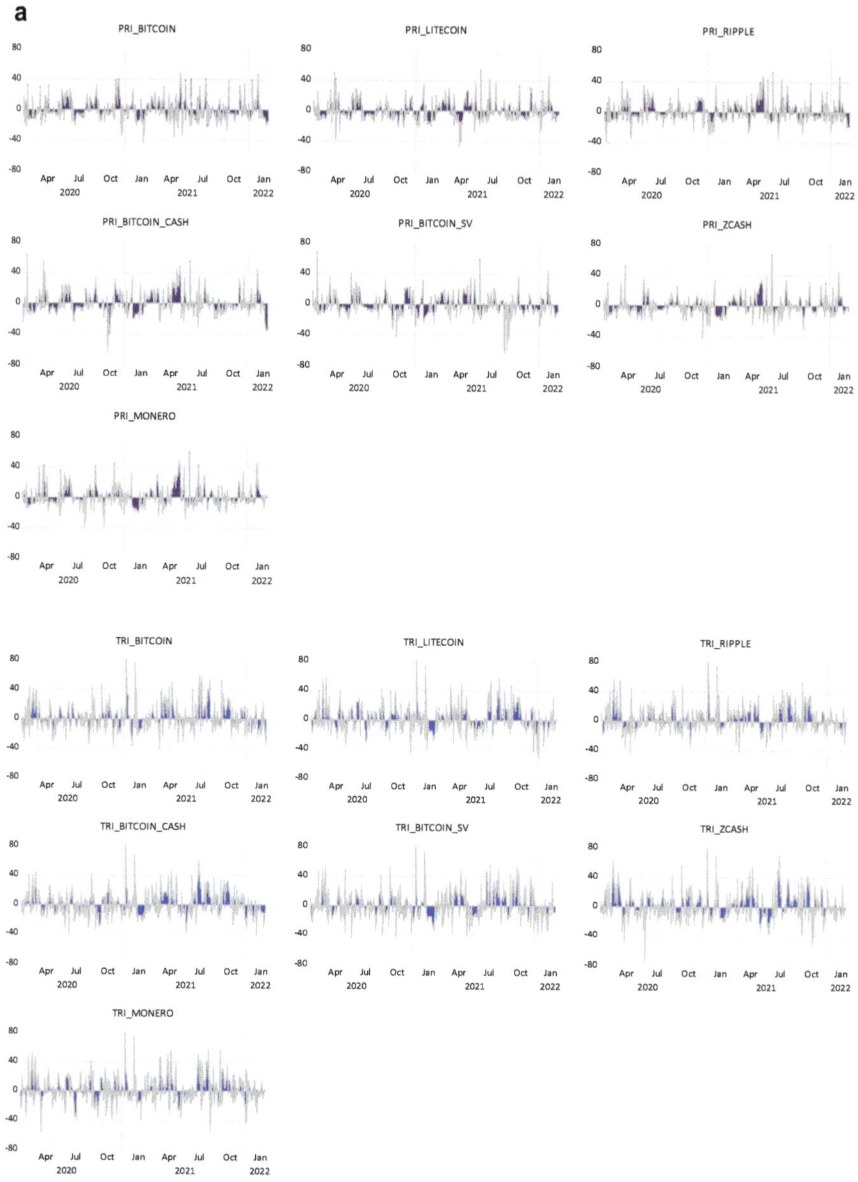

Fig. 2 **a** Net pairwise directional connectedness between **i** PRI and **ii** TRI and cryptocurrencies at the 5*th* quantile during the Covid-19 **b** Net pairwise directional connectedness between **i** PRI and **ii** TRI and cryptocurrencies at the 50*th* quantile during the Covid-19 c. Net pairwise directional connectedness between **i** PRI and **ii** TRI and cryptocurrencies at the 95*th* quantile during the Covid-19. *Note* The label on each graph informs about the pair of variables investigated regarding their net pairwise directional connectedness. The level of net pairwise directional connectedness is displayed on the vertical axis while time is shown on the horizontal axis. The environmental risk being the net source of spillovers is revealed by net directional connectedness lying on the positive area while net connectedness lying in the negative area informs that the cryptocurrency is a net generator of effects

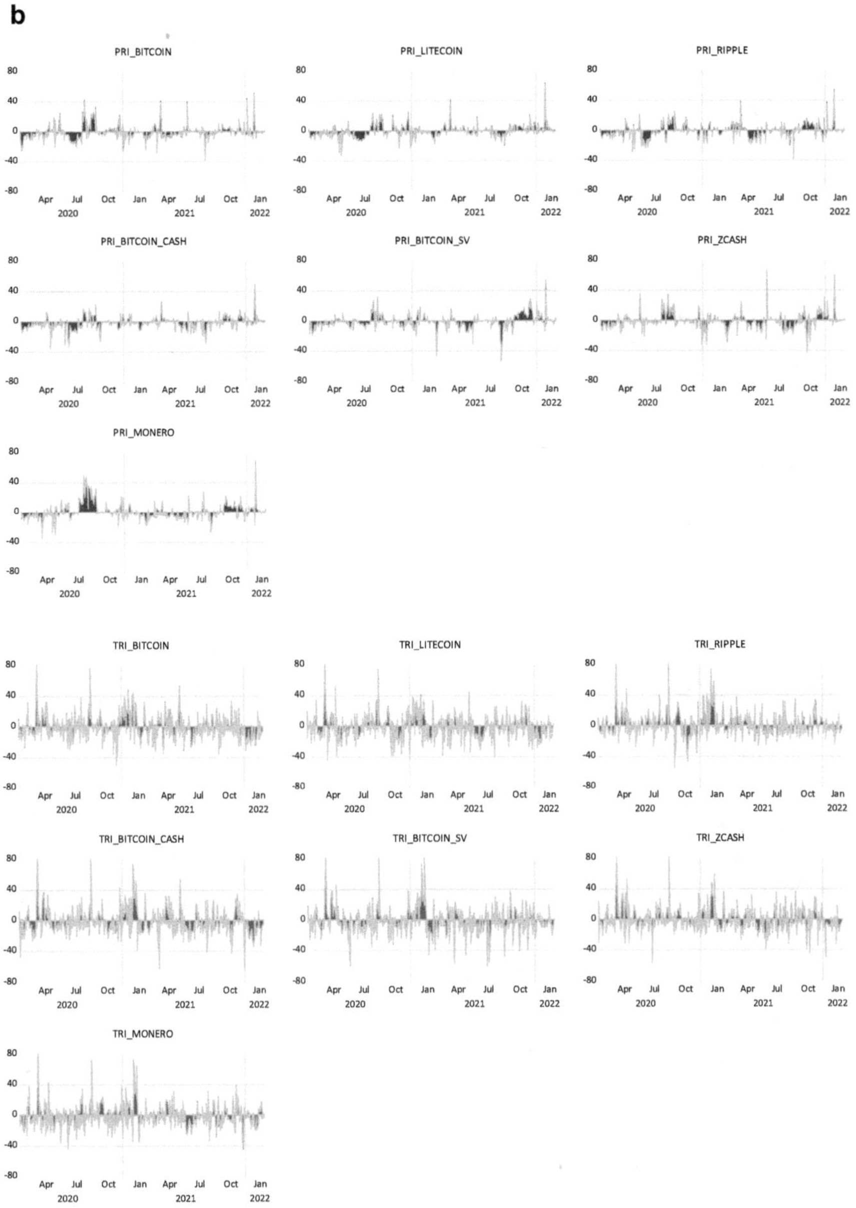

Fig. 2 (continued)

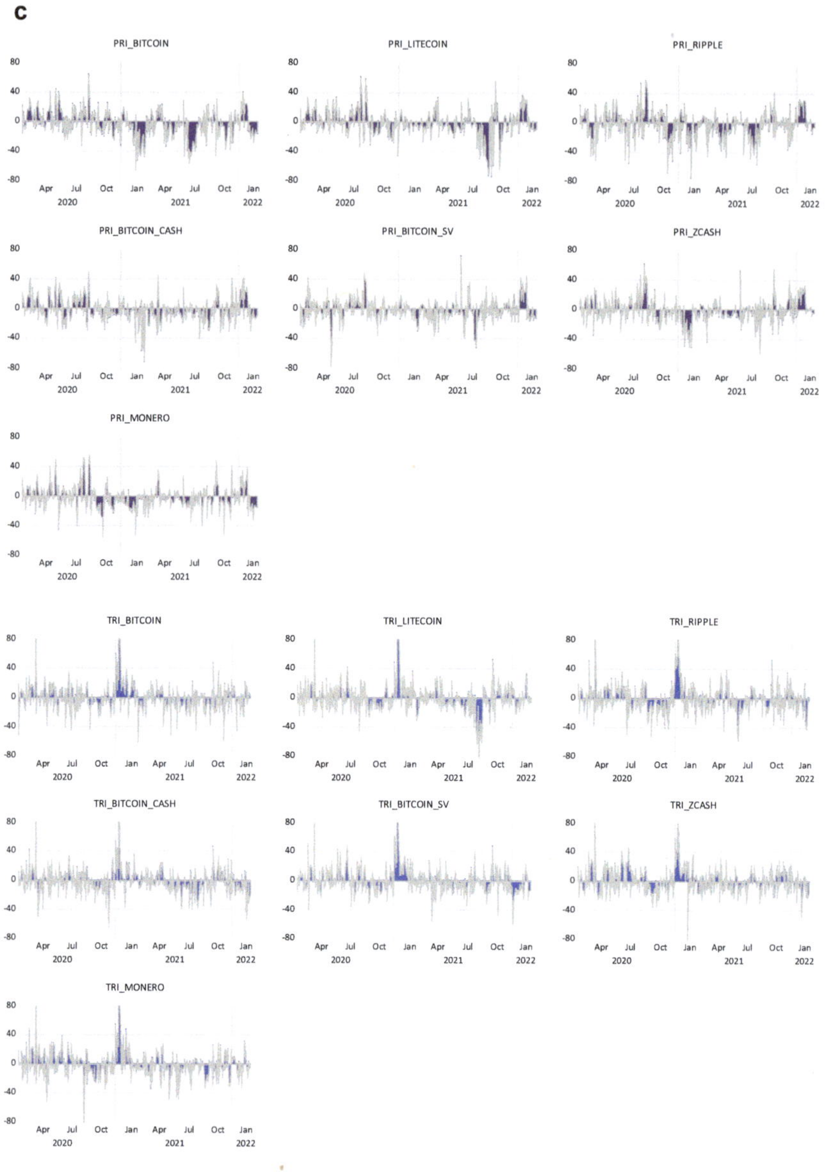

Fig. 2 (continued)

between the TRI and major cryptocurrencies present in normal market conditions reveals that new forms of liquidity are indeed affected by the costs for 'greening' even when no extreme climate events occur. This is important as it adds another influential factor to the set of determinants of money forms. This also explains why green investments are more popular. The green label constitutes an attractive feature that increases the market value of financial assets. Cryptocurrencies being able to acquire this characteristic could serve for their inclusion to the group of acceptable money forms that could enjoy the 'legal tender'.

When it comes to estimations at the upper quantile (Fig. 2c), mixed findings are laid out about the direction of causality between each of the environmental indices and cryptocurrencies. It is worth mentioning though that in a significant number of small periods the PRI serves as a receiver rather than as a source of spillovers. It is of great interest to see that in late 2020—when good news about the possibility of fighting the disease due to initial vaccine creation take place—transition risk becomes a great source of causality towards cryptocurrency markets. This shows that optimism feeds the effect of environmental risk on digital currency markets and the weakening of other sources of turmoil strengthens the footprint of environmental transition costs on the modern financial outlook. This is valid in bull markets and high environmental risk as good cryptocurrency performance is fed by investor optimism but also the threat of high climate costs. Nevertheless, this impact fades out in later stages.

These findings reinforce the notion that the environmental footprint on digital currency markets can be substantial and lead to over-enthusiasm of investors that seek to increase their wealth. In other words, elevated climate uncertainty can display a temporary but substantial effect on aggressive cryptocurrency markets. This can lead to fluctuations in consumption by making more important the pseudo-wealth component of wealth. This phenomenon cannot constitute a norm for financial markets but could prove to be the fuel for remarkable alterations in optimal asset allocation in extreme events like inflationary crises.

Moreover, Figs. 3a, b and c lay out the network connectedness between (i) the PRI and (ii) the TRI vis-à-vis major cryptocurrencies at the lower, middle and upper quantiles, correspondingly. Bitcoin, Bitcoin SV, Bitcoin Cash, Zcash and Ethereum are the sources of spillovers at the lower quantile while physical risk is not a net generator of impacts. This comes in contrast with the case of the transition risk index, which serves as the strongest net source of systemic spillovers and Bitcoin Cash, Bitcoin SV, Zcash and Monero follow. Interestingly, Bitcoin loses its major generator role when the TRI is included in estimations. It is worth noticing that in both cases Ripple functions as the largest net absorbers of spillovers.

When it comes to estimations at the middle quantile, Litecoin and Bitcoin Cash take the lead as net spillover sources and Bitcoin follows. This indicates that the absence of extreme levels of environmental risk or extreme parts of return distributions do not favourite Bitcoin acting as a driver of cryptocurrency markets. Remarkably, both the environmental indices function as net receivers of effects from major cryptocurrencies.

Notably, indices remain net absorbers at the upper quantile. Intriguingly, Ripple, which functions as the strongest net receiver in bear market conditions switches into the most powerful net source of spillovers in bull market conditions and very high-risk levels. This is of particular importance as the levels of environmental risk and the levels and direction of market sentiment are found to serve as determinants concerning which cryptocurrency could constitute the leading driver of

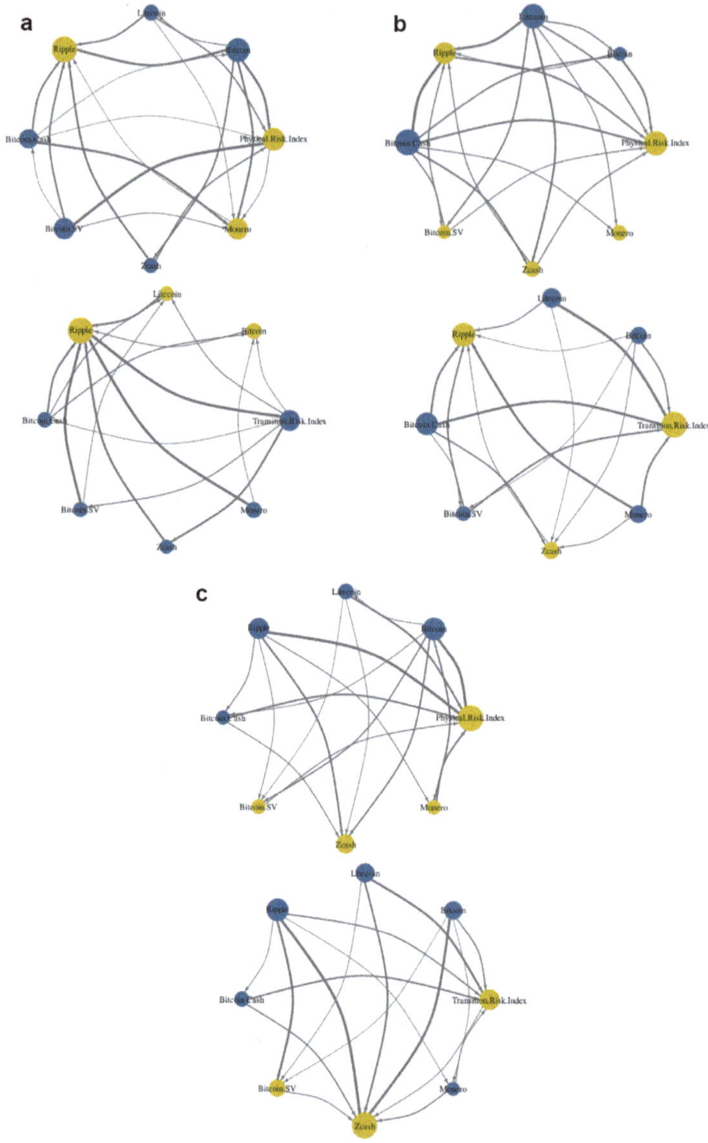

Fig. 3 **a** Network connectedness between **i** PRI and **ii** TRI and cryptocurrencies at the 5th quantile during the Covid-19 **b** Network connectedness between **i** PRI and **ii** TRI and cryptocurrencies at the 50th quantile during the Covid-19 c. Network connectedness between **i** PRI and **ii** TRI and cryptocurrencies at the 95th quantile during the Covid-19. *Note* Big blue node stands for variables that function as strong net sources of spillover effects while small blue node represents weak net sources, In contrast, big yellow node reveals that the variable works as strong net spillover receiver whereas small yellow node presents its functioning as a weak net spillover receiver

cryptocurrency markets. Ripple becoming the net leading force in bull cryptocurrency markets and intense environmental risk paves the way for better understanding whether Bitcoin's dominance in the cryptocurrency markets could be threatened and by which extent environmental issues could prove to be crucial in this evolution.

4.2 *Outcomes About the Russia-Ukraine Conflict Period*

Econometric outcomes about net pairwise directional connectedness during the Russia-Ukraine conflict at the lower, middle and upper quantiles are displayed in Fig. 4a, b and c, respectively. By focusing on the dynamic interplay of the environmental risk indices with cryptocurrencies in bear markets, it is clear that physical risk has a larger net effect on cryptocurrencies in the first semester of 2023 while net impacts continue being strong in the second semester of 2023. On the other hand, the transition risk has a higher net impact on digital money in late 2022. Notably, physical risk is found to be substantially more influential (in net terms) than the transition risk overall. This comes into contrast with findings in the Covid-19 crisis when transition risk was clearly more influential. These findings give credence to the notion that the source of each inflationary crisis is crucial for determining which type of environmental risk can better serve as driving force of the cryptocurrency markets.

When analysis in conducted at the middle quantile during geopolitical tensions, outcomes do not differ much than in the Covid-19 case. Normal levels of environmental risk and normal market conditions translate into weak net dynamic pairwise connectedness while some sporadic negative spikes emerge in the case of physical risk. Notably, the physical risk becomes somewhat more influential on Monero in the later war phase while the same holds about the transition risk impacts mostly on Ripple but also on Monero and Zcash.

Remarkably, the net spillover impact of both environmental indices on cryptocurrencies is obviously larger at high levels of climate risk and bull cryptocurrency markets. Physical risk is more influential than the transition risk on these digital assets, abiding by findings at the lower quantile in war periods but being in stark contrast with results in the pandemic. The late summer and autumn of 2022 are the periods when these net effects are more pronounced. In general, mixed signs of pairwise causality are traced, which impedes from extracting safe results about the net sources and receivers of spillovers overall.

Network connectedness at the 5th, 50th and 95th quantiles are displayed in Fig. 5a, b and c, respectively. Interestingly, both environmental indices are net sources of spillovers—with physical risk being a stronger generator than transition risk—at the lower quantile. Moreover, the Zcash cryptocurrency becomes a considerably strong settler of systemic causality in the cryptocurrency markets and this is clearer in the TRI case. Bitcoin remains a driver of causality but plays a secondary role.

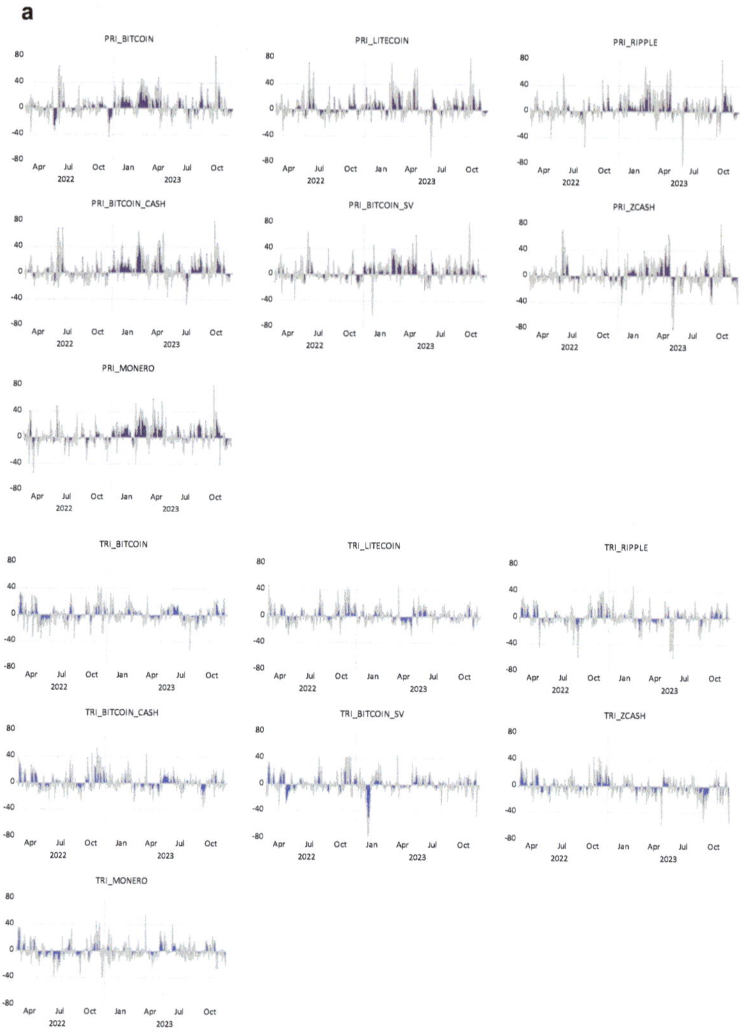

Fig. 4 **a** Net pairwise directional connectedness between **i** PRI and **ii** TRI and cryptocurrencies at the 5*th* quantile during the Russia-Ukraine conflict **b** Net pairwise directional connectedness between **i** PRI and **ii** TRI and cryptocurrencies at the 50*th* quantile during the Russia-Ukraine conflict c. Net pairwise directional connectedness between **i** PRI and **ii** TRI and cryptocurrencies at the 95*th* quantile during the Russia-Ukraine conflict. *Note* The label on each graph informs about the pair of variables investigated regarding their net pairwise directional connectedness. The level of net pairwise directional connectedness is displayed on the vertical axis while time is shown on the horizontal axis. The environmental risk being the net source of spillovers is revealed by net directional connectedness lying on the positive area while net connectedness lying in the negative area informs that the cryptocurrency is a net generator of effects

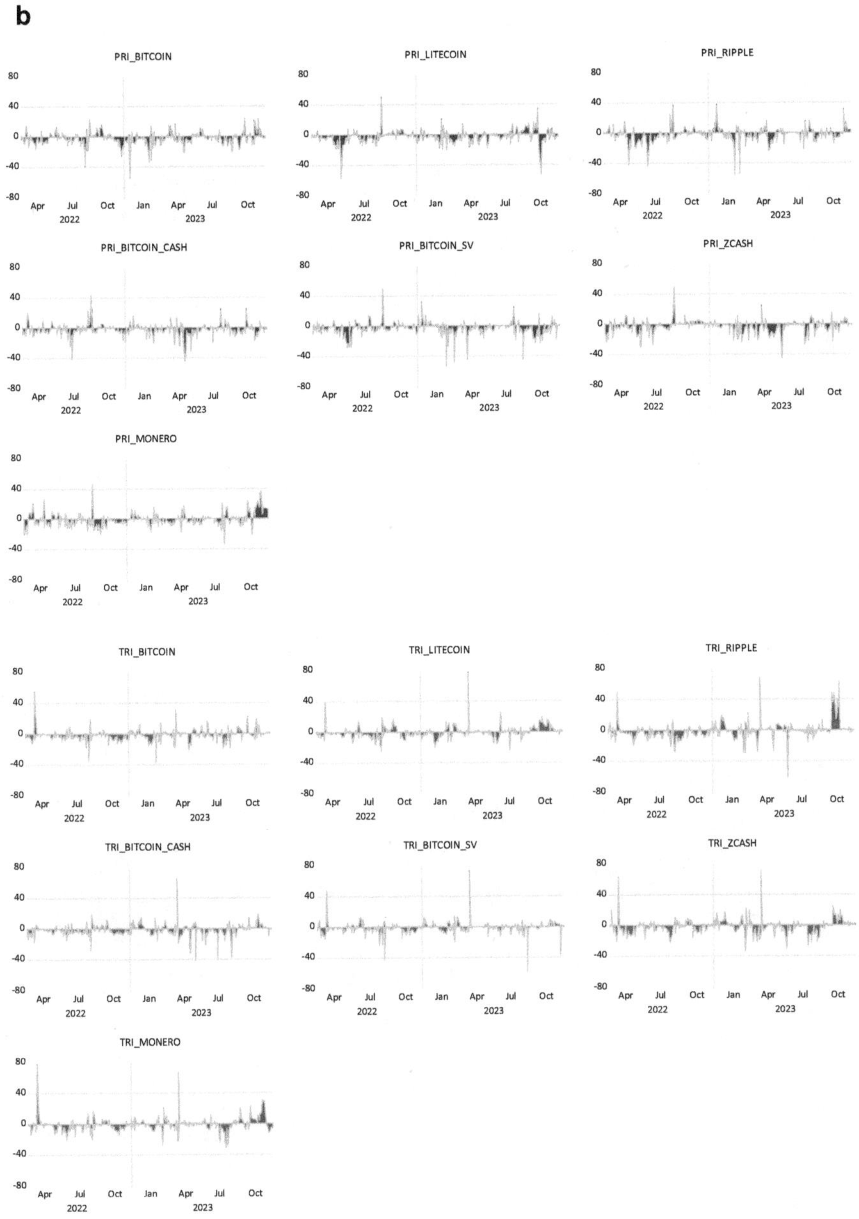

Fig. 4 (continued)

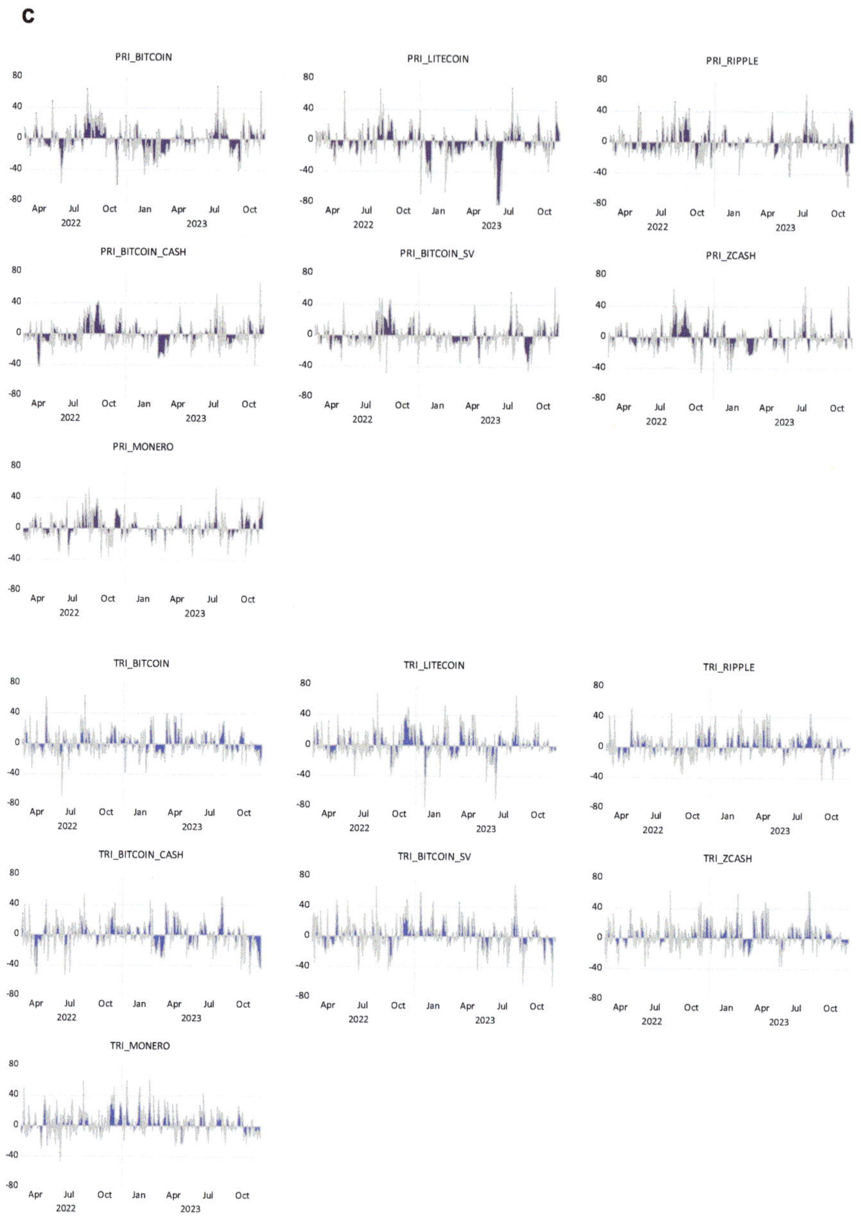

Fig. 4 (continued)

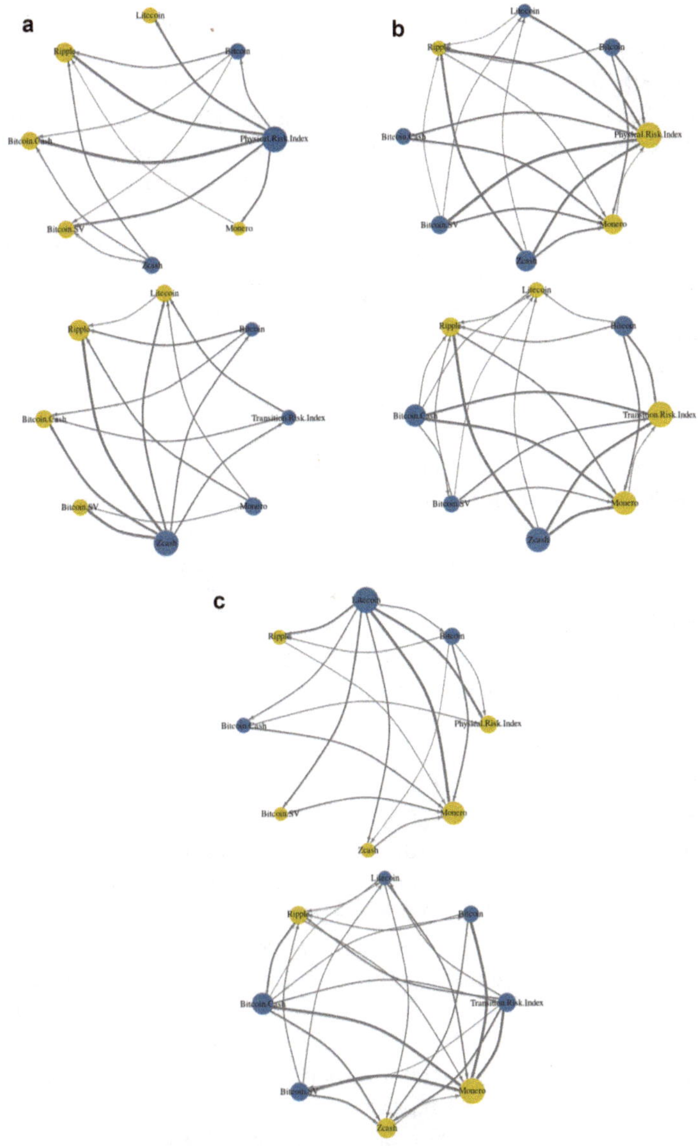

Fig. 5 **a** Network connectedness between **i** PRI and **ii** TRI and cryptocurrencies at the 5*th* quantile during the Russia-Ukraine conflict **b** Network connectedness between **i** PRI and **ii** TRI and cryptocurrencies at the 5*th* quantile during the Russia-Ukraine conflict. *Note* Big blue node stands for variables that function as strong net sources of spillover effects while small blue node represents weak net sources, In contrast, big yellow node reveals that the variable works as strong net spillover receiver whereas small yellow node presents its functioning as a weak net spillover receiver

Apart from these, in normal risk levels and normal market conditions, both environmental indices become net receivers of spillovers. Notably, Zcash remains important systemically and constitutes the main determinant of systemic risk along with Bitcoin and Bitcoin Cash. Litecoin and Bitcoin SV also serve as net sources of impacts. This makes more solid the viewpoint that alternative risk and market conditions could promote the emergence of new leading forces in the cryptocurrency markets.

It is really interesting that in high environmental risk levels and bull cryptocurrency markets, Litecoin takes the lead when physical risk index is considered in the examination. In contrast, Bitcoin Cash, Bitcoin, Bitcoin SV and Litecoin are the main causal sources along with the transition risk that act as net settlers when the TRI case is examined. These outcomes underline the differences as concerns the effects of alternative types of environmental risk. Outcomes underscore how physical environmental risk differs from transition environmental risk regarding their potential to bring to the forefront new leaders of cryptocurrency markets. It is evident that the transition risk leaves space for less concentration of power in these markets as the role of driver is split into an array of major currencies with lower-cap candidates also being promising leaders. On the other hand, high physical risk in inflationary war periods favourites Litecoin—one of the established vice-leaders of cryptocurrencies—to take charge of systemic risk causality. Thereby, higher risk of physical destruction boosts the evolution of the modern financial outlook in the highly anticipated manner. In contrast, higher transition risk—in the form of costs for greening firms—urges towards decentralization of net causal power in the cryptocurrency markets.

Overall, outcomes indicate that the type of crisis that functions as a source for inflationary phenomena is crucial for determining the leaders in the cryptocurrency markets and the driving forces of the transition to the new financial landscape. Moreover, the type of environmental risk serves as an important settler of systemic influence that each digital promising successor of traditional money exerts on the monetary outlook. Bitcoin and Bitcoin-related cryptocurrencies as well as Litecoin—which is constructed to have many similarities with Bitcoin—are the indisputable leaders in the race for driving the cryptocurrency markets. Notably, a major cryptocurrency with low nominal value but significant levels of volatility take the lead in bull markets when disease-caused inflation prevails. At the same time, it is physical risk—which reflects costs for repairing from natural disasters—that acts as the environmental settler of the digital economy markets. These findings partly come into contrast with previous research that indicates transition risk being clearly more influential than physical risk in financial markets [56] but abide by studies stressing the great role of transition risk in economies [8].

Notably, results in war-induced inflation periods fertilize the ground for further contemplation on the dynamics of money evolution. Zcash—which presents lower market capitalization than Bitcoin, Bitcoin-related cryptocurrencies and Litecoin—emerges as a protagonist. This currency along with the transition risk—which illustrates the cost for greening financial markets—are the most influential in cryptocurrency markets. This indicates that the reformation of the existing monetary and financial status quo could be paved by less popular but volatile digital money able to finance the 'environmentally friendly' transition. This is in accordance with earlier research that indicates modern digital money forms constituting potential competitors of traditional money in international markets [47]. This evolution would be pushed by strong climate degradation and should prove capable of alleviating from the burden of the reduced purchasing capacity that traditional money devaluation imposes to consumers. It is noteworthy that although transition risk favours Litecoin, one of the four historic pillars of the cryptocurrency markets, to take the lead and act as a settler of this emerging money type, physical risk increases lower concentration with regard to the primary causal sources of cryptocurrency markets. Moreover, it should indispensably be underlined that these findings provide new insights into Gresham's law as they corroborate that 'bad money' in terms of volatility and store-of-value capacity drive out 'good money' in terms of international prestige and the potential to be the successors of the US dollar in international transactions.

5 Conclusions

This paper examines the influential impacts of two distinct types of environmental risk—the physical risk and the transition risk—on the markets of seven major large-cap cryptocurrencies separately in the Covid-19 pandemic and the Russia-Ukraine war inflationary crises. This investigation examines the dynamics of modern money forms through the lens of i) the environmental risk effects that can work for the advancement of the existing monetary and financial landscape and ii) the boost that inflationary pressures bring about to innovative liquidity assets. Differently said, this study provides insights into the environmental alongside with the monetary forces of major cryptocurrencies that aspire to constitute successors of international fiat currencies.

Daily data and the innovative Quantile-VAR methodology expressed by three advanced specifications—the net pairwise directional connectedness, the net extended joint connectedness and the network connectedness—are employed for estimations. Econometric outcomes reveal that Bitcoin, Bitcoin Cash and Bitcoin SV serve as leaders in the cryptocurrency markets in the Covid-19 period, especially in bear market conditions, while Ripple gains prominence in bull markets. Zcash becomes influential mostly in war times. Moreover, Litecoin—one of the most prominent vice-leaders of Bitcoin—is the main driver in bull markets during the war. Notably, the transition risk has a larger footprint on cryptocurrencies in the Covid-19 but this relationship changes in the war period. Interestingly, physical risk enhances lower concentration regarding the leading causal sources of cryptocurrency markets while transition risk favourites Litecoin—one of the four traditional pillars of the cryptocurrency markets—to take the lead and serve as a settler of this upcoming money type.

Overall, findings indicate that environmental risk functions as a driver of digital currency markets in specific monetary and market conditions. This implies that the environment and climate features belong to the group of sources that create uncertainty and push towards changes in investor behaviour and the evolution of acceptable money forms. This improves our understanding of the factors that feed over-optimism of investors and strengthen the pseudo-wealth component of wealth as developed by Guzman et al. [26].

This study advances the understanding of how environmental risks alongside with differently-sourced inflationary phenomena render impactful on innovative money forms and contribute towards the sophistication and modernization of the existing monetary and financial landscape. These findings provide a compass for policy-makers and investors for enabling better decision-making in turbulent eras with natural disasters and financial instability. Potential avenues for further research could include the examination of the market price crash risk of modern financial assets by the use of advanced econometric methodologies. Incorporating a spectrum of uncertainties into these estimations would enable to distinguish and quantify the factors that fuel the evolution of the current financial outlook. This would fertilize the ground for a brave new era in the field of monetary economics and finance.

6 Disclosure of Interests:

The authors have no competing interests to declare that are relevant to the content of this article.

Acknowledgements The author is grateful to Prof. Dr. George Halkos and the participants of the 10th Conference on 'Economics of Natural Resources and the Environment' (ENVECON 2024) for their useful remarks that contributed to the significant amelioration of this study.

References

1. Alessi, L., Di Girolamo, E.F., Pagano, A., Giudici, M.P.: Accounting for climate transition risk in banks' capital requirements. J. Financ. Stab. **73**, 101269 (2024)
2. Alessi, L., Ossola, E., Panzica, R.: What greenium matters in the stock market? The role of greenhouse gas emissions and environmental disclosures. J. Financ. Stab. **54**, 100869 (2021)
3. Aliyev, F., Eylasov, N.: The impact of Nasdaq100, US Dollar index and commodities on Cryptocurrency: new evidence from augmented ARDL approach. Econ. Lett. 112191 (2025)
4. Bauer, M.D., Huber, D., Rudebusch, G.D., Wilms, O.: Where is the carbon premeium? Global performance of green and brown stocks. J. Clim. Financ. **1**, 100006 (2022)
5. Bazán-Palomino, W., Svogun, D.: On the drivers of technical analysis profits in cryptocurrency markets: a Distributed Lag approach. Int. Rev. Financ. Anal. **86**, 102516 (2023)
6. Bernanke, B.S.: The new tools of monetary policy. Am. Econ. Rev. **110**(4), 943–983 (2020)
7. Bitros, G.C., Economou, E.M.L., Kyriazis, N.C.: Democracy and money: Lessons for today from Athens in Classical times. London and New York: Routledge (2020). https://www.routledge.com/Democracy-and-Money-Lessons-for-Today-from-Athens-in-Classical-Times/Bitros-Economou-Kyriazis/p/book/9780367509170

8. Bolton, P., Kacperczyk, M.: Global pricing of carbon-transition risk. J. Financ. **78**(6), 3677–3754 (2023)
9. Bua, G., Kapp, D., Ramella, F., Rognone, L.: Transition versus physical climate risk pricing in European financial markets: a text-based approach. Eur. J. Financ. 1–35 (2024)
10. Böhme, R., Christin, N., Edelman, B., Moore, T.: Bitcoin: Economics, technology, and governance. J. Econ. Perspect. **29**(2), 213–238 (2015)
11. Caferra, R., Tedeschi, G., Morone, A.: Bitcoin: Bubble that bursts or Gold that glitters? Econ. Lett. **205**, 109942 (2021)
12. Chatziantoniou, I., Gabauer, D., Stenfors, A.: Interest rate swaps and the transmission mechanism of monetary policy: a quantile connectedness approach. Econ. Lett. **204**, 109891 (2021)
13. Corbet, S., Lucey, B., Yarovaya, L.: Datestamping the Bitcoin and Ethereum bubbles. Financ. Res. Lett. **26**, 81–88 (2018)
14. Crosignani, M., Faria-e-Castro, M., Fonseca, L.: The (unintended?) consequences of the largest liquidity injection ever. J. Monet. Econ. **112**, 97–112 (2020)
15. Cunado, J., Chatziantoniou, I., Gabauer, D., de Gracia, F.P., Hardik, M.: Dynamic spillovers across precious metals and oil realized volatilities: Evidence from quantile extended joint connectedness measures. J. Commod. Mark. **30**, 100327 (2023)
16. Dimitriadis, K.A., Koursaros, D., Savva, C.S.: Evaluating the sophisticated digital assets and cryptocurrencies capacities of substituting international currencies in inflationary eras. Int. Rev. Financ. Anal. **96**, 103693 (2024)
17. Dimitriadis, K.A., Koursaros, D., Savva, C.S.: The influence of the "environmental-friendly" character through asymmetries on market crash price of risk in major stock sectors. J. Clim. Financ. **9**, 100052 (2024)
18. Dimitriadis, K.A., Koursaros, D., Savva, C.S.: The influential impacts of international dynamic spillovers in forming investor preferences: a quantile-VAR and GDCC-GARCH perspective. Appl. Econ. 1–21 (2024b)
19. Dimitriadis, K.A., Koursaros, D., Savva, C.S.: Exploring the dynamic nexus of traditional and digital assets in inflationary times: the role of safe havens, tech stocks, and cryptocurrencies. Econ. Model.107195 (2025)
20. Economou, E.M.L., Kyriazis.: Daily Life in Classical Athens: A Socioecononomic Analysis. Cham: Springer Nature (2024). https://link.springer.com/book/9783031585401
21. Economou, E.M.L.: The Economy of Classical Athens Organization, Institutions and Society. London and New York: Routledge (2023). https://www.routledge.com/The-Economy-of-Classical-Athens-Organization-Institutions-and-Society/Economou/p/book/9781032561547
22. Ferrari, A., Nispi Landi, V.: Will the green transition be inflationary? Expectations matter. IMF Econ. Rev 1–64 (2024)
23. Gerritsen, D.F., Lugtigheid, R.A., Walther, T.: Can bitcoin investors profit from predictions by crypto experts? Financ. Res. Lett. **46**, 102266 (2022)
24. Gkargkavouzi, A., Halkos, G., Matsiori, S.: How do motives and knowledge relate to intention to perform environmental behavior? Assessing the mediating role of constraints. Ecol. Econ. **165**, 106394 (2019)
25. Gkargkavouzi, A., Halkos, G.: The psychology of non-market environmental valuation: an integrative review of research evidence, theoretical insights, and policy considerations. J. Environ. Psychol. 102510 (2024)
26. Guzman, M., Stiglitz, J.E.: Pseudo-wealth and consumption fluctuations. Econ. J. **131**(633), 372–391 (2021)
27. Halkos, G., de Alba, J.M., Todorov, V.: Economies' inclusive and green industrial performance: An evidence based proposed index. J. Clean. Prod. **279**, 123516 (2021)
28. Halkos, G.E., Aslanidis, P.S.C.: Addressing multidimensional energy poverty implications on achieving sustainable development. Energies **16**(9), 3805 (2023)
29. Halkos, G.E., Aslanidis, P.S.C.: Green energy pathways towards carbon neutrality. Environ. Resource Econ. **87**(6), 1473–1496 (2024)

30. Halkos, G., Managi, S., Tsilika, K.: Ranking countries and geographical regions in the international green bond transfer network: a computational weighted network approach. Comput. Econ. **58**(4), 1301–1346 (2021)
31. Halkos, G., Zisiadou, A.: Relating environmental performance with socioeconomic and cultural factors. Environ. Econ. Policy Stud. **20**, 69–88 (2018)
32. Halkos, G., Zisiadou, A.: An overview of the technological environmental hazards over the last century. Econ. Disasters Clim. Chang. **4**, 411–428 (2020)
33. Halkos, G., Gkargkavouzi, A.: Environmentalism in the light of behavioral economics (No. 120752). University Library of Munich, Germany (2024)
34. Hsu, S.H., Sheu, C., Yoon, J.: Risk spillovers between cryptocurrencies and traditional currencies and gold under different global economic conditions. N. Am.n J. Econ. Financ. **57**, 101443 (2021)
35. Kang, K.Y., Lee, S.: Money, bitcoin, and monetary policy. J. Money, Credit, Bank. **56**(1), 225–256 (2024)
36. Karau, S.: Monetary policy and Bitcoin. J. Int. Money Financ. **137**, 102880 (2023)
37. Kyriazis, N.A.: Is Bitcoin similar to gold? An integrated overview of empirical findings. Journal of Risk and Financial Management **13**(5), 88 (2020)
38. Kyriazis, N.A.: The effects of gold, stock markets and geopolitical uncertainty on Bitcoin prices and volatility. Glob. Econ. J. **20**(04), 1–15 (2020)
39. Kyriazis, NA.: The effects of geopolitical uncertainty on cryptocurrencies and other financial assets. SN Business & Economics **1**(1), 5 (2020)
40. Kyriazis, N.A.: The nexus of sophisticated digital assets with economic policy uncertainty: A survey of empirical findings and an empirical investigation. Sustainability **13**(10), 5383 (2021)
41. Kyriazis, N.A.: Trade policy uncertainty effects on macro economy and financial markets: an integrated survey and empirical investigation. J. Risk Financ. Manag. **14**(1), 41 (2021)
42. Kyriazis, N.A., Economou, E.M.: Unveiling the Impacts of Geopolitical Risk on the Transition to the Decentralized Financial Landscape. Peace Science and Public Policy, Peace Economics (2025)
43. Kyriazis, N., Papadamou, S., Corbet, S.: A systematic review of the bubble dynamics of cryptocurrency prices. Res. Int. Bus. Financ. **54**, 101254 (2020)
44. Kyriazis, N., Papadamou, S., Tzeremes, P., Corbet, S.: The differential influence of social media sentiment on cryptocurrency returns and volatility during COVID-19. Q. Rev. Econ. Finance **89**, 307–317 (2023)
45. Kyriazis, N., Papadamou, S., Tzeremes, P., Corbet, S.: Examining spillovers and connectedness among commodities, inflation, and uncertainty: A quantile-VAR framework. Energy Economics **133**, 107508 (2024)
46. Kyriazis, N. A.: Herding behaviour in digital currency markets: an integrated survey and empirical estimation. Heliyon **6**(8) (2020a)
47. Kyriazis, N.A., Fassas, A., Papadamou, S.: Can metaverse coins crowd out traditional money forms when inflation surges? Financ. Res. Open 100008 (2025)
48. Li, Q., Shan, H., Tang, Y., Yao, V.: Corporate climate risk: Measurements and responses. Rev. Financ. Stud. **37**(6), 1778–1830 (2024)
49. Majumder, S.B.: Is cryptocurrency a new digital gold? Evidence from the macroeconomic shocks in selected emerging economies. J. Econ. Stud. **52**(1), 88–103 (2025)
50. Nouir, J.B., Hamida, H.B.H.: How do economic policy uncertainty and geopolitical risk drive Bitcoin volatility? Res. Int. Bus. Financ. **64**, 101809 (2023)
51. Oad Rajput, S.K., Soomro, I.A., Soomro, N.A.: Bitcoin sentiment index, bitcoin performance and US dollar exchange rate. J. Behav. Financ. **23**(2), 150–165 (2022)
52. Papadamou, S., Kyriazis, N.A., Tzeremes, P.G.: Non-linear causal linkages of EPU and gold with major cryptocurrencies during bull and bear markets. N. Am.n J. Econ. Financ. **56**, 101343 (2021)
53. Podhorsky, A.: Bursting the bitcoin bubble: Do market prices reflect fundamental bitcoin value? Int. Rev. Financ. Anal. **93**, 103158 (2024)

54. Raya, J.M., Vargas, C.: How to become a cashless economy and what are the determinants of eliminating cash. J. Appl. Econ. **25**(1), 543–562 (2022)
55. Rebonato, R., Kainth, D., Melin, L.: The link between physical and transition risk (2024). https://climateimpact.edhec.edu/sites/ercii/files/ercii_wp_link_between_physical_transition_risk_april_2024.pdf
56. Salisu, A.A., Ndako, U.B., Vo, X.V.: Transition risk, physical risk, and the realized volatility of oil and natural gas prices. Resour. Policy **81**, 103383 (2023)
57. Su, C.W., Yang, S., Qin, M., Lobonţ, O.R.: Gold vs bitcoin: Who can resist panic in the US? Resour. Policy **85**, 103880 (2023)

A Time Series Analysis Comparing the Main Air Pollutants of Athens and Salonica in Greece

Christos Kitsos and Constantinos-Symeon Nisiotis

Abstract This paper examines the air pollutants CO, NO, NO_2, SO_2, and O_3 comparing the two most economical significant areas of Greece: Athens and Salonica. Therefore, we are working to investigate similarities and differences within each city—working on two areas of interest for each city, and between the cities, for each pollutant. The different obtained models are compared, but one of the main characteristics of our research is based on EDA. Certainly there are turning points, but such research needs particular investigation. The magnitude and complexity of the data necessitate the application of advanced time series models for accurate characterization. Nevertheless, the overarching pattern remains consistent across the various pollutants, exhibiting a gradual decline over time—albeit at differing rates—in both the densely impacted central areas (Aristotelous and Patission) and the peripheral zones (Kalamaria and Peristeri) of the two cities.

Keywords Pollutants · ARMA · AR(p) · MA(q)

1 Introduction

The effects of climate change are very strong and visible, for example, in the Netherlands, farmers have to reduce their herds so that to reduce nitrogen [44, 50] or in France where large lots had to leave 4% fallow [43].

For farmers, the new EU measures included reducing the use of fertilizers and pesticides, while simultaneously reducing the area of productive land used by farms [41].

C. Kitsos (✉)
University of West Attica, Egaleo, Athens, Greece
e-mail: xkitsos@uniwa.gr

C.-S. Nisiotis
University of the Aegean, University Hill, Mytilene, Greece
e-mail: csnisiotis@aegean.gr

G. Halkos et al. (eds.), *Environmental Risk Analysis*, Springer Proceedings in Mathematics & Statistics 540, https://doi.org/10.1007/978-3-032-24155-9_8

According to the plan in the Netherlands, by 2030, nitrogen emissions should be reduced on average by 50% and in Natura areas by 70%. That is, to reduce dramatically, up to 50%, the numbers of animals on Dutch farms [20].

Air consists mainly of nitrogen N_2 in 78,08% v/v and oxygen O_2 in 20,95% v/v. The CO_2 is in 0,03% v/v, it has been increased at a surprisingly rapid rate the last 100 years, while Ar is in 0,93% v/v [26]. Therefore, violation for these limits might be considered dangerous.

The continuous addition of CO_2 to the atmosphere causes an increased activity in terms of the "greenhouse effect" on the Earth. Therefore, an increase in Earth's temperature has been noticed. Then, warming the oceans would cause an increased flow of CO_2 from them to the atmosphere. This would result in the greenhouse effect and a further increase in the planet's temperature. Such a version could be repeated. However, from the temperature trends observed today, no such increase in temperature can be found [10, 27, 37]. Statistical figures for this kind of Environmental data are not unique and can be approached in different ways. A historical development can be traced in Kandel [29]. We note that women, who make up the majority of the world's poor, are particularly vulnerable to the effects of climate change, see Nyumayer and Plomber [38] among others.

The effect of the consequences of the increase of CO_2 in the atmosphere is compensated by the addition of various particles that cut off solar radiation, Ahmad et al. [1]. Thus, the warming of the earth is delayed, reduced somehow. However, we believe, we should control the CO_2 content of the atmosphere on a "global scale"—this might be more important than the globalization of the Economy, Asumadu-Sarkodie et al. [2], Balcilar et al. [3].

Agriculture and animal husbandry emit about 25% of the world's pollutants. Agriculture takes up about 50% of land use and 70% of fresh water and ultimately causes about 78% of fresh water and ocean pollution. Such results are not permanent in time, and are depending more on policies and serve as indicators, rather than solid statistical figures.

There is a question concerning the existence of interactions between the components—pollutants in a mixture consideration, but we would like to emphasize that,

1. The action of the chemicals in the mixture, under the Statistical and Toxicological point of view
   (a) Follow some mode of action of the chemicals
   (b) Follow different mode of action of the chemicals
2. The effects in the exposed individuals, either animal or human
   (a) Effect on the same target organ
   (b) Effect on different target organs

It has been emphasized, Kitsos and Edler [32] that when synergy or antagonistic exists this is NOT equivalent to an additive response. Such health-influencing problems are beyond the target of this paper.

The main target of this paper is to handle in a Statistical way of thinking, a large data set of Environmental information, on the main pollutants, for the main two cities of Greece, were almost 70% of the Greek population is living. For this reason, Statistical methodology has been presented, a theoretical framework has been tackled and applied obtaining a number of models for the existed data set. Surprising enough there was no such an analysis before, although both cities, Athens and Salonica are considered polluted. The results obtained, even in this first level analysis, were too many, and only a portion is presented here. Moreover in another paper, Kitsos et al. [35], where a probability—oriented data analysis has been performed, and hopefully more results will be presented on the subject. We were surprised when we realized, that although data exist, a published Statistical analysis does not! So, we shall try to provide particular studies on this Environmental data set, concerning mainly Greece, to attract interest on the subject, under the Tokey [49], Hackl [11], Halkos [14] line of thought.

The article proceeds as follows. Section 2 outlines the theoretical background and reviews relevant literature. Section 3 describes the conceptual framework that defines the criteria and rationale guiding our analysis. Section 4 presents the data and Sect. 5 reports the empirical results. Section 6 concludes and discusses our findings.

2 Background

The problem of Economy and Environment has been extensively discussed by Halkos [13, 14], Halkos and Paizanos [15], Halkos and Kitsos [16] among others. There are certainly excellent papers on the Climate Change, most of them due to Intergovernmental Panel on Climate Change, IPCC, see IPCC [26] among others, who supported the correlation between temperature and pollutants like CO_2. Moreover, statistical techniques have been proposed for the evaluation of hazard functions for air pollution problems, Kitsos and Trandafir [33]. For the change point methods, discussed in Sect. 3, some statistical methods are proposed to be applied when the change of climate is to be under research consideration, Hackl [11], Darling [8], Carlstein [6], among others. This paper is adopting the Exploratory Data Analysis (EDA), an analysis identifying the main hidden patterns in the data, introduced by Tukey [49], and can be adopted to a number of applications, Behrens [4], adopting computational techniques, see the early work of Kitsos [31], while the time series models, Halkos [12] (Chaps. 9 and 11), Box et al. [5], are the main Statistical framework on which this paper is based.

China is the biggest methane polluter with 57×10^6 tons, South Asia with 43×10^6 tons, North America with 35×10^6 tons and the European Union (EE) with 23×10^6 tons. Although methane has only been used in recent decades, it contributes 80 times more than carbon to global Earth warming.

Roughly speaking since 1850 to today, the total emissions of carbon dioxide (CO2), from the USA, China, and Russia are ranked as the first three. But in the measurement of carbon dioxide emissions per capita, Saudi Arabia (18.1 tons/capita) and China (16.6 tons/capita) are ahead. China is currently the largest producer of CO2. But, it should be remembered that many American and European industries have settled in China, because they can pollute more easily! Hence the utilization of the measurement is difficult.

The most common air pollutants are the following:

1. **Carbon monoxide (CO)**
   In principle, we would say that carbon monoxide is a very poisonous gas that you can't see, taste, or smell. Therefore, some devices should never be used (heaters, barbeques, tools with gasoline engines, etc.). The main symptoms of the health risk due to CO poisoning include headache, nausea, vomiting, dizziness, among others, and in some instances, permanent brain injury or death.
2. **Nitric oxide (NO)**
   Nitrogen or nitric oxide (or nitrogen monoxide) reacts with the hydroperoxyl radical HO_2 to form nitrogen dioxide. NO_2, which then can react with a hydroxyl radical (OH) to produce nitric acid, HNO_3. Liquid nitrogen oxide is very sensitive to detonation even in the absence of fuel. Nitric oxide should not be confused with nitrogen dioxide (NO_2), a brown gas and major air pollutant.
3. **Nitrogen dioxide (NO_2)**
   Nitrogen dioxide is a highly reactive gas. It is formed by emissions mainly from motor vehicles, industry, that is why high concentrations are near busy roads and indoors where unflued gas heaters are in use.
   Outdoors, NO_2 contributes to the formation of ground-level ozone, O_3, as well as particulate matter pollution. The most potential health effect from exposure to nitrogen dioxide is to increase susceptibility to lung infections in people with asthma, as well as the Airway inflammation in healthy people—which can be take place in hospitals.
4. **Sulfur dioxide (SO_2)**
   Sulfur dioxide is highly reactive gas with a pungent irritating smell. It is mainly formed by fossil fuel combustion at power plants and other huge industrial installations. The most common potential health effect to humans from exposure to SO_2 is narrowing of the airways leading to wheezing, chest tightness, and shortness of breath.
   In general, Shipping emits 2.7% of global pollutants. But in Shipping, the International Maritime Organization (IMO) requested the reduction of sulfur dioxide (SO_2) emissions to 0.1%. Ship-owners accepted it, but no more fuel is known to be suitable for reducing pollutants.
5. **Ozone (O_3)**
   Ozone is found in the upper atmosphere as well as at ground level. O_3, in the upper atmosphere protects Life, by filtering out damaging ultraviolet radiation from the Sun. But, at ground level, acts as a high-risk pollutant, damaging human's health.

It is the main component of smog and is the product of the interaction between sunlight and emissions from sources (motor vehicles and industry).
In principle, O_3 can travel long distances and accumulate to high concentrations far away from the sources of the original pollutants. This creates food for thought to work on transfer entropy influence by this particular air pollutant.

In the present paper, interest is focused on pollutants (i) to (v), as above, for the areas of our interest: Metropolitan Areas of Athens and Salonica.

For the time series observed data Y_t, $t = 1, \dots, n$, we briefly review the main points for this paper. Let us consider the Autocorrelation Function (ACF) and particularly the sample one r_k estimating the ACF, ρ_k, as

$$r_k = \frac{\sum_{t=k+1}^{n}(Y_t - \bar{Y})(Y_{t-k} - \bar{Y})}{\sum_{t=1}^{n}(Y_t - \bar{Y})^2}, \qquad k = 1, 2, \dots \tag{1}$$

It is crucial to know that for the stationary ARMA, Halkos [12] among others, process and for fixed in the joint distribution of $[\sqrt{n}(r_i - \rho_i)]$ $i = 1, 2, \dots, m$ as $n \to \infty$ is $N(0, \Sigma)$, i.e., normal with zero means and $\Sigma = (\sigma_{ij}) \in \mathbb{R}^{m \times m}$ covariance matrix with

$$\sigma_{ij} = \sum_{k=-\infty}^{\infty} \rho_{k+i}\rho_{k+j} + \rho_{k-i}\rho_{k+j} - (2\rho_j\rho_k\rho_{k+i} + 2\rho_i\rho_j\rho_k) \tag{2}$$

Moreover,

$$r_k \sim N(\rho_k, \frac{c_{kk}}{n}), \quad Corr(r_k, r_j) = \frac{\sigma_{kj}}{\sqrt{\sigma_{kk}\sigma_{ij}}} \tag{3}$$

If Y_t follows an $AR(1)$ model, with $|\phi| < 1$, then

$$Var(r_1) = \frac{1 - \phi^2}{n} \tag{4}$$

with ϕ close to ± 1

$$Var(r_k) \approx \frac{1}{n}\left(\frac{1 + \phi^2}{1 - \phi^2}\right) \quad \text{with } k >> \tag{5}$$

while for large $\rho_k \approx \phi^k$ the correlation is

$$Corr(r_1, r_2) \approx 2\phi \left[\frac{1 - \phi^2}{1 + 2\phi^2 - 3\phi^4}\right] \tag{6}$$

If we consider Y_t $t = 1, 2, \dots$ to follow a $MA(1)$ model, then

$$Var(r_k) = \frac{1}{n}\sigma_{kk}, \quad Corr(r_k, r_j) \approx \frac{\sigma_{kj}}{\sqrt{\sigma_{kk}\sigma_{ij}}} \tag{7}$$

$$\text{with } \sigma_{kk} = 1 + 2\rho_1^2,\ \sigma_{11} = 1 - 3\rho_1^2 + 4\rho_1^4,\ \sigma_{12} = 2\rho_1(1 - \rho_1^2)$$

When lag k, say, is present the Partial Autocorrelation function (PACF) between Y_t and Y_{t-k} is considered after removing the effect of the intervening variables (i.e., $Y_{t-1}, Y_{t-2}, \ldots, Y_{t-k+1}$).

For the PACF of an $AR(p)$ model, with $k > p$, it holds that $\phi_{kk} = 0$ with

$$\phi_{kk} = Corr\left[\left(Y_t - \sum_{j=1}^{p} \phi_j Y_{t-j}\right), \left(Y_{t-p} - \hat{Y}_{t-p}\right)\right] \tag{8}$$

with $\hat{Y}_i$ being the estimation of Y_i.

For the autoregressive model of order p, $AR(p)$, (autoregressive model with p lags) as denoted already in (9), with white noise (with mean zero and variance = constant) then the Yule–Walker equation for the $AR(p)$ model are of the form

$$p = P\phi^* \tag{9}$$

with P being the autocorrelation matrix

$$P = \begin{pmatrix} p(0) & \ldots\ p(p-1) \\ p(1) & \ldots\ p(p-2) \\ \vdots & \\ p(p-1)\ \ldots & p(0) \end{pmatrix} \tag{10}$$

and ϕ^* the vector of AR coefficients $\phi^* = (\phi_1, \ldots, \phi_p)'$, and p the autocorrelation vector $p = (p(1), \ldots, p(p-1))^T$.

Example 8.1 For the $AR(1)$ model, simultaneous Equation (10) are reduced to

$$p(1) = p(0)\phi_1 \Rightarrow \phi_1 = \frac{p(1)}{p(0)}$$

For the $AR(2)$ model holds

$$\begin{pmatrix} p(1) \\ p(2) \end{pmatrix} = \begin{pmatrix} p(0) & p(1) \\ p(1) & p(2) \end{pmatrix} \begin{pmatrix} \phi_1 \\ \phi_2 \end{pmatrix}$$

thus $p(1) = \frac{\phi_1}{1-\phi_2}$, $p(2) = \frac{\phi_1^2 - \phi_2^2 + \phi_2}{1-\phi_2}$.

Eventually, the Yule–Walker Equation (11) can be solved and provide

$$\phi_{kk} = \frac{p_k - \sum_{j=1}^{k-1} \phi_{k-1,j}\, p_{k-j}}{1 - \sum_{j=1}^{k-1} \phi_{k-1,j}\, p_j}$$

Table 1 AR(1) simulation study, with small sample size, n where α: nominal level, MSE: mean square error, Sk: average skewness, Ku: average kurtosis, $\bar{\phi}$: average value of estimates

n	ϕ	α	MSE	Sk	Ku	$\bar{\phi}$
10	−1.5	.919	0.03	3.88	21.07	−1.44
	−1.0	.947	0.09	1.39	5.47	−0.84
	−0.5	.970	0.09	0.63	3.27	−0.42
	0.0	.967	0.10	−0.06	2.56	0.00
	0.5	.965	0.10	−0.58	3.01	0.41
	1.0	.955	0.09	−1.29	5.45	0.85
	1.5	.928	0.03	−3.96	22.21	1.45
5	−1.5	.968	0.27	1.13	5.79	−1.29
	−1.0	.969	0.27	0.08	10.00	−.82
	−0.5	.979	0.25	−0.09	6.05	−0.39
	0.0	.970	0.25	0.22	4.92	0.02
	0.5	.973	0.33	−0.58	4.95	0.79
	1.0	.975	0.61	−0.14	4.98	0.38
	1.5	.957	0.29	−1.23	6.02	1.29

with

$$\phi_{k,j} = \phi_{k-1,j} - \phi_{kk}\phi_{k-1,k-j} \qquad j = 1, 2, \ldots, k-1$$

In practice, ACF and PACF can be proved useful for $AR(p)$ or $MA(2)$ models, see Table 1, simulation Study for the AR(1) model with n being the sample size, MSE the evaluated Mean Square Error, Sk the evaluated average Skewness, Ku the evaluated average kurtosis, and α the chosen nominal level, see Kitsos [30].

The $AR(1)$ model of the form $Y_{t+1} = \phi Y_t + \epsilon_{t+1}, t = 1, 2, \ldots, n$ model has been considered by Kitsos [30], in his Ph.D. thesis, Glasgow University, as a sequential design, with white noise. The sample information is $\frac{1}{\sigma^2} I_n$, with $I_n = \sum_t Y_t^2$. The $AR(1)$ was also discussed under stochastic process theory by Lai and Siegment [36] who proved that
$I_n^{1/2}(\hat{\phi}_n - \phi) \xrightarrow{\mathcal{L}} N(0, \sigma^2)$, $|\phi| < 1$, with $\mathcal{L}$ mean "convergences in Law."

Fisher's information was evaluated as the "input" and "output" to be fixed, despite their time series or sequential design or stochastic process manner hidden by the model, while the optimal design approach and its geometrical interpretation it was discussed by Papakyriazis [39].

Therefore, an estimate of the variance can be theoretically provided through Residual Sum of Squares (RSS)

$$RSS = \sum Y_t^2 - \hat{Y} \sum Y_{t+1} Y_t$$

and

$$\hat{\sigma}^2 = \frac{1}{n-1} RSS$$

Therefore, confidence intervals can be constructed for ϕ by the usual way. As it was pointed out under 1000 simulations, considering $\epsilon_t \sim N(0, 1)$, and $Y_1 = 0.0$ and nominal level $a = 0.05$ the obtained estimates are in Table 1. By a, the estimated coverage is presented, S_k the average skewness, Ku the Average Kurtosis of the estimates and $\bar{\phi} = \frac{1}{1000} \sum_{t=1}^{1000} \hat{\phi}_t$. The question investigated is "how well behaves the asymptotic result with a small number of observation, i.e., sample size $n = 10$ or 5." The answer is: quite well. As always everything close to infinite is perfect but not that bad with small sample size.

3 The Change Point Approach

When a series of observations are received, such as from an industrial process, or from a series of daily observations, we are interested in investigating cases where a specific event occurred and the process has been violated. In such cases, we are referred that the process is out of control, for the Industrial Statistics or that a violation of the weather took place in Environmental or Economical Statistics, Hinkley [21], Horvath [22].

When a time series of observations are examined, such as from the market, or for the weather characteristics, there is still significant interest in examine, or in statistical terms in test, if at a certain point a violation of the time series process took place. In this, latter case, the change point exists and remains, in principle, while in the Industrial case, the change point can be detected, and the process return to its typical characteristics.

Therefore, we examine statistical tests where the null hypothesis H_0 states that the process is stationary, while the alternative H_1 status that the process is non-stationary since the change point.

The simplest presentation of this change point idea is to consider the standardized variables Y_i, $i = 1, 2, \ldots, n$, $E(Y_i) = 0$, $E(Y_i^2) = \sigma^2$, with errors e_i, say, $i = 1, 2, \ldots, n$. At the null hypothesis H_0, we consider Y_i, $i = 1, \ldots, n$ to remain with the errors e_i, when in the alternative H_1, there is a change point occurred at point k_0, i.e.,

$$H_0 : Y_i = e_i, i = 1, 2, \ldots, n \qquad H_1 : \exists\, k_0 = 1, \ldots, n-1 \text{ with}$$

$$Y_i = \begin{cases} e_i, & i = 1, \ldots, k_0 \\ \mu + e_i, & i = k_0 + 1, \ldots, n \end{cases} \tag{11}$$

with e_i, $i = 1, 2, \ldots, n$ being independent identically distributed (iid) errors. When inference is required, we assume that $e_i \sim N(0, 1)$, $i = 1, 2, \ldots, n$ with $N(0, 1)$ the standard normal distribution with density function $\phi(x)$ and distribution function $\Phi(x)$, as usually. The k is known as change point.

The average of the observations at the first and second part of the observations, divided by the change point m, are, respectively,

$$\bar{Y}_k = \frac{1}{k}\sum_{i=1}^{n} Y_i \text{ and } \bar{Y}'_k = \frac{1}{n-k}\sum_{i=k+1}^{n} Y_i \tag{12}$$

We mention that $\bar{Y}'_k$ is the LSE of the unknown $\mu \in \mathbb{R}$. Moreover, it holds

$$(n-k)^{1/2}\bar{Y}'_k \sim N(0,1) \tag{13}$$

and the k_0 value is needed to be evaluated, through the chosen k.

As test statistic is considered the statistic

$$T_{n,k} = \max_{0 \le k \le n-1} \left\{ \frac{1}{\sqrt{n-k}} \sum_{i=k+1}^{n} Y_i \right\} \tag{14}$$

and therefore, the null hypothesis is rejected for a two-sided test, when

$$|T_{n,k}| > z_{1-a/2} \tag{15}$$

with $z_{1-a/2}$ the $(1-a/2)100\%$ quantile of the standard Normal $N(0,1)$.

Indeed: As Y_i, $i = 1, 2, \ldots, n$ are iid $N(0,1)$, then

$$T_{n,k} = \frac{1}{\sqrt{n-k}} \sum_{i=k+1}^{n} Y_i \sim N(0,1) \tag{16}$$

Thus for k known and equal to k, we would reject H_0 at significant level a if

$$\left| \frac{1}{\sqrt{n-k}} \sum_{i=k+1}^{n} Y_i \right| > z_{1-a/2} \tag{17}$$

and therefore, this is true also for $|T_{n,k}|$.

Although the exact distribution of $|T_{n,k}|$ it is not an easy task one could evaluate

$$Corr(T_{n,k}, T_{n,q}) = \sqrt{n-q/n-k} \qquad k \le q \tag{18}$$

see Hawkins [18]. Under H_0 a very nice asymptotic result, for n large was proposed by Darling and Erdos [8]. Let us define the quantities
$\gamma_n = \sqrt{2\log\log n}$, $\delta_n = \gamma_n^2 + \frac{1}{2}\xi_n - \frac{1}{2}\log\pi$

Then for n large and $x \in \mathbb{R}$

$$P\left[|T_{n,k}| > \frac{x+\delta_n}{\gamma_n}\right] \approx 1 - exp[-e^(- x)] \quad (19)$$

Now we shall refer to another problem. Recall (1), the hypothesis definition of the testing. This can be changed to

$$H_0 : Y_1, \ldots, Y_n \sim N(\mu, \sigma^2) \qquad H_1 : \exists k_0 \in \{2, \ldots, n-2\} \text{ with}$$

$$Y_1, \ldots, Y_{k_0} \sim N(\mu_1, \sigma_1^2) \quad (20)$$

$$Y_{k_0+1}, \ldots, Y_n \sim N(\mu_2, \sigma_2^2)$$

with $\mu_1 \neq \mu_2 \in \mathbb{R}$ and $\sigma_1^2 \neq \sigma_2^2 \in \mathbb{R}$. For the test statistic

$$\Psi_k^2 = n\log\left[\frac{1}{n}S\left(\bar{Y}_n\right)\right] - k\log\left[\frac{1}{k}S\left(\bar{Y}_k\right)\right] - (n-k)\log\left[\frac{1}{n-k}S\left(\bar{Y}_k'\right)\right] \quad (21)$$

with $S(A) = \sum_{i=1}^{n}(Y_i - A)^2$, recall (2), it can be proved, Horvath [22], that $P\left[\max_{1\leq k\leq n-1} \Psi_k > \frac{x+\theta_n}{\gamma_n}\right] \approx 1 - exp[-2e^{-x}]$, $x \in \mathbb{R}$, $\theta_n = a_n^2 + \xi_n$, recall (10).

We shall try to work on this framework completing the Explanatory Data Analysis (EDA), mainly graphically, to get the time series features of the data, and work with ACF and PACF, for the data which we collected and we tried to analyze in this present paper. Therefore, the target of this paper is to provide a complete presentation of the data under EDA principles, see Appendix, compact which supports the decision.

4 Data and Results

There are a number of different databases concerning different areas in the world. We are focused, in this study, to compare the air pollution between the most industrialized areas in Greece, Athens and Salonica, as in Sect. 3.

The data consist of long time series for various pollutants (among which, the five pollutants of great interest, CO, NO, NO_2, O_3, and SO_2), in many cities, in most of the cases covering the same period. We ultimately focused on two specific areas (Patisia, Peristeri) in the metropolitan area of Athens, and on two areas (Aristotelous, Kalamaria) in the metropolitan city of Salonica.

In detail, out of the four areas, three (Aristotelous, Patisia, and Peristeri) have data for each of the following six time periods: 1984–2000, 2001–2005, 2006–2010, 2011–2015, 2016–2020, and 2021. Kalamaria has no data for 1984–2000, lacks CO, NO, and NO_2 measurements for 2006–2010, and has CO data only for 2011–2015. For the remaining periods, data for Kalamaria are complete. CO data for Aristotelous

Table 2 ACF and PACF compact interpretation of time series

	Model	$AR(p)$	$MA(q)$	$ARMA(p, q)$
Plot	ACF	T_O	$S(p, q)$	T_O
	PACF	$S(p, q)$	T_O	T_O

are available only for 1984–2000, and there are no O_3 records. The remaining three pollutants are complete for this area, except SO_2 for 2006–2010. Patisia has records of all pollutants for all periods except 2021, in which only SO_2 has been recorded. Finally, Peristeri is complete except for 2016–2021, for which CO and SO_2 are missing.

Pollutant concentrations are measured continuously throughout the entire 24-hour period. The response time of the automated analyzers is on the order of one minute, meaning that each analyzer provides approximately one measurement per minute. Using a microprocessor installed at each automatic monitoring station and connected to the analyzers, mean hourly pollutant concentrations are calculated. The analysis is based on daily mean pollutant concentrations recorded at 9 a.m. (see Appendix Tables 3, 4, 5, 6, and 7).

For the time series models we worked with, such as AR, MA, and ARMA. Interest is focused on the ACF and PACF, i.e., the Autocorrelation Function and the partial autocorrelation function, respectively, plots are essential. An ACF plot indicates the existing correlation between a time series and lagged versions of itself. Recall that autocorrelation exists at zero lag, equivalently is the correlation of the time series with itself and therefore results in a correlation of 1. A PACF plot indicates the correlation between a time series and its lagged values after accounting for the correlations of shortest lags. Actually, the partial autocorrelation at lag q is the autocorrelation between X_t and X_{t-q} that is not accounted for by lags 1 through $q - 1$. Presented plots, following the augmented Dickey–Fuller test, Dickey and Fuller [9], for stationarity of the series, are implemented for AR and MA ordering and produced for all the above-mentioned (i–v) pollutants.

The following diagram offers a compact interpretation of both ACF and PACF for the given models (Table 2).
where, T_O: Tails off (Geometric delay), $S(t, r)$: Significant at each lag t/cuts off after lag r.

Extrapolatory Data Analysis (EDA) was introduced by Turkey [49] in order to analyze data set from different areas of interest, to obtain their main characteristics (i.e., statistical functions, test) and visualize the data sets with graphical methods, see also Behrens [4].

A sort manual of statistical packages was discussed by Kitsos [31], while a detailed approach with a number of applications with a recent software to the Time Series problem see Hyndman et al. [25], Hyndman and Khandakan [24], Halkos [12].

5 Empirical Results

5.1a NO in the areas of Athens

5.1a.1 Pollutant NO, Peristeri

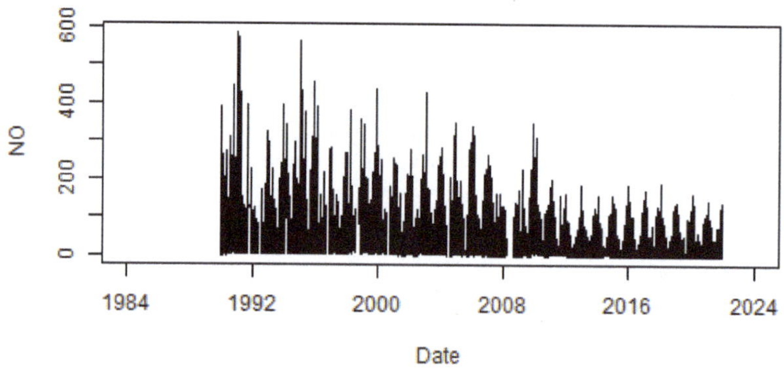

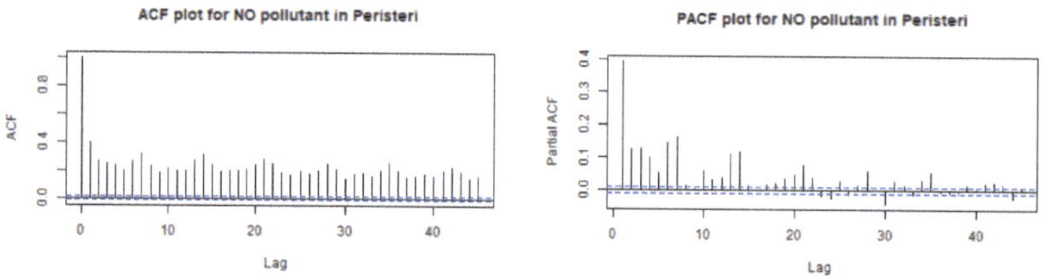

In the presented figures, the history of nitric oxide, NO, pollutant for the areas discussed above are indicatively represented and the corresponding ACF and PACF plots are considered, Kitsos and Nisiotis [34]. In detail:

5.1a.2 Pollutant NO, Patision

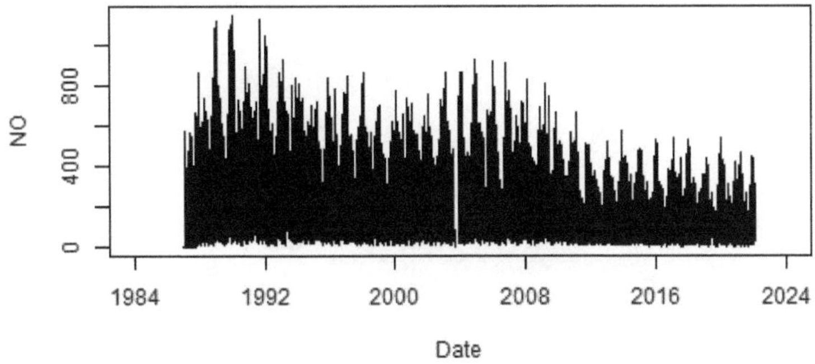

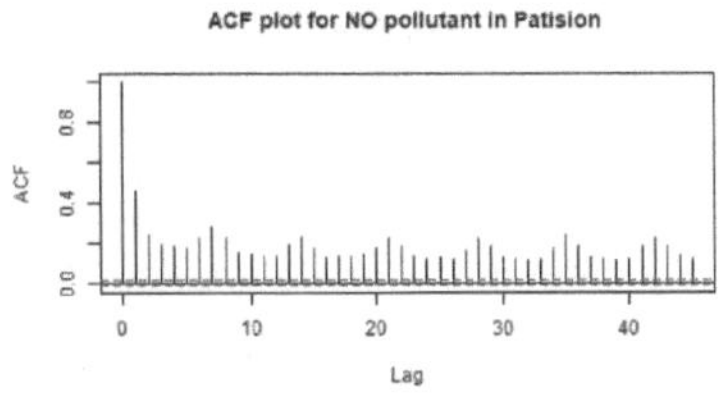

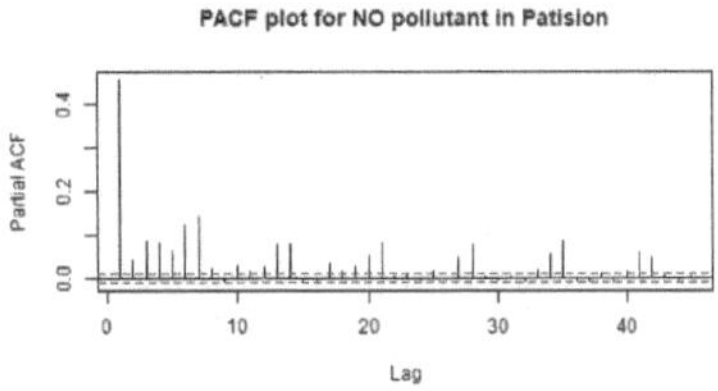

5.1b NO in the areas of Salonica

5.1b.1 Pollutant NO, Aristotelous

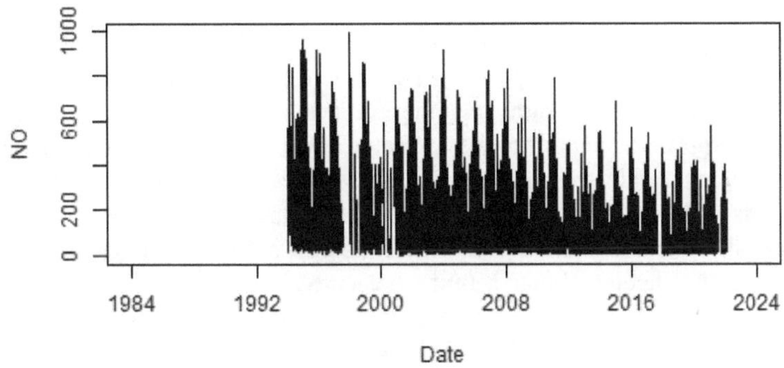

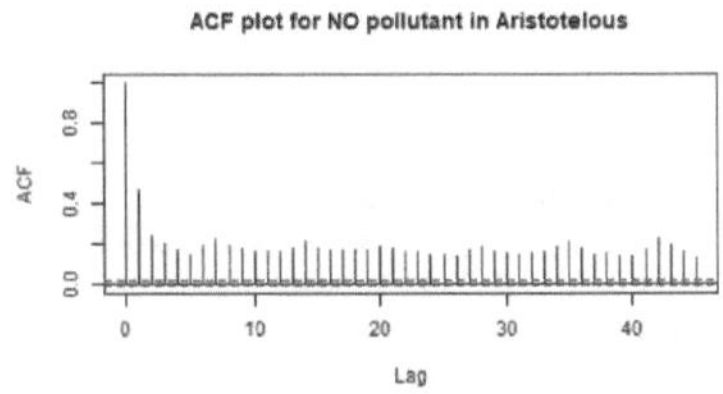

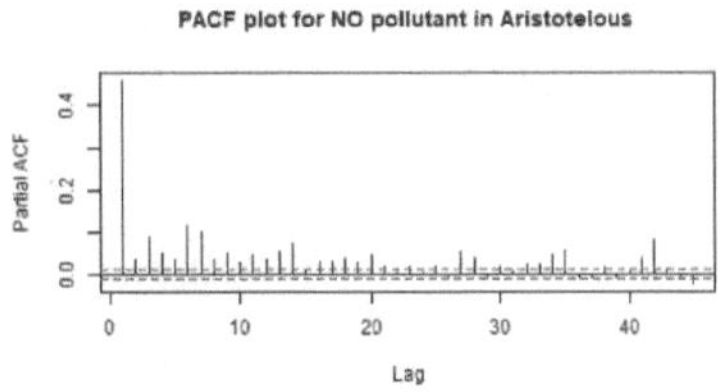

5.1b.2 Pollutant NO, Kalamaria

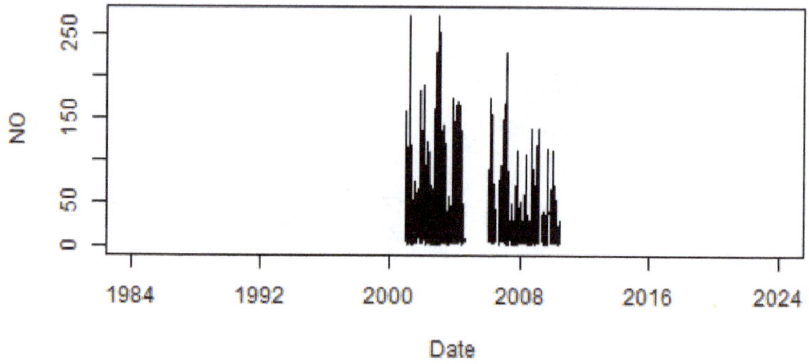

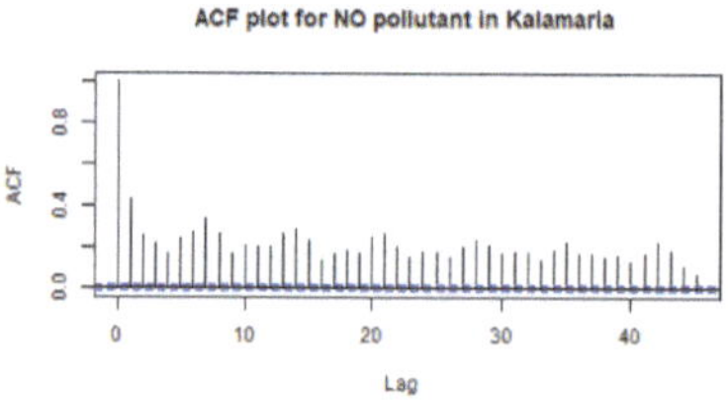

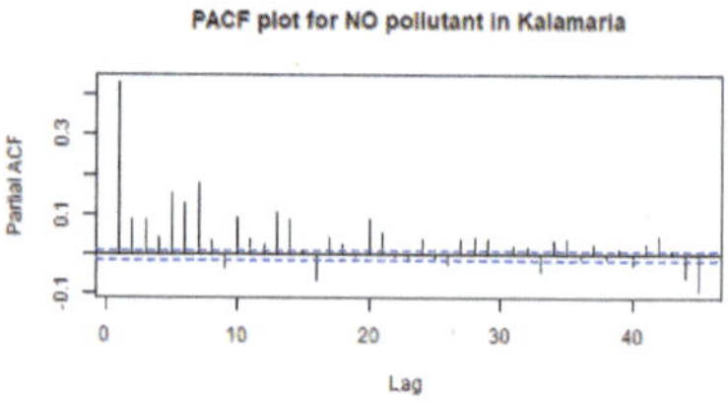

The ACF plot indicates long-term decreasing autocorrelation. The gradual decay suggests non-stationarity, and the presence of seasonality is evident from the repeating pattern at lag 7. The data does not appear to be white noise.

The PACF plot shows significant spikes at first six lags, with a gradual decay, suggesting a potential complex process. The presence of a spike at lag 6 may indicate seasonality. ACF and PACF plots, together, suggest an ARIMA (p, d, q) model.
A similar interpretation applies to the ACF and PACF plots of the remaining regions, given the evident tendency for these to align with each other. Note, however, that time periods of available data differ among the four areas.

For the above models and the presented figures, a number of ARIMA models, Halkos [12], have been estimated. In the presented Tables 8, 9, 10, 11, and 12, the appropriate model for estimating the pollutants under consideration for Athens and Salonica, per particular area under investigation are presented.

Table 3 Pollutant **CO**

		AR(p)	ar(1)	ar(2)	ar(3)	ar(4)	ar(5)	AIC
Athens	Patision	Coefficient	−0, 4677	−0, 4347	−0, 3441	−0, 2676	−0, 2333	66364, 8
		(SE)	(0, 0001)	(0, 0001)	(0, 0001)	(0, 0001)	(0, 0001)	
	Peristeri	Coefficient	−0, 6093	−0, 5154	−0, 3788	−0, 2730	−0, 2282	25005, 0
		(SE)	(0, 0001)	(0, 0002)	(0, 0002)	(0, 0002)	(0, 0001)	
Salonica	Aristotelous	Coefficient	−0, 5497	−0, 4418	−0, 3644	−0, 3324	−0, 2765	9509, 3
		(SE)	(0, 0005)	(0, 0006)	(0, 0006)	(0, 0006)	(0, 0005)	
	Kalamaria	Coefficient	−0, 5399	−0, 4853	−0, 3906	−0, 3013	−0, 2377	4339, 2
		(SE)	(0, 0002)	(0, 0003)	(0, 0003)	(0, 0003)	(0, 0002)	

Table 4 Pollutant **NO**

		AR(p)	ar1	ar2	ar3	ar4	ar5	AIC
Athens	Patision	Coefficient	−0, 5102	−0, 4682	−0, 3710	−0, 2609	−0, 2064	159508, 8
		(SE)	(0, 0001)	(0, 0001)	(0, 0001)	(0, 0001)	(0, 0001)	
	Peristeri	Coefficient	−0, 6211	−0, 5254	−0, 4037	−0, 2775	−0, 2236	106322, 4
		(SE)	(0, 0001)	(0, 0001)	(0, 0001)	(0, 0001)	(0, 0001)	
Salonica	Aristotelous	Coefficient	−0, 4867	−0, 4454	−0, 3338	−0, 2419	−0, 2008	113000, 4
		(SE)	(0, 0001)	(0, 0001)	(0, 0001)	(0, 0001)	(0, 0001)	
	Kalamaria	Coefficient	−0, 4341	0, 1716				22889, 7
		(SE)	(0, 0554)	(0, 0047)				
		MA(q)	**ma1**	**ma2**	**ma3**			
		Coefficient	−0, 2380	−0, 6887	0, 0193			
		(SE)	(0, 0558)	(0, 0312)	(0, 0055)			

Table 5 Pollutant **NO_2**

		AR(p)	ar1	ar2	ar3	ar4	ar5	AIC
Athens	Patision	Coefficient	−0, 5028	−0, 4566	−0, 3650	−0, 2674	−0, 2161	127159, 4
		(SE)	(0, 0001)	(0, 0001)	(0, 0001)	(0, 0001)	(0, 0001)	
	Peristeri	Coefficient	−0, 5487	−0, 4851	−0, 3704	−0, 2889	−0, 2364	92492, 2
		(SE)	(0, 0001)	(0, 0001)	(0, 0001)	(0, 0001)	(0, 0001)	
Salonica	Aristotelous	Coefficient	−0, 5015	−0, 4618	−0, 3467	−0, 2702	−0, 2363	81659, 0
		(SE)	(0, 0001)	(0, 0001)	(0, 0001)	(0, 0001)	(0, 0001)	
	Kalamaria	Coefficient	0, 9290					20636, 1
		(SE)	(0, 0001)					

The estimated ARIMA (p, d, q) models for the available data sets are similar, and the following comments can be applied across them. A substantial number of AR coefficients are statistically significant ($p < 0.001$)—only the first five are presented here—suggesting a meaningful autocorrelation structure. Data richness justifies the adoption of higher-order time series models to capture potentially complex lag structures and autocorrelations. Almost all AR coefficients are negative, indicating a complex oscillatory pattern involving multiple past values. All series required differencing (parameter d of the ARIMA model) to achieve stationarity. All estimated parameters satisfy the stationarity and invertibility conditions.

Only the nitric oxide (NO) data set from Kalamaria includes statistically significant MA coefficients, similar to the ozone (O_3) data set from Peristeri. In contrast, the ozone data set from Patision does not support any significant AR coefficients. Overall, the ozone data sets appear to deviate from the pattern observed in the other pollutants. It is also noted that ozone data are not available for Aristotelous.

We briefly refer that AIC is the Akaike information criterion (AIC): For the given statistical model, describing a data set, with p the number of the estimated parameters we let the maximized value of the likelihood function to be $L*$. Then the AIC value of the model is $AIC = 2[p - ln(L*)]$. We note that SE is the standard error, as usual and the fitted models.

Table 6 Pollutant SO_2

		AR(p)	ar1	ar2	ar3	ar4	ar5	AIC
Athens	Patision	Coefficient	−0, 4604	−0, 4223	−0, 3430	−0, 2047	−0, 1883	123301, 7
		(SE)	(0, 0001)	(0, 0001)	(0, 0001)	(0, 0001)	(0, 0001)	
	Peristeri	Coefficient	−0, 5899	−0, 4515	−0, 3521	−0, 2691	−0, 1595	58946, 2
		(SE)	(0, 0001)	(0, 0002)	(0, 0002)	(0, 0002)	(0, 0001)	
Salonica	Aristotelous	Coefficient	−0, 5265	−0, 4346	−0, 3737	−0, 3115	−0, 2259	45005, 5
		(SE)	(0, 0002)	(0, 0002)	(0, 0002)	(0, 0002)	(0, 0002)	
	Kalamaria	Coefficient	−0, 6095	−0, 4710	−0, 3349	−0, 3226	−0, 1673	7451, 1
		(SE)	(0, 0011)	(0, 0014)	(0, 0016)	(0, 0015)	(0, 0012)	

Table 7 Pollutant O_3

		MA(q)	ma1	ma2	ma3			AIC
Athens	Patision	Coefficient	−0, 6501	−0, 1735	−0, 0753			97437, 2
		(SE)	(0, 0001)	(0, 0001)	(0, 0001)			
	Peristeri	Coefficient	−0, 9336					87495, 9
		(SE)	(0, 0000)					
		AR(p)	ar1	ar2	ar3	ar4	ar5	
		Coefficient	0, 3185	−0, 0286	−0, 0016			
		(SE)	(0, 0001)	(0, 0001)	(0, 0001)			
Salonica	Aristotelous	Coefficient						
		(SE)	no data available					
	Kalamaria	Coefficient	−0, 5321	−0, 3240	−0, 2380	−0, 2155	−0, 1855	10524, 8
		(SE)	(0, 0008)	(0, 0010)	(0, 0010)	(0, 0010)	(0, 0008)	

The definition of ARMA model used here is

$$X_t = a_1 X_{(t-1)} + \cdots + a_p X_{(t-p)} + e_t + b_1 e_{(t-1)} + \cdots + b_q e_{(t-q)}$$

The differenced series for ARIMA models with differencing follows a zero-mean ARMA model. The exact likelihood was chosen for parameters' estimation in ARIMA process. Model selection was based on AIC value, through the stepwise algorithm outlined in Hyndman and Khandaakar [24] over possible models within the five order constraints provided. The variance matrix of the estimates is found from the Hessian of the log-likelihood. All analyses were carried out in the R environment, using appropriate packages (stats, forecast).

6 Conclusions

There are different considerations and extensions of the ideas we discussed in this paper. The autoregression has been in relation to causality, Toda and Phillips [45], on the distribution of the estimators for the autoregressive model, Dickey and Fuller [9], Lai and Siegmud [36] Kitsos [30], to use the canonical analysis, so useful to optimal experimental design theory, Kitsos [30], to time series, Tsay and Tiao [48]. There is an interest in working with the change-point theory in Environmental Problems and estimating the number of such change points in the time series under consideration, Yao [51].

For all pollutants, apart from O_3, and for all areas, their concentrations seem to be eliminated during the time, while the compact results for an EDA analysis are in Appendix so that the interest researcher to work on them. ACF plots reveal a tiny, steady weekly periodicity, observed optically to values recorded through time. In most of the cases, an ARIMA (p, 1, 0) is proposed—manifesting a consistent dependency on previous values, and only for the cases of NO [ARIMA (2, 1, 3)] and NO_2 [ARIMA (1, 0, 0)] in Kalamaria, O_3 [ARIMA (3, 1, 1)] in Peristeri and [ARIMA (0, 1, 3)] in Patision the proposed model change. Almost all models have first-order differencing to stationarize the series. Coefficients in Tables 3, 4, 5, 6, and 7 are estimated for the majority of the models for up to five AR terms—i.e., an ARIMA (5, d, q) model. All coefficients for the ARIMA (p, 1, 0) are negative, while all coefficients for all the models presented are statistically significant (Tables 3, 4, 5, 6, and 7).

Disclosure of Interests. The authors declare no conflicts of interest.

Acknowledgements We express our gratitude to the anonymous reviewers for their valuable and insightful comments of this work.

7 Appendix: Main EDA Parameters for Pollutants

See (Tables 8, 9, 10, 11, and 12).

Table 8 Pollutant **CO**

	ARI	KAL	PAT	PER
n	1847	4116	13376	7539
Mean	4.516	0.67	4.51	1.698
Median	3.8	0.5	3.3	1.1
Trimmed mean	4.22	0.616	4.164	1.51
Geometric mean	3.292	0.531	0	0
Harmonic mean	1.69	0.419	1.68	0.67
1st quantile	2.2	0.3	1.7	0.5
3rd quantile	5.9	0.9	6.4	2.3
Range	24	4.3	25.5	16.1
IQR	3.7	0.6	4.7	1.8
Variance	11.178	0.249	14.557	2.722
Std deviation	3.343	0.499	3.815	1.65
Skewness	1.524	2.123	1.372	1.964
Kurtosis	3.304	7.074	1.86	5.43

Table 9 Pollutant $\mathbf{NO_2}$

	ARI	KAL	PAT	PER
n	9022	2375	12403	9963
Mean	127.532	28.181	211.681	39.466
Median	76.5	14	176	16
Trimmed mean	110.736	23.602	198.984	31.946
Geometric mean	69.724	14.91	0	0
Harmonic mean	32.333	8.314	68.337	4.819
1st quantile	30	6	78	4
3rd quantile	177.75	37	307	55
Range	991	270	1144	586
IQR	147.75	31	229	51
Variance	19315.668	1175.412	27555.987	3019.196
Std deviation	138.981	34.284	166	54.947
Skewness	1.948	2.282	1.104	2.588
Kurtosis	4.487	6.633	1.326	9.932

Table 10 Pollutant O_3

	ARI	KAL	PAT	PER
n		1249	12361	9528
Mean		20.09	14.459	32.314
Median		11	10	22
Trimmed mean		17.624	12.727	30.139
Geometric mean		0	8.195	18.253
Harmonic mean		3.498	4.165	7.881
1st quantile		1	4	7
3rd quantile		32	20	51
Range		124	153	127
IQR		31	16	44
Variance		520.149	223.646	854.002
Std deviation		22.807	14.955	29.223
Skewness		1.535	1.85	0.934
Kurtosis		2.444	4.403	–0.107

Table 11 Pollutant NO_2

	ARI	KAL	PAT	PER
n	9026	2378	13410	9971
Mean	71.722	46.073	92.522	49.346
Median	68	45	90	49
Trimmed mean	69.78	45.649	91.259	48.27
Geometric mean	65.083	0	85.396	39.701
Harmonic mean	57.119	35.62	77.025	27.296
1st quantile	51	32	67	28
3rd quantile	87	59	113	67
Range	293	120	387	223
IQR	36	27	46	39
Variance	984.196	337.297	1251.256	761.009
Std deviation	31.372	18.366	35.373	27.586
Skewness	1.21	0.346	0.68	0.527
Kurtosis	3.286	-0.209	1.578	0.506

Table 12 Pollutant SO_2

	ARI	KAL	PAT	PER
n	5488	908	13010	7490
Mean	17.517	19.438	36.012	12.187
Median	7	12	20	6
Trimmed mean	13.869	17.182	30.832	10.038
Geometric mean	8.677	0	19.806	0
Harmonic mean	5.079	7.959	10.786	5.577
1st quantile	3	7	8	5
3rd quantile	21	25	50	14
Range	260	183	464	205
IQR	18	18	42	9
Variance	609.819	390.412	1713.357	225.728
Std deviation	24.695	19.759	41.393	15.024
Skewness	3.194	2.134	2.371	4.273
Kurtosis	14.777	7.136	8.538	30.543

References

1. Ahmad, N., Du, L., Lu, J., Wang, J., Li, H.-Z., Hashmi, M.Z.: Modelling the CO_2 emissions and economic growth in Croatia: Is there any environmental Kuznets curve? Energy **123**, 164–172 (2017)
2. Asumadu-Sarkodie, S., Owusu, P.A., Asumadu-Sarkodie, S., Owusu, P.A.: Carbon dioxide emissions, GDP per capita, industrialization and population: an evidence from Rwanda. Environ. Eng. Res. **22**, 116–124 (2016)
3. Balcilar, M., Ozdemir, Z.A., Arslanturk, Y.: Economic growth and energy consumption causal nexus viewed through a bootstrap rolling window. Energy Econ. **32**, 1398–1410 (2010)
4. Behrens, J.T.: Principles and procedures of exploratory data analysis. Psychol. Methods **2**(2), 131 (1997)
5. Box, G.E.P., Jenkins, G.M., Reinsel, G.C.: Time Series Analysis: Forecasting and Control. 3d Edn., JRSS, B, vol. 26, pp. 211–243 (1994)
6. Carlstein, E.: Nonparametric change-point estimation. Ann. Stat. **16**(1), 188–197 (1988)
7. Chernoff, H., Zacks, S.: Estimating the current mean of a normal distribution which is subjected to changes in time. Ann. Math. Stat. **35**(3), 999–1018 (1964)
8. Darling, D.: A limit theorem for maximum of normalized sums of independent random variables. Duke Math. J. **23**, 143–155 (1956)
9. Dickey, D.A., Fuller, W.A.: Distribution of the estimators for autoregressive time series with a unit root. J. Am. Stat. Assoc. **74**(366a), 427–431 (1979)
10. Gruber, N., Bakker, D., DeVries, T., Gregor, L., Hauck, J., Landschützer, P., Mckinley, G., Müller, J.: Trends and variability in the ocean carbon sink. Nat. Rev. Earth Environ. **4**, 1–16 (2023)
11. Hackl, P. (Ed.).: Statistical Analysis and Forecasting of Economic Structural Change. Springer Science & Business Media (2013)
12. Halkos, G.E.: Econometrics, Theory and Practice. 2d Edn. DISIGMA Pub (2019). (in Greek)
13. Halkos, G.E.: Modeling energy–environment–economy interrelations. Energies **17**(24), 6343 (2024)

14. Halkos, G.E.: Economy and Environment. Liberal Books (2013). (in Greek)
15. Halkos, G.E., Paizanos, E.A.: Environmental macroeconomics: economic growth, fiscal spending and environmental quality. Int. Rev. Environ. Resour. Econ. **9**(3–4), 321–362 (2016)
16. Halkos, G.E., Kitsos, C.P.: Uncertainty in environmental economics: the problem of entropy and model choice. Econ. Anal. Policy **60**(2018), 127–140 (2018)
17. Halkos, G.E., Gkampoura, E.C.: Examining the linkages among carbon dioxide emissions, electricity production and economic growth in different income levels. Energies **14**(6), 1682 (2021)
18. Hawkins, D.M.: Testing a sequence of observations for a shift in location. J. Am. Stat. Assoc. 180–186 (1977)
19. He, J., Richard, P.: Environmental Kuznets curve for CO2 in Canada. Ecol. Econ. **69**, 1083–1093 (2010)
20. de Heer, M., Roozen, F., Maas, R.: The integrated approach to nitrogen in the Netherlands: a preliminary review from a societal, scientific, juridical and practical perspective. J. Nat. Conserv. **35**, 101–111 (2017)
21. Hinkley, D.V.: Inference about the change-point in a sequence of random variables. Biometrika **57**(1), 1–1 (1970)
22. Horváth, L.: The maximum likelihood method for testing changes in the parameters of normal observations. Ann. Stat. 671–680 (1993)
23. Horváth, L.: Detecting changes in linear regressions. Stat. J. Theor. Appl. Stat. **26**(3), 189–208 (1995)
24. Hyndman, R.J., Khandakar, Y.: Automatic time series forecasting: the forecast package for R. J. Stat. Softw. **26**(3) (2008)
25. Hyndman, R., Athanasopoulos, G., Bergmeir, C., Caceres, G., Chhay, L., O'Hara-Wild, M., Petropoulos, F., Razbash, S., Wang, E., Yasmeen, F.: Forecast: Forecasting Functions for Time Series and Linear Models (2023)
26. IPCC: Scientific Assessment of Climate Change: Report Prepared for IPCC by Working Group 1. World Meteorological Organization and United Nations Environmental Program. New York (1990)
27. IPCC: Climate Change 2021: The Physical Science Basis. Contribution of Working Group I to the Sixth Assessment Report of the Intergovernmental Panel on Climate Change[Masson-Delmotte, V., P. Zhai, A. Pirani, S.L. Connors, C. Péan, S. Berger, N. Caud, Y. Chen, L. Goldfarb, M.I. Gomis, M. Huang, K. Leitzell, E. Lonnoy, J.B.R. Matthews, T.K. Maycock, T. Waterfield, O. Yeleki, R. Yu, and B. Zhou (eds.)]. Cambridge University Press, Cambridge, United Kingdom and New York, NY, USA (2021)
28. Jarušková, D.: Testing appearance of linear trend. J. Stat. Plan. Infer. **70**(2), 263–276 (1998)
29. Kandel, R.: Le réchauffement climatiques: Le grand risque, Presses Universitaire de France, Collection Que sais-je, 127 (2002)
30. Kitsos, C.P.: Design and inference in Nonlinear Problems. Ph.D thesis, University of Glasgow (1986). https://thesis.gla.ac.uk
31. Kitsos, C.P.: Computational Statistics. New Technologies Pub (1996). (in Greek)
32. Kitsos, C.P., Edler, L.: Cancer risk assessment for mixtures. In: Edler, L., Kitsos, C. (eds.) Recent Advances in Quantitative Methods in Cancer and Human Health Risk Assessment, pp. 281–298. Wiley, Chichester, England (2004)
33. Kitsos, P.C., Trandafir, P.-C.: Hazard function of $N_\gamma(\mu, \sigma^2)$ distribution for environmental pollutants. In: ENVECON **8**, 2–3 Dec 2022, Department of Public and Community Health Athens, Book of Abstracts, Session 2, pp. 65–75 (2022)
34. Kitsos, P.C., Nisiotis, C.S.: Comparing Air Pollution levels in Greece: the case of Athens and Salonica. In: ENVECON **7**, 8–9 Dec 2023, Book of Abstracts, Session 4, Climate Change and Urban Environment (2023)
35. Kitsos, P.C., Nisiotis, C.S., Stamatiou I.S.: The γ-order generalized distributions for environmental economics applications: a compact review. Submitted (2025)
36. Lai, T.L., Siegmund, D.: Fixed accuracy estimation of an autoregressive parameter. Ann. Stat. 478–485 (1983)

37. Mesick, K.E., Feldman, W.C., Coupland, D.D.S., Stonehill, L.C.: Benchmarking Geant4 for simulating galactic cosmic ray interactions within planetary bodies. Earth Space Sci. **5**, 324–338 (2018)
38. Neumayer, E., Plomber, T.: The Gender Nature of Natural Disasters: the impact of catastrophic events on the genter gapin the expectancy 1981–2002. London School of Economics, U. of Essex, Max Plank Institute for Economics, London, UK (2007)
39. Papakyriazis, P.A.: Optimal experimental design in econometrics: the time series problem. J. Econ. **7**(3), 351–372 (1978)
40. R Core Team: R: A language and environment for statistical computing. R Foundation for Statistical Computing, Vienna, Austria (2023). https://www.R-project.org/
41. Silva, V., Yang, X., Fleskens, L., Ritsema, J.C., Geissen, V.: Environmental and human health at risk Scenarios to achieve the Farm to Fork 50% pesticide reduction goals. Environ. Int. **165** (2022)
42. Sims, C.A., Stock, J.H., Watson, M.W.: Inference in linear time series models with some unit roots. Econometrica **58**, 113–144 (1990)
43. Skevas, I., Zhu, X., Shestalova, V., Emvalomatis, G.: The impact of Agri-environmental policies and production intensification on the environmental performance of Dutch dairy farms. J. Agric. Resour. Econ. **43**, 423–440 (2018)
44. Stokstad, E.: Nitrogen crisis threatens Dutch environment and economy. Science **366**, 1180–1181 (2019). https://doi.org/10.1126/science.366.6470.1180
45. Toda, H.Y., Phillips, P.C.: Vector autoregressions and causality. Econom. J. Econ. Soc. **61**, 1367–1393 (1993)
46. Toda, H.Y., Phillips, P.C.: Vector autoregression and causality: a theoretical overview and simulation study. Econ. Rev. **13**, 259–285 (1994)
47. Toda, H.Y., Yamamoto, T.: Statistical inference in vector autoregressions with possibly integrated processes. J. Econ. **66**, 225–250 (1995)
48. Tsay, R.S., Tiao, G.C.: Use of canonical analysis in time series model identification. Biometrika **72**(2), 299–315 (1985)
49. Tukey, J.W.: Exploratory Data Analysis. Addison-Wesley Series in Behavioral Science: Quantitative Methods (1977)
50. Upham, P., Scholte, R., Feron, S.: Social representations of injustice in sustainability transitions: the case of farmers and Dutch agricultural nitrogen policy. Front. Polit. Sci. **7** (2025)
51. Yao, Y.C.: Estimating the number of change-points via Schwarz' criterion. Stat. Probab. Lett. **6**(3), 181–189 (1988)

Identifying Impacts and Causalities Regarding Air Quality, Environment, and Tourism in the Eurozone Economic Space: A Panel Data Analysis

George Ekonomou and George Halkos

Abstract Air pollution remains an important issue in academic and empirical research since it is connected with climate change issues and quality of life. Given its importance in implementing climate resilience policies, the present paper discusses the Environmental Kuznets Curve hypothesis in the Eurozone countries using a less visible set of variables in the relevant literature. It includes methane emissions exclusively released from the energy sector and carbon dioxide intensity as environmental degradation proxies, whereas it uses population rates and Gross Domestic Product (GDP) to define growth. Furthermore, it investigates the impact of the tourism sector on the air degradation variables to test the EKC curve. Research findings confirm the EKC in all econometric model specifications, whereas causality tests suggest two unidirectional relationships between growth variables and air pollution proxies. Practical implications call for increasing energy efficiency rates and adopting eco-friendly practices to establish sustainability options in socio-economic systems.

Keywords Environmental degradation · Air pollution · Economic growth · Tourism

1 Introduction

Climate change harms the quality of natural resources, whereas climate crises threaten the well-being of all living and nonliving things. Climate change and air pollution concerns should not be addressed jointly; instead, they should focus on protecting people's health, strengthening human capital, and advancing the quality of

G. Ekonomou (✉)
Department of Planning and Regional Development, School of Engineering, University of Thessaly, Volos, Greece
e-mail: oikonomoug@aua.gr

G. Halkos
Laboratory of Operations Research, Department of Economics, University of Thessaly, Volos, Greece
e-mail: halkos@econ.uth.gr

G. Halkos et al. (eds.), *Environmental Risk Analysis*, Springer Proceedings in Mathematics & Statistics 540, https://doi.org/10.1007/978-3-032-24155-9_9

life [1]. Environmental degradation is a challenging research issue since it concerns the quality and quantity of ecosystem services, the severity, and the intensity of economic activities interacting with socio-ecological systems. In open economies, environmental quality status is not static—in contrast, it undergoes dynamic changes due to economic development [2]. A major concern in these efforts remains the struggle to achieve sustainable growth patterns and secure the provision of ecological benefits, nature grants to our society without abolishing economic goals and objectives. Practically, it is essential to establish a balance between economic growth and natural resource exploitation. Interestingly, as indicated by [3], at least in the short term, economic growth cannot be experienced without threatening the environmental quality status [4–6].

Recognizing research gaps and priorities is essential to promote sustainability science and move toward a sustainable future [7], whereas multiple actors are involved in meeting the combined goals of a low-carbon, resource-circulating, nature-harmonious society [8]. To this end, the EKC hypothesis can help define whether economic growth degrades the environment (e.g., air quality) or the economy "de-couples" from environmental quality status. Upon confirmation of the EKC, an inverted U-shaped relationship exists between the tested growth variable (e.g., income per capita) and environmental degradation proxy (e.g., pollution). EKC indicates that environmental degradation rises in an economy's early growth phase [9, 10]. However, after a specific (turning) point in this process (e.g., rising per capita income), environmental degradation decreases [9, 10]. This upward trend in the early growth phases reverses, taking on a downward direction after the turning point, since more resources are allocated to improve environmental quality [3, 11].

This study aims to identify impacts and causalities between air pollution and growth variables in the context of the EKC hypothesis in Eurozone countries. The authors conceptualize air pollution as methane emissions explicitly released from the power sector and as the carbon dioxide intensity of GDP. Furthermore, model specifications include a country's GDP and population rates as growth variables. This study's environmental degradation proxies are often overlooked or underutilized when testing the EKC hypothesis. Moreover, one particular contribution of this study is the inclusion of population rates as a growth variable to justify the U-shaped curve of the EKC hypothesis. Despite its significance in a country's growth process, such a variable is unobserved in relevant literature. To augment our scientific approach to the EKC hypothesis, the authors test the impacts of high-leverage sectors such as tourism by including "tourism spending" variables, namely capital investments and internal consumption in the travel and tourism sector. This research effort views holistically the EKC hypothesis, encompassing high-impact variables, and investigates relevant causalities to address useful, practical implications toward sustainable growth patterns.

The World Bank [1] reports that the cost of the health damage generated by air pollution amounts to $8.1 trillion a year, equivalent to 6.1% of global GDP. Additionally, it stresses the importance of prioritizing air pollution amid competing development challenges in designing and implementing policies and interventions to improve air quality. Moreover, the power industry is the largest emitter of carbon dioxide; CO_2 emissions constitute the principal greenhouse gas (GHG) contributing to global warming [12]. Also, 34% of the global power supply is produced by coal, which is responsible for almost 70% of emissions linked to electricity [12].

The structure of this paper is as follows: The theoretical background discusses the concept of the EKC hypothesis and presents this study's contributions to the literature. The Methodological section analyzes the econometric approach of panel data analysis and includes panel variable interpretation. The Results section discloses the research findings, whereas the Discussion section concerns practical implications. Finally, the last section concludes the results.

2 Theoretical Framework

One of the most demanding and contemporary issues across the scientific community is the struggle to foster sustainability in the economic system. [13] argues that economic growth impacts environmental quality, particularly at the beginning stages of output. Increasing pressure on the environment urges the world to structure and implement policies toward sustainable development [14]. Such an approach profoundly requires well-defined and detailed research efforts that generate meaningful results to confront climate change conditions effectively (e.g., applicable policies that create measurable results) and efficiently (e.g., resource efficiency, energy efficiency). The EKC hypothesis advances the meaning of sustainable growth rates, indicating the core issue of establishing equilibrium between the economy and the environment.

Following [15] approach, the EKC hypothesis can be explained by the scale, composition, and technique effects. The scale effect indicates that as per capita income grows, the demand for natural resources increases, too. This means that as the size of the economy increases, the extraction and use of natural resources generate pollution, which drives environmental degradation [2]. Practically, as economic growth accelerates, in the early stages, environmental degradation levels increase because of economic growth's scale effects [16]. As a country's economy expands, production and consumption increase, creating the need for more resource extraction and waste generation. This can degrade the country's environmental quality levels [17]. After the scale effect, the composition effect takes place. This effect suggests that a country's industrial structure starts to shift due to income increases, indicating changes in that country's economic structure [4]. In this phase, shifts are experienced regarding the type of production, indicating a transition in the economic structure from polluting industries to cleaner sectors [17]. These shifts include eco-friendly practices (e.g., cleaner activities) that limit pollution levels [11]. In the inverted U-shaped curve, the technique effect follows the composition effect. Such an effect implies that as the economy continues to increase, technological progress reduces environmental degradation and generates environmental improvements [11]. This reduction results from cleaner technologies replacing dirtier ones in production [18]. For instance, technological innovation, efficient production processes, and less polluting technologies favor environmental improvements [17]. Composition and technique effects generate the "pollution-income" trajectory downward of the inverted U-shaped curve [2].

When relevant econometric analysis is processed, the EKC hypothesis is justified by an inverted U-shaped curve. However, many research efforts report variations on differently shaped curves regarding the EKC. Indicatively, [19] evidenced an N-shaped curve between natural resource rents and economic variables. Furthermore, [20] confirmed an N-shaped curve for all tested income group variables, except for countries that belong to the upper-middle-income level. Practically, when an N-shaped curve is confirmed, it indicates that the U-shaped curve suggested by the EKC will not continue to exist with a long-term perspective. As soon as economic growth overcomes (passes) a certain point, an increase in income level may generate a positive relationship again between economic growth and environmental degradation [21]. As indicated by AlKhars et al. [4, 22], processed a review and elaborated on 168 studies concerning the EKC hypothesis in the context of air quality (CO_2 emissions). The period was from 1991 to 2017. The review advanced the power of income in the adopted model specifications (e.g., quadratic or cubic power), the type of econometric analysis, namely panel data or time series analysis, and the shape of the curve justified. For instance, [4] mention variations in the justified curves. They mention a linear, U-shaped, or inverted U-shaped curve, a N-shaped or inverted N-shaped curve, or no justification of the EKC hypothesis. Another interesting issue in confirming the EKC concerns the turning point after which the growth variables do not degrade the environment. From this perspective, [23] noted that different pollutant determinants have different turning points. Understandingly, studies concerning various countries suggest that the turning point of the inverted U-shaped curve, where degradation begins to decrease, differentiates significantly depending on the country's developmental status, policy frameworks, and industrial composition [3].

Despite its broad applicability and contribution to defining links and trajectories between the environment and the economy, the ECK hypothesis and the implied inverted U-shaped curve are criticized across academia. [17] mention some critical points of criticism. Their study indicates that the EKC tends to oversimplify the complex linkages between environmental outcomes and economic growth [24]. An additional criticism argument is that the EKC does not address the diversity of environmental problems since various forms of pollution may have distinct relationships (associations) with income levels [25]. Moreover, [17] mention normative implication issues when applying the EKC hypothesis. In this context, [25] notes that economic growth unavoidably creates environmental degradation, while the solution is grounded in reaching a particular income level rather than radically implementing policies to protect environmental quality. One way to address these issues is to implement the EKC hypothesis under different econometric schemes, account for the heterogeneous nature of the economic system, including well-defined environmental proxies, and enhance the predictive power of control variables in model specifications.

By nature, the supply–demand interaction in the economic system reflects a dynamic system that varies in time and space while new developments and trends arise. Multiple drivers can be considered to test their predictive power on energy—and environment-related variables. This approach highlights the multifaceted scientific nature of the economy and the natural environment. This study's research approach is grounded on the EKC hypothesis to test whether adopted growth variables impact environmental quality levels. The EKC hypothesis has been tested extensively in relevant literature. Despite these outstanding research efforts, the present study adds to the literature in the following ways: It uses environment degradation proxies rarely met in previous research attempts. Specifically, it uses methane emissions exclusively released from the power industry (energy) to focus on how this air quality determinant interacts with the economic sector (e.g., a nation's GDP). Furthermore, the authors extend their approach by testing carbon dioxide emissions in the EKC framework. This proxy stands for the carbon intensity of GDP, which remains a challenging research indicator for structuring resilient economies. In particular, investigating linkages between environmental degradation levels and macroeconomic variables is significant and highly informative in implementing policies, such as fiscal policies on CO_2 emissions [26]. The current scientific discussion focuses on reducing carbon dioxide releases without considerably narrowing growth rates, especially for emerging and developing economies [27].

Second, this study uses the population rates across the Eurozone countries as a growth variable to evidence the inverted U-shaped curve in the EKC hypothesis [28]. Claim that population, in parallel with other social determinants, remains essential for processing research to report the human dimensions of global environmental change. Population as a growth proxy is unobserved in relevant research efforts despite its significance and contribution to energy needs, environmental issues, and consumption rates in the economic system. Researchers have paid attention to population issues and environmental quality. For instance, [29] investigated the impact of population on carbon emissions by implementing cross-national panel analysis from 86 countries. They concluded that regardless of the growth rates within an economy, the population still correlates strongly and positively with carbon dioxide emissions. Supportively, [28] explored the relationship between population factors and environmental degradation. They evidenced a positive correlation among OECD countries. Furthermore, [30] concluded that population growth leads to environmental degradation in the long run when Panel Vector Autoregressive (PVAR) and Fully Modified OLS (FMOLS) approaches are applied from 1990 to 2014 in 28 European countries.

Third, the Eurozone economic space is less investigated in empirical studies when elaborating on the EKC hypothesis. As a result, an additional research opportunity appears to investigate impacts and causalities based on the EKC hypothesis. One particular issue that calls for our scientific attention is how high-leverage tourism development indicators impact environmental quality. Interestingly, capital investment spending and internal travel and tourism consumption are included in this study's model specification. The purpose behind such an approach is twofold: first, to test how impactful economic sectors are on environmental quality, and second, to highlight the deeper meaning of "spending," considering it not only as a monetary variable but also as a core determinant in formulating pro-environmental behavior and sustainable consumptive patterns. Moreover, maintaining a balance between tourism expansion and natural resource preservation is essential to avoid environmental degradation [31]. The adopted groups of variables structure a logical and compelling research approach scarcely met in relevant research efforts. Nevertheless, investigating the EKC hypothesis has raised concerns from several viewpoints, indicating a question that still needs an answer: "Can economic growth be part of the solution rather than the cause of the environmental problem?" [11]. Consequently, this study holistically views the context of the EKC hypothesis to generate contributions in the scientific community, providing justified answers to the following research questions: Is there any justified causality linkage between a country's population rate and methane emissions? Is there any justified causality linkage between a country's GDP and carbon dioxide intensity? Is there any impact of capital investment spending and internal travel and tourism consumption on the air pollutants adopted in this study? What are the directions, if evidenced, of these causality relationships? If any causality exists, what is the impact of such potential relationships? How much does a change in one act upon the other? These questions need well-defined answers based on statistically significant and interpretable research findings from robust econometric approaches.

3 Methodology

This study elaborates on the interaction of air quality and economic growth in the context of the EKC hypothesis. The author's approach encompasses the countries' GDP and population as growth variables against methane emissions and carbon dioxide intensity of GDP. To augment this perspective, the authors include tourism-related variables as regressors to explore how these high-leverage and profitable sectors interrelate with air pollution. To this end, panel data analyses are processed across countries in the Eurozone's economic space between 1996 and 2019. Data were derived from the World Bank and World Travel and Tourism Council (WTTC), whereas the time frame of the analysis was based on data availability. All panel variables were ln-transformed. As a result, regression coefficients can be considered as elasticities. In Table 1, descriptive statistics of the panel variables are presented. Specifically, the following model specifications were developed to meet this study's research objectives.

Model Specification 1

$$\text{lnmethane} = \text{f}\,(\text{lnpop},\ \text{lnpopsq},\ \text{lngdp}, \text{lninvest}) \tag{1}$$

Model Specification 2

$$\text{lndioxideint} = \text{f}\,(\text{lngdp},\ \text{lngdpsq}, \text{lnpop},\ \text{lnittc}) \tag{2}$$

Table 1 Descriptive statistics

	Pop	GDP	Invest	Ittc	Methane	Dioxide int
Mean	17,434,652,70	573,115,846,282,28	5,396,430,462,72	46,666,865,013,16	0,0,616,162	0,000,000,455
Median	5,421,225,00	200,158,758,897,60	1,773,450,000,00	14,837,450,000,00	0,02,275	0,000,000,393
Maximum	83,092,962,00	3,597,317,337,494,24	39,040,900,000,00	334,476,000,000,00	0,4549	0,000,002,190
Minimum	379,905,00	4,835,164,585,22	50,000,000,00	623,236,000,00	0,0003	0,000,000,146
Std. Dev	24,296,225,97	862,960,745,315,64	7,967,391,029,16	77,621,093,636,05	0,0,878,492	0,000,000,275
Skewness	1,56	1,81	3,03	2,25	2,44	2,43
Kurtosis	0,97	2,22	5,396,430,462,72	4,47	6,72	7,88
Observations	456	456	456	456	456	456

In the above models, lnmethane stands for methane emissions from the power industry (energy) measured in Mt of CO_2 equivalent. Lndioxideint denotes the carbon intensity of GDP measured in kg of CO_2 equivalent per constant 2015 US dollars of GDP. Lngdp is the Gross Domestic Product measured in constant 2015 US dollars. Lngdpsq is the squared ln-transformed GDP. Lnpop denotes the population of the Eurozone countries. Lnpopsq is the squared ln-transformed population of the country of interest. Lninvest represents capital investment spending by all industries directly related to travel and tourism. Additionally, this panel variable concerns investment spending by other industries on particular tourism assets in real prices in US dollars. It includes spending on accommodation and passenger transport equipment, leisure facilities for specific tourism reasons, and restaurants. Lnittc denotes the total revenue generated by industries that deal directly with tourists, including visitor exports, domestic spending, and government individual spending at real prices in US dollars.

3.1 Cross-Section Dependence (CD) Tests

Attention should be paid to testing whether cross-section phenomena appear in panel data series. If present, CD phenomena may create biases in error terms, providing misleading results. Two main reasons CD is frequently present in panel data analysis are spatial dependence across tested countries, such as economic distance [32] and policy and social distance [33]. For instance, CD indicates unobserved components in the disturbance term [34]. These dependencies are rooted in open economies, relevant trade or commodity exchanges, resource mobility, and border abolition. CD phenomena are primarily featured in panel analysis when countries of reference, such as the Eurozone member states, adopt similar economic schemes or follow commonly agreed directives and commercial deals.

The present study processes [35] test to investigate if the CD is present (Eq. 3). In Eq. 4, T stands for the time interval, N denotes the cross-section units, whereas $\widehat{p_{ij}}$ provides the pair-wise correlation between cross-sections. The null hypothesis in [35] CD test suggests cross-section independence.

$$CD = \sqrt{\frac{2T}{N(N-1)}} \sum_{i=1}^{N-1} \sum_{j=i+1}^{N} \widehat{p_{ij}} \tag{3}$$

3.2 Unit Root Tests

Unit root tests indicate whether the mean and variance across panel variables remain constant over the analysis's time span. If unit roots are not present, meaning that panel variables are stationary, then the mean and variance are constant. To test the presence of unit roots, the present study adopts [36] test (Eq. 4). Grounded on the augmented IPS test (CIPS), in Eq. 4, N represents the number of cross-section units, whereas T concerns the time-series dimension. Moreover, $t_i(N, T)$ stands for the cross-sectionally augmented Dickey-Fuller statistic for the i th cross-section unit. Importantly, this test effectively treats CD phenomena identified in the previous step [36].

$$CIPS = N^{-1}\sum_{i=1}^{N} t_i(N, T) \tag{4}$$

3.3 *Driscoll-Kraay's Standard Errors Regression*

To identify whether the EKC is confirmed regarding this study's model specifications (Eqs. 1 and 2), we employed Driscoll-Kraay's standard errors regression, based on Ordinary Least Square with pooled effects procedure [37]. This test accommodates CD phenomena (e.g., spatial and temporal dependence), is used for panel samples with small data properties, and treats heteroscedasticity and autocorrelated standard errors, generating robust results [38].

3.4 *Ganger Non-Causality Tests*

This study implements [39] Granger non-causality procedure to explore the causal relationships between pair tests of panel variables. The null hypothesis of this test suggests that no causality exists between the tested pairs of panel variables. The test relies on the Wald statistic calculated of Eq. 5. This equation generates the average statistic $W_{N,T}^{H_{nc}}$ concerning the null Homogeneous Non-Causality (HNC) hypothesis. In Eq. 5, $W_{i,T}$ denotes the individual Wald statistics for the i th cross-section unit corresponding to the individual test $H_o : \beta_\iota = 0$.

$$W_{N,T}^{H_{nc}} = \frac{1}{N}\sum_{i=1}^{N} W_{i,T} \tag{5}$$

Table 2 CD test results

CD test formodel specification of Eq. 1	
Dependent variable: lnmethane	
Explanatory variables: lnpop, lnpopsq, lngdp, lninvest	
Pesaran's test of cross-sectional independence = 9.605	p-value = 0.000
CD test for model specification of Eq. 2	
Dependent variable: lndioxideint	
Explanatory variables: lngdp, lngdpsq, lnpop, lnittc	
Pesaran's test of cross-sectional independence = 24.680	p-value = 0.000

4 Results

Following concrete and well-defined methodological steps, the present study's panel analysis provides statistically significant, meaningful, and interpretable results. These results, in turn, provide useful and helpful inputs directly connected with physical implications and policy formulation, as indicated in the following sections.

4.1 Results of Cross-Section Dependence Tests (CD)

This study adopted [35] procedure to test whether CD is present in panel data. Test results reject the null hypothesis of cross-sectional independence (Table 2). The CD is usually met in panel data analysis, and tests that accommodate it should be implemented to receive dependable results. The reasons for the CD presence across the panel series are that countries in the Eurozone share and follow similar economic schemes, whereas they support open economies and capital flows or exchange of goods and services to create relevant dependencies. Furthermore, one interesting point is that spatial and spatial distance can enhance such dependencies.

4.2 Results of Unit-Root Tests

Unit roots indicate that the mean and variance across panel variables are not constant over time. [36] adopted unit root test indicates that this study's panel variable carries no unit roots at the first difference level, I(1). The panel variable lnpop is stationary at the 10% significance level, whereas all other variables are stationary at the 1% significance level (Table 3). Significantly, this test treats the CD phenomenon identified in the previous step and is based on the cross-sectionally augmented IPS test.

Table 3 Unit root test results

Variables	CIPS test
Levels	
lnmethane	−1.692
lndioxideint	−2.124
lngdp	−1.200
lnpop	−1.249
lninvest	−2.276
lnittc	−1.910
First Difference	
Δ.lnmethane	−4.267
Δ.lndioxideint	−4.431
Δ.lngdp	−3.048
Δ.lnpop	−2.148
Δ.lninvest	−4.669
Δ.lnittc	−3.862

Pesaran Panel Unit Root Test with cross-sectional and first difference mean
Δ: is the first difference operator
Ho: homogeneous non-stationary, p value < 0.005: reject *Ho*

Critical values: 10%: −2.07, 5%: −2.15, 1%: −2.32

4.3 *Results of* Driscoll and Kraay's Standard Error Regression

The authors used the pooled OLS [37] standard error procedure to test the EKC hypothesis concerning air pollution. This study conceptualized air pollution using methane emissions from the power industry (energy) and the carbon dioxide intensity of GDP. Furthermore, it adopted the GDP and population rates of the Eurozone member states as proxies for growth patterns. The EKC hypothesis is confirmed in all cases, and test results provide useful insights for formulating sustainable practices in socio-economic systems across the Eurozone. In this study, the authors also tested the tourism sector's predictive power and relevant impacts, stressing its importance in air quality levels.

Interestingly, in Model 1 (Table 4), where the population rates are the growth variable, the results indicate that a one-unit increase in GDP will increase methane emissions by 0.80%. In contrast, increased capital investment spending will decrease methane emissions by 0.36%.

Moreover, the results are on the same wavelength in Model 2, where the GDP is the growth variable (Table 5). Research findings suggest that an increase in population will increase the carbon dioxide intensity of GDP by 0.55%. In contrast, an increase

Table 4 Driscoll and Kraay's standard error regression results with the dependent variable methane emissions

Dependent variable	Explanatory variables	Coefficients	Drisc/ Kraay Std. Err	t	P > |t|	95% Conf. Interval	
lnmethane	lnpop	2.902242	0.1840439	15.77	0.000	2.51558	3.288904
	lnpopsq	−0.0781619	0.0055031	−14.2	0.000	−0.08972	−0.066600
	lngdp	0.8092699	0.0505039	16.0	0.000	0.703165	0.915375
	lninvest	−0.3621627	0.0708505	−5.1	0.000	−0.51101	−0.213311
	Constant	−43.15446	1.439705	−29.9	0.000	−46.17917	−40.12975

Table 5 Driscoll and Kraay's standard error regression results with the dependent variable carbon dioxide intensity of GDP

Dependent variable	Explanatory variables	Coefficients	Drisc/ Kraay Std. Err	t	P > |t|	95% conf. interval	
lndioxideint	lngdp	0.6308388	0.1535548	4.11	0.001	0.3082322	0 0.953446
	lngdpsq	−0.0229021	0.0031925	−7.17	0.000	−0.02961	−0.01619
	lnpop	0.5536316	0.0315586	17.54	0.000	0.487329	0.619934
	lnittc	−0.0570079	0.0317511	−1.80	0.089	−0.123715	0.0096987
	Constant	−23.02986	2.211423	−10.4	0.000	−27.67588	−18.3838

in internal travel and tourism consumption will decrease this air pollution proxy by 0.05%.

These results indicate the need, or even better, the necessity, to target eco-friendly practices in the growth process. The evidence of the U-shaped curve of the EKC hypothesis implies that air quality will not be degraded after a certain (turning) point. After this point, growth variables generate environmental improvements, and economic benefits related to clean air will continue to advance the quality of life. Moreover, high-leverage and impactful sectors, such as tourism, enter the equation to preserve good air quality since their dynamic and multifaceted nature are considered core determinants in experiencing sustainable growth patterns with a long-term perspective.

4.4 Results of Granger Non-Causality Tests

This study investigated causality relationships by implementing pair-wise tests across panel variables of reference. [39] adopted procedure revealed various causalities with interpretable results (Table 6). The authors processed the Schwarz criterion (BIC) to determine the optimum number of lags.

In Model 1, the first unidirectional relationship runs from population rates to methane emissions, indicating that an increase in population generates an increase in methane emissions rates. Moreover, a unidirectional hypothesis is evidenced between GDP and methane emissions. An additional identified causality stems from methane emissions to capital investment spending. Such a result indicates that capital investment spending does not generate methane emissions.

In Model 2, a unidirectional causality runs from GDP to the carbon dioxide intensity of GDP. Moreover, a feedback hypothesis has been confirmed between population rates and carbon dioxide intensity of GDP, highlighting the importance of sustainable practices in living, consuming, and acting. Additionally, internal travel and tourism consumption justify a second feedback hypothesis with carbon dioxide

Table 6 Granger non-causality test results

Null hypothesis	Z-bar Tilde (Wald statistic)	P values	*The Decision for* H_0
lnpop does not Granger-cause lnmethane	6.9206	0.000	*Reject*
lnmethane does not Granger-cause lnpop	1.0574	0.2903	*Accept*
lngdp does not Granger-cause lnmethane	4.9108	0.0000	*Reject*
lnmethane does not Granger-cause lngdp	-0.1236	0.9017	*Accept*
lninvest does not Granger-cause lnmethane	1.5754	0.1152	*Accept*
lnmethane does not Granger-cause lninvest	2.3161	0.0206	*Reject*
lngdp does not Granger-cause lndioxideint	3.0428	0.0023	*Reject*
lndioxideint does not Granger-cause lngdp	0.4511	0.6519	*Accept*
lnpop does not Granger-cause lndioxideint	3.0794	0.0021	*Reject*
lndioxideint does not Granger-cause lnpop	1.6906	0.0909	*Reject*
lnittc does not Granger-cause lndioxideint	3.9179	0.0001	*Reject*
lndioxideint does not Granger-cause lnittc	4.6013	0.0000	*Reject*

P values < 0.10 indicate rejection of H_0 *at a 10% significance level*
P values < 0.05 indicate rejection of H_0 *at a 5% significance level*
P values < 0.01 indicate rejection of H_0 *at a 1% significance level*

intensity of GDP, underscoring the need for sustainable consumption rates in the tourism sector.

These Granger non-causality test results communicate the importance of including sustainability characteristics in the growth process across the Eurozone and match them with economic viability patterns at the interface of the environment (air quality) and relevant socio-economic systems (population-economy). Mainly, these causalities directly connect profitable and high-impact economic sectors, such as tourism, as core determinants in formulating sustainable growth patterns.

5 Discussion

Environmental studies investigating impacts and causalities between the economic system (e.g., growth) and pollutants (e.g., air pollutants) remain challenging in the literature. These studies provide empirical evidence regarding sustainable economic growth and whether identified growth patterns can generate environmental improvements in the long run. The present study evidences the EKC hypothesis regarding methane emissions from the power sectors and the carbon dioxide intensity of GDP in the Eurozone countries. The tested growth variables were the country's GDP and pollution rates. Research findings revealed meaningful results that transformed them into practical and applicable implications. Then, in turn, these test results can provide inputs to inform robust decision-making processes for establishing sustainability in the Eurozone's economic system. Notably, [40] stress the importance of how mitigation policies at a global level interrelate with urban political–ecological transformations. Consequently, all relevant interested parties and stakeholders can form strategies and policies to establish resilience in the economic system, especially in light of climate change and increased environmental degradation rates. [41] claim that markets should have environmental policy orientations (e.g., environmental policy instruments) providing, at the same time, strategies for governance. Enhancing more sustainable and ecological development patterns is highly important,thus, financial and economic policies should target this direction accordingly [42].

The evidence of a U-shaped curve in the case of methane emissions of the power industry sector calls for managing the air quality advantageously, avoiding market failure phenomena from uncontrolled economic activities that create negative externalities widely affecting society and well-being status. This is evident since the theoretical background of the EKC communicates that growth patterns can create environmental improvement after a certain point in the economic process. Consequently, for large economies in the Eurozone countries, this issue requires serious attention to integrating sustainability patterns in the struggle for economic expansion and viability concerns. This is especially the case for methane emissions. Methane emissions are the second most devastating air pollutant after carbon dioxide releases. Interestingly, one-third of methane emissions come from the energy sector, whereas the European Union [43] has structured a methane strategy to reduce these emissions by incorporating efficiency and technology issues. Specifically, the European

Union's Methane Regulation targets the drastic reduction of devastating releases due to fossil fuel use in Europe and abroad. According to this new regulation, Europe's fossil gas, oil, and coal industries have to measure, monitor, report, and verify their methane emissions following the highest monitoring standards and take immediate action to limit them [44]. This study's research results are aligned with these well-defined and well-detailed approaches to reducing harmful emissions and "de-growth" the economic output from environmental quality issues. The Green Deal Agreement also highlights this perspective, which calls for a "green" transition toward climate-neutral economies by 2050. Notably, the Green Deal Agreement supports resource-efficient and competitive economies by meeting targets of no net emissions of greenhouse gases by 2050 and experiencing economic growth decoupled from resource use (European [45]). In the energy sector, systemic changes are required, such as drastically reducing fossil fuel use and switching to renewable energy sources [46]. Interestingly, one option could be a recycling-based perspective to increase sustainability in urban–rural areas and cities [47].

One particular contribution of this study is including population rates to test the EKC hypothesis. This approach remains less visible in relevant literature despite the population's predictive power on economic activities. Consumption rates can potentially be affected by population and relevant income, spending habits, lifestyle, mindset, and living standards, especially in developed economies across the Eurozone economic space, given households' mean income and economic power. According to the Statista Research Department [48], in 2022, the mean net earnings in the European Union were 26,136 € for a single person who has no children. The average net earnings for a couple with children who both worked were estimated at 55,573€. Whatever the population consumes depends on various factors, so energy must be produced and distributed. Thus, if fossil fuels generate this energy, harmful emissions are released, devastating the air quality. Interestingly, energy consumption can be linked with energy justice issues, suggesting an orientation to equitable sustainability [49], stimulating relevant opportunities for future research.

The authors of this study also discuss the role of high-leverage economic sectors in achieving advanced air quality and becoming more sustainable. Understandingly, they used capital investment spending and internal travel and tourism consumption as core determinants of tourism demand to underscore the sector's potential and predictive power. Judging from the pooled OLS Driscoll and Kraay's Standard Error Regression, both tourism proxies decrease methane emissions and carbon dioxide intensity of GDP, providing evidence of a sustainable tourism sector in the growth process. Consequently, tourism's contribution to air quality improvements should be monitored and controlled in the long run, especially for tourism-dependent economies across the Eurozone countries. Recent empirical research has shown that core tourism market segments support the EKC hypothesis. Indicatively, business and leisure tourism spending evidence the EKC hypothesis across the Eurozone countries [50], whereas the total contribution of tourism to a country's GDP also confirms the EKC hypothesis for the Eurozone member states [51]. In these studies, authors investigated these impactful tourism proxies against air quality indicators (e.g., greenhouse gas emissions). Arguably, the tourism sector reflects a representative example of integrating sustainable practices and considers the concept of "spending" not only a monetary variable to calculate the economic output of tourism to a nation's economy but also a vehicle to experience responsible and eco-friendly consumptive and buying behavior. This approach is more evident if we consider the multiplicative effects of tourism on other economic sectors, advancing its multifaceted character and relevant interactions with sustainability patterns.

As is the case with every empirical research, results vary based on the research purpose and econometric procedures used, methodology and data availability issues, the set of panel growth variables, pollutant indicators, and time spans to process the analysis. This is the case for many research endeavors that seek to define causality and impacts and transform them into practical implications. Notwithstanding, what is crucial in research endeavors is to conclude research findings, gather all existing knowledge, and provide opportunities for future research. From this perspective, research in the context of the EKC hypothesis should test additional growth and environmental degradation variables by targeting discrete economic sectors and market segments that primarily affect a nation's economic output or GDP. Furthermore, despite the predictive power of a nation's GDP, research can focus on economic growth determinants that carry on sustainable characteristics beyond its economic dimensions. Future research can employ the Index of Sustainable Economic Welfare (ISEW) that has been tested, for instance, in robust research efforts for the energy-growth nexus discussion [52, 53]. Last, the value of a more diversified ecosystem extends nature's potential to the human world, benefiting each individual in the system, securing a safe and sustainable environment [54].

6 Conclusions

The present study examined whether economic growth in the Eurozone economic space degrades environmental quality in the EKC framework. To this purpose, the authors determined economic growth in terms of a country's GDP and population rates. Furthermore, two less visible environmental degradation proxies were used in the relevant literature: methane emissions from the power sector (energy) and the carbon intensity of GDP. To strengthen the adopted approach, the authors included high-leverage tourism proxies in this study's model specifications: capital investment spending and internal travel and tourism consumption.

A balanced panel data analysis was processed for the Eurozone member states from 1996 to 2019. All panel tests, regression analysis, and causality evidence provided meaningful and interpretable results. In both econometric models, the growth variables confirmed the EKC hypothesis, implying the inverted U-shaped curve. Moreover, the Granger non-causality pairwise tests revealed two unidirectional relationships running from the growth variables to this study's air quality pollutants. Interestingly, the pooled OLS Driscoll and Kraay's Standard Error Regression results revealed that the two tourism-related proxies decrease environmental degradation levels. This result indicates that tourism spending can be considered not only a monetary variable but also a vehicle to depict pro-environmental behavior.

This study investigated the nonlinear nature of the economy and the environment regarding quality levels and sustainability concerns with a long-run perspective. It can motivate future research efforts and test the economy's heterogeneity against an additional set of environmental degradation proxies in different groups of countries.

In conclusion, continuous research is required to address issues and concerns at the interface of the economy and socio-ecological systems. Such an attempt requires thoroughly examining the multifaceted and multidisciplinary character of the economy and relevant causalities with environmental improvement efforts, especially in the face of climate change and crisis conditions.

Acknowledgements This paper cordially appreciates the editor's and anonymous reviewers' respected comments.

Funding The authors declare that no funds, grants, or other support were received during the preparation of this manuscript.

Data Availability The datasets generated during and/or analyzed during the current study are available from the corresponding author upon reasonable request.

Conflict of Interest The authors declare no competing interests.

References

1. World Bank: The Global Health Cost of PM2.5 Air Pollution: A Case for Action Beyond 2021. International Development in Focus;. © Washington, DC: World Bank. http://hdl.handle.net/10986/36501 License: CC BY 3.0 IGO." (2022)
2. Guo, X., Shahbaz, M.: The existence of environmental Kuznets curve: Critical look and future implications for environmental management. J. Environ. Manage. **351**, 119648 (2024). https://doi.org/10.1016/j.jenvman.2023.119648
3. Shahbaz, M., Nejati, M., Du Min, A., Jia, Z.: A recursive dynamic CGE approach to investigate environmental Kuznets curve. J. Environ. Manage. **370**, 122359 (2024). https://doi.org/10.1016/j.jenvman.2024.122359
4. Shahbaz, M., Sinha, A. (2019). Environmental Kuznets curve for CO2 emissions: a literature survey. J. Econ. Stud. **46(1),** 106–168 (2024). https://doi.org/10.1108/JES-09-2017-0249
5. Danish Baloch, M.A., Mahmood, N., Zhang, J.W.: Effect of natural resources, renewable energy and economic development on CO2 emissions in BRICS countries. Sci. Total. Environ. **678**, 632–638 (2019). https://doi.org/10.1016/j.scitotenv.2019.05.028
6. Jahanger, A., Yu, Y., Hossain, M.R., Murshed, M., Balsalobre-Lorente, D., Khan, U.: Going away or going green in NAFTA nations? Linking natural resources, energy utilization, and environmental sustainability through the lens of the EKC hypothesis. Resour. Policy **79**, 103091 (2022). https://doi.org/10.1016/j.resourpol.2022.103091
7. Sahle, M., Lahoti, S.A., Lee, S.Y., et al.: Revisiting the sustainability science research agenda. Sustain. Sci. **20**, 1–19 (2025). https://doi.org/10.1007/s11625-024-01586-3

8. Takeuchi, K.: Rebuilding the relationship between people and nature: the Satoyama Initiative. Ecol. Res. **25**, 891–897 (2010). https://doi.org/10.1007/s11284-010-0745-8
9. Halkos, G.E.: Environmental Kuznets Curve for sulfur: evidence using GMM estimation and random coefficient panel data models. Environ. Dev. Econ. **8**(4), 581–601 (2023). https://doi.org/10.1017/S1355770X0300317
10. Halkos, G.: Exploring the economy–environment relationship in the case of sulphur emissions. J. Environ. Planning Manage. **56**(2), 159–177 (2013). https://doi.org/10.1080/09640568.2012.657756
11. Dinda, S.: Environmental kuznets curve hypothesis: a survey. Ecol. Econ. **49**, 431–455 (2004). https://doi.org/10.1016/j.ecolecon.2004.02.011
12. Statista. Global electric power industry emissions - statistics & facts. Published by Ian Tiseo. (2023). https://www.statista.com/topics/11057/power-sector-emissions-worldwide/#topicOverview
13. Stern, D.I.: The rise and fall of the environmental Kuznets curve. World Dev. **32**(8), 1419–1439 (2004). https://doi.org/10.1016/j.worlddev.2004.03.004
14. Tian, X.-L., Belaid, F., Ahmad, N.: Exploring the nexus between tourism development and environmental quality: Role of Renewable energy consumption and Income. Struct. Chang. Econ. Dyn. **56**, 53–63 (2021). https://doi.org/10.1016/j.strueco.2020.10.003
15. Grossman, G.M., Krueger, A.B.: Environmental impacts of a north american free trade agreement. NBER Working Paper, p. 3914 (1991)
16. Tsurumi, T., Managi, S.: Decomposition of the environmental Kuznets curve: scale, technique, and composition effects. Environ. Econ. Pollut. Stud. **11**, 19–36 (2010). https://doi.org/10.1007/s10018-009-0159-4
17. Horobet, A., Belascu, L., Radulescu, M., Balsalobre-Lorente, D., Botoroga, C.-A., Negreanu, C.-C.: Exploring the nexus between greenhouse emissions, environmental degradation and green energy in europe: a critique of the environmental kuznets curve. Energies **17**, 5109 (2024). https://doi.org/10.3390/en17205109
18. Hussen, A.M.: Principles of environmental economics and sustainability: an integrated and ecological approach. In: Air Pollution and Per Capita Income, vol. 84. Working Paper Series (2005)
19. Shehzad, K., Zeraibi, A., Zaman, U.: Testing the N-shaped environmental Kuznets Curve in Algeria: An imperious role of natural resources and economic globalization. Resour. Policy **77**, 102700 (2022). https://doi.org/10.1016/j.resourpol.2022.102700
20. Numan, U., Ma, B., Saeed Meo, M., Dino Bedru, H.: Revisiting the N-shaped environmental kuznets curve for economic complexity and ecological footprint. J. Clean. Prod. **365**, 132642 (2022). https://doi.org/10.1016/j.jclepro.2022.132642
21. De Bruyn, S.M., van den Bergh, J.C.J.M., Opschoor, J.B.: Economic growth and emissions: reconsidering the empirical basis of environmental Kuznets curves. Ecol. Econ. **25**, 161–175 (1998). https://doi.org/10.1016/S0921-8009(97)00178-X
22. Alkhars, M.A., Alwahaishi, S., Fallatah, M.R.: Kayal, A: A literature review of the environmental kuznets curve in GCC for 2010–2020. Environ. Sustain. Indic. **14**, 100181 (2016). https://doi.org/10.1016/j.indic.2022.100181
23. Dai, Y., Zhang, H., Cheng, J., Jiang, X., Ji, X., Zhu, D.: Whether ecological measures have influenced the environmental Kuznets curve (EKC)? An analysis using land footprint in the Weihe River Basin. China. Ecological Indicators **139**, 108891 (2022). https://doi.org/10.1016/j.ecolind.2022.108891
24. Barbier, E.B.: Introduction to the environmental Kuznets curve special issue. Environ. Dev. Econ. **2**(4), 369–381 (1997). https://doi.org/10.1017/S1355770X97000193
25. Magnani, E.: The environmental Kuznets Curve: development path or policy result? Environ Model Softw. **16**, 157–165 (2001). https://doi.org/10.1016/S1364-8152(00)00079-7
26. Halkos, G.E., Paizanos, E.A.: The effects of fiscal policy on CO2 emissions: evidence from the USA. Energy Policy **88**, 317–328 (2016). https://doi.org/10.1016/j.enpol.2015.10.035
27. Ma, X., Ahmad, N., Oei, P.-Y.: Environmental Kuznets curve in France and Germany: Role of renewable and nonrenewable energy. Renewable Energy **172**, 88–99 (2021). https://doi.org/10.1016/j.renene.2021.03.014

28. Menz, T., Welsch, H.: Population aging and carbon emissions in OECD countries: accounting for life-cycle and cohort effects. Energy Economics **34**, 842–849 (2012). https://doi.org/10.1016/j.eneco.2011.07.016
29. Jorgenson, K.A., Clark, B.: Assessing the temporal stability of the population/ environment relationship in comparative perspective: a cross-national panel study of carbon dioxide emissions, 1960–2005. Popul. Environ. **32**, 27–41 (2010). https://doi.org/10.1007/s11111-010-0117-x
30. Pham, N.N, Huynh, T.L.D., Ali Nasir, M.: Environmental consequences of population, affluence and technological progress for European countries: A Malthusian view. J. Environ. Manag. **260**, 110143 (2020). https://doi.org/10.1016/j.jenvman.2020.110143
31. Wilson, C., Tisdell, C.: Coastal development, coral reefs and marine life in Asia: Tourism's double-edged sword. In: Managi, S (Ed.), The Routledge HandBook of Environmental Economics in Asia. In: Routledge International Handbooks, Routledge, United Kingdom, pp. 379–400 (2015)
32. Conley, T.G.: GMM estimation with cross sectional dependence. J. Econ. **92**(1), 1–45 (1999). https://doi.org/10.1016/S0304-4076(98)00084-0
33. Conley, T.G., Topa, G.: Socio-Economic distance and spatial patterns in unemployment. J. Appl. Economet. **17**, 303–327 (2002)
34. De Hoyos, R., Sarafidis, V.: Testing for cross-sectional dependence in panel-data models. Stand. Genomic Sci. **6**(4), 482–496 (2006)
35. Pesaran, M. H.: General diagnostic tests for cross section dependence in panels. Cambridge Working Papers in Economics. No: 0435, Faculty of Economics, University of Cambridge (2004)
36. Pesaran, M.H.: A simple panel unit root test in the presence of cross-section dependence. J. Appl. Economet. **22**(2), 265–312 (2007). https://doi.org/10.1002/jae.951
37. Driscoll, J.C., Kraay, A.C.: Consistent covariance matrix estimation with spatially dependent panel data. Rev. Econ. Stat. **80**(4), 549–560 (1998). https://doi.org/10.1162/003465398557825
38. Hoechle, D.: Robust standard errors for panel regressions with cross-sectional dependence. Stand. Genomic Sci. **7**(3), 281–312 (2007)
39. Dumitrescu, E.I., Hurlin, C.: Testing for Granger noncausality in heterogeneous panels. Econ. Model. **29**(4), 1450–1460 (2012). https://doi.org/10.1016/j.econmod.2012.02.014
40. Ernstson, H., Swyngedouw, E.: Wasting CO_2 and the clean development mechanism: The remarkable success of a climate failure. Environ. Plan. E: Nat. Space **7**(2), 654–680 (2024). https://doi.org/10.1177/25148486231196677
41. Robertson, M., Lave, R., Doyle, M.: Making a market in environmental credits II: Watershed moments. Environ. Plan. E: Nat. Space **6**(4), 2452–2471 (2023). https://doi.org/10.1177/25148486231151807
42. Naz, F., Karim, S., Zahra, K.: Carbon footprints, dynamic capabilities, and financial inclusion in G7 and E7 nations. Sustain. Futur. **8**, 100337 (2014). https://doi.org/10.1016/j.sftr.2024.100337
43. European Union.: Methane Emissions. (2024a). https://energy.ec.europa.eu/topics/carbon-management-and-fossil-fuels/methane-emissions_en (Accessed November 29, 2024)
44. European Union.: New EU Methane Regulation to reduce harmful emissions from fossil fuels in Europe and abroad. (2024b). https://energy.ec.europa.eu/news/new-eu-methane-regulation-reduce-harmful-emissions-fossil-fuels-europe-and-abroad-2024-05-27_en (Accessed November 29, 2024).
45. European Commission.: The European Green Deal. (2021). https://commission.europa.eu/strategy-and-policy/priorities-2019-2024/european-green-deal_en (Accessed November 29, 2024).
46. European Council. European Commission.: Fit for 55: shifting from fossil gas to renewable and low-carbon gases. (2023). https://www.consilium.europa.eu/en/infographics/fit-for-55-hydrogen-and-decarbonised-gas-market-package-explained/ (Accessed December 2, 2024)
47. Takeuchi, K., Hara, Y.:Recycling-oriented society and urban-rural sustainability in Asian megacities. Assoc. Rural. Plan. **25(3),** 201–205 (2006). https://doi.org/10.2750/arp.25.201

48. Statista. Annual average net earnings in Europe by country in 2022. Statista Research Department. (2024). https://www.statista.com/statistics/1449039/average-net-annual-earnings-europe-country/ (Accessed November 29, 2024)
49. Sen, K.K., Hosan, S., Karmaker, S.C., Chapman, A.J., Sha, B.B.: Clarifying the linkage between renewable energy deployment and energy justice: Toward equitable sustainability. Sustain. Futur. **8**, 100236 (2024). https://doi.org/10.1016/j.sftr.2024.100236
50. Halkos, G., Ekonomou, G.: Can business and leisure tourism spending lead to lower environmental degradation levels? research on the eurozone economic space. Sustain. **15**, 6063 (2023). https://doi.org/10.3390/su15076063
51. Ekonomou, G., Halkos, G.: Can tourism growth drive environmental improvements in the Eurozone economic space: A panel data analysis. World Dev. Sustain. **4**, 100139 (2024). https://doi.org/10.1016/j.wds.2024.100139
52. Menegaki, A., Tugcu, C.T.: Energy consumption and Sustainable Economic Welfare in G7 countries; A comparison with the conventional nexus. Renew. Sustain. Energy Rev. **69**, 892–901 (2017). https://doi.org/10.1016/j.rser.2016.11.133
53. Menegaki, A., Tugcu, C.T.: Two versions of the Index of sustainable economic welfare (ISEW) in the energy-growth nexus for selected Asian countries. Sustain. Prod. Consum. **14**, 21–35 (2018). https://doi.org/10.1016/j.spc.2017.12.005
54. Zhang, Y., Xue, Y., Guo, K., et al.: Biodiversity and happiness: exploring the value of nature. Sustain. Sci. (2024). https://doi.org/10.1007/s11625-024-01599-y

Global Bioeconomy: History, Visions, and Limitations

Panagiotis Koronaios, Panos Kalimeris, and Georgios Maroulis

Abstract The present paper aims to shed light in the interdisciplinary field of bioeconomy. Bioeconomy is represented by key stakeholders as a prominent alternative paradigm of socioeconomic organization. However, bioeconomy is still a work in progress. As a result, there are diversified perceptions about the definition, the visions, and the goals which the proposed bioeconomy strategies aim to address. A detailed literature review, through a bibliometric analysis, was conducted in order to depict the evolution of bioeconomy, as a concept. The evolutionary process of the definition led to the creation of the three main visions, namely: biotechnology, bioresource, and bioecology. Furthermore, the paper aims to identify the major constraints of implementing bioeconomy in practice, emerging as obstacles in the acceleration of the transformational process toward forming bio-economies worldwide.

Keywords Bioeconomy · Bioeconomy visions · Bioeconomy strategies · Sustainable development

JEL Classification Q57 · Q56 · Q43 · Q2

1 Introduction

The bioeconomy has emerged as a significant alternative economic framework for both the production (supply) and consumption (demand) sides, aligning with the principles of the circular economy and potentially promoting sustainable development,

P. Koronaios (✉) · P. Kalimeris · G. Maroulis
Institute of Urban Environment and Human Resources, Department of Economic and Regional Development, Panteion University of Social and Political Sciences, Athens, Greece
e-mail: koronaiospanagiotis@gmail.com

P. Kalimeris
e-mail: p.kalimeris@panteion.gr

G. Maroulis
e-mail: g.maroulis@panteion.gr

G. Halkos et al. (eds.), *Environmental Risk Analysis*, Springer Proceedings in Mathematics & Statistics 540, https://doi.org/10.1007/978-3-032-24155-9_10

under certain conditions. At its core, bioeconomy seeks to substitute non-renewable fossil-based resources with renewable alternatives in order to fulfil economy's needs, such as food, feed, materials, and energy [1, 2]. As a rapidly evolving multidisciplinary field, the bioeconomy characterized by an increasing attention within the scientific community, evidenced by a rising number of publications focused on terms such as "bioeconomy" or "bio-based economy" [2, 3]. However, its definition has continually evolved, incorporating new perspectives, principles, and objectives to reflect the interests of various stakeholders. The concept of "bioeconomics" was initially introduced in the 1920s by the Russian biologist F. I. Baranoff, who used it to describe optimal management of fish resources. By the 1950s, its scope broadened to include the sustainable utilization of renewable resources, with a focus on conservation to prevent depletion [2, 4]. In the 1970s, economist Nicholas Georgescu-Roegen redefined the bioeconomy as an economic framework which is functioning under physical restrictions imposed by nature, underscoring the profound interdependence between human societies and the environment. From this viewpoint, permanent economic growth is seen as a utopic scenario [5]. These early interpretations closely align with the "bioecology" vision, which emphasizes to the ecological aspects of economic processes, considering the importance of natural limits and the planetary boundaries. The technological advancements of the late twentieth century led to the emergence of biotechnology which became key component regarding the bioeconomy perception. The beginning of the twenty-first century marked with an increasing interest for bringing bio-based models into the political arena. As the definition evolved through time, new aspects emerged leading to the creation of two new visions which influenced the way that policy makers perceive bioeconomy, the "bio-technology" perspective, which advocates for economic growth driven by biotechnology without focusing on sustainability related issues, and the "bio-resource" perspective, which prioritizes the replacement of non-renewable resources with renewable ones to harmonize economic growth with sustainability objectives [6, 7]. Although bio-resource and bio-technology visions seem to have many differences, they share the use of technology and innovation, as a core, in the economic process. However, each vision is characterized by diversified goals regarding environment. Evidently, the first definitions emphasized in the ecological perspective of the economic process, considering the importance of natural limits and the planetary boundaries. These early attempts are currently enhanced with new approaches and visions, such as bio-technology and bio-resource approaches. Nowadays the existing bioeconomy strategies worldwide appear to incorporate elements from different visions. The purpose of the study is to present the whole spectrum of bioeconomy as a scientific field, as well as to explore bioeconomy potential as an alternative socioeconomic framework, aiming to address societal needs and issues regarding environmental problems.

2 Methodology and Data

Both quantitative and qualitative approaches were employed in the present literature review. The initial step involved narrowing the scope of analysis. The year 2008 was chosen as the starting point due to a significant increase in relevant publications after that period. Additionally, 2008 marks the beginning of major efforts by national and international authorities to develop bioeconomy strategies. The review was conducted primarily by using Google Scholar, with search results filtered to include studies published from 2008 onward. As of November 2024, the term "Bioeconomy" yielded 45,700 references. To refine the results, additional keywords were used, including "Bioeconomy," "Bioeconomy Visions," "Bioeconomy Strategies," "Bioeconomy Perceptions," "Bioeconomy Definitions," and "Bio-based Economy." Specifically, searches for "bioeconomy visions" returned 17,400 references, "bioeconomy strategies" yielded 19,500, "bioeconomy perceptions" resulted in 18,300, "bioeconomy definitions" produced 17,600, and "bio-based economy" generated 17,400 references (see Fig. 1). A qualitative classification process was then applied to exclude studies focusing on highly specialized technologies, micro-sectoral implementations, product-level case studies, or broader scientific fields unrelated to the analysis. Further refinement was based on citation counts, ensuring that less influential material was filtered out. A final screening was conducted to remove duplicate studies. In total, 145 studies were included in the bibliometric analysis. Due to space limitations, the complete dataset can be provided by the authors upon reasonable request. The final selection of studies was then analyzed using the VOSviewer online tool [8] for an in-depth statistical process based on bibliometric attributes. The scope of the bibliometric analysis was to identify patterns in the evolution of the field throughout a period of 17 years (2008–2024). The data were extracted from the scientific papers and sorted under certain criteria in order to investigate the interaction of bioeconomy with different concepts and environmental related topics.

3 Results

The bibliometric analysis of the scientific literature led to the creation of two bibliometric maps. We choose to depict our results in a format of a co-occurrence network, based on bibliographic and text data. The co-occurrence network based on bibliographic data, includes information related with two or more concepts that tend to appear together. The map depicts the concepts and their links produced by the bibliometric data of the scientific papers, e.g., the author's names, the title of the paper, the publisher, the scientific journal, and the keywords of the paper. The co-occurrence network based on text data was used to depict the interconnection between different terms, based on the data extracted from the title and the abstract of the scientific papers. Firstly, the co-occurrence network based on the bibliographic data is presented (Fig. 2). The co-occurrence network was created by using a threshold

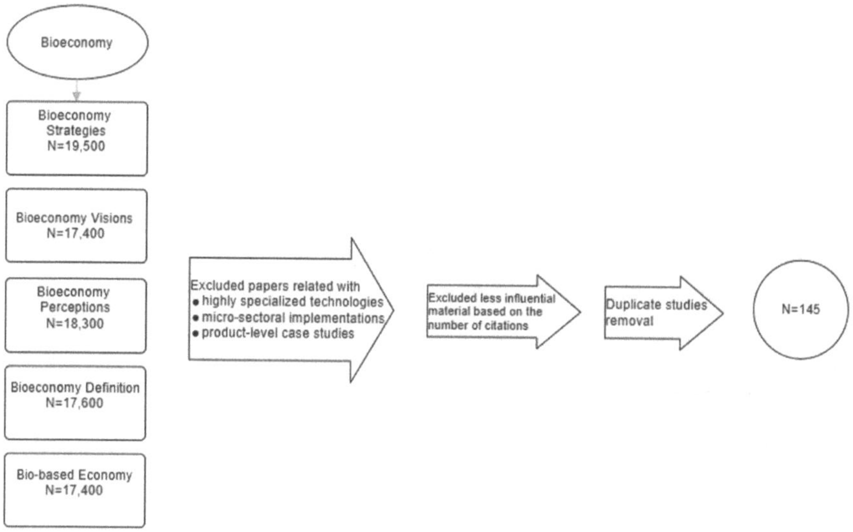

Fig. 1 Selection process of papers for conducting the bibliometric analysis

of minimum 5 times appearance of a term in order to be depicted in the map. The network indicates the relationship between bioeconomy and different concepts. As it is observable, bioeconomy is related with many concepts and scientific fields, such as biology, philosophy, natural resource economics, business, geography, law, sociology, sustainable development, and economics. The evolution of the field is gradually shifting from general fields to more specialized ones, as it is indicated by the yellow nodes. After 2020 there is identified a coupling between bioeconomy and environmental related topics, as well as an emphasis in regionality.

Figure 3 presents a co-occurrence map based on text data extracted from the title and the abstract of the papers. The largest nodes depict the most related terms based on appearance and link strength. The most commonly used terms related with bioeconomic research appeared to be growth, innovation, perspective, sustainable bioeconomy, circular economy, and bioeconomy strategy. A large number of scientific papers investigating issues concerning visions, perspectives, bioeconomy concepts, and definitions, are indicating an emphasis to a theoretical framework. A shift toward more practical concerns is also observed, as indicated by the color of the nodes. The scientific community's recent efforts emphasize in research related issues with terms such as implementation, bioeconomy development, and bioeconomy strategy.

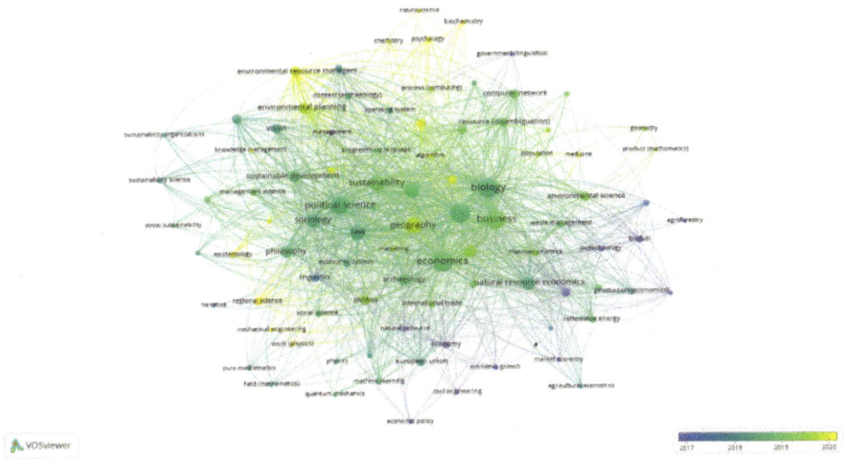

Fig. 2 Co-occurrence network based on bibliographic data

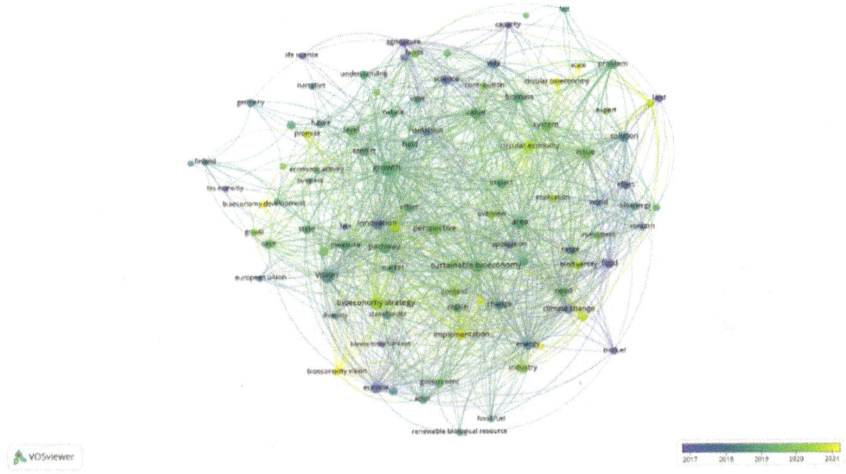

Fig. 3 Co-occurrence network based on text data

4 Bioeconomy Visions

This section describes and analyses the existing bioeconomy visions. We follow the classification of visions by Bugge et al. [6] which is the most comprehensive framework based on the bibliometric analysis of the literature. The bioecology vision, also referred as the agro-ecological approach in the work of Hausknost et al. [9], aligns closely with Georgescu's concept of bioeconomy [4]. This perspective prioritizes environmental sustainability over economic growth [10, 11]. Bioecology seeks to achieve sustainability by ensuring the optimal use of natural resources and emphasizes in the ecosystem conservation and biodiversity protection as core principles [11, 12]. Moreover, minimizing waste and aspects of circular economy could be observed in areas with great bioecological focus [13]. The strategies involved in bioecological perception and could observed globally are mainly focusing on the regional level following a bottom-up approach [11].

The biotechnology vision centers on applying scientific knowledge and integrating innovation into economic systems by utilizing advancements in natural and biological sciences [6, 14]. Rather than transforming the existing linear production model, this approach largely aligns with business-as-usual frameworks [6, 14]. Consequently, basic goals of biotechnology are unemployment decrease by creating job opportunities and a consistent economic growth [10]. Biotechnology is further categorized based on its areas of application. White Biotechnology focuses on industrial products, such as chemicals, Green Biotechnology applies to agriculture, including genetically modified crops and fertilizers, and Blue Biotechnology involves innovations derived from marine biodiversity [15]. Additionally, Red Biotechnology pertains to advancements in medicine, such as the development of new drugs and therapies [16]. In biotechnology perspective the emphasis is in the creation of a global value chains, therefore the importance of the regions is limited [13]. The bioresource vision seeks to move away from the current linear economic system by incorporating renewable and biological resources in the economic process in order to shift in a bio-based economy [6, 14]. This approach is emphasizing both in sustainability related aspects as well as to economic growth [10]. As its name suggests, the bioresource vision is centered around natural resources and especially biomass. Shifting from fossil-based to renewable resources plays a crucial role in reducing carbon emissions and promoting economic decarbonization [13]. This vision is closely linked to optimizing land use, expanding biofuel utilization, and implementing waste management policies [6]. While innovation and technology are essential components of the bioresource approach, its key distinction from biotechnology lies in its strong emphasis on biomass as the foundation of the economic model [4, 9]. Lastly, similar to the bioecology perspective, the bioresource vision adopts a regional approach of implementation.

5 Bioeconomy Limitations

5.1 Sustainability Issues

Bioeconomy does not inherently ensure sustainability despite using renewable resources [17, 18]. Sustainable bioeconomy embodies sustainability in the context of economic processes and consumption patterns, as well as sustainable use of natural resources [17]. The relationship between bioeconomy and sustainability varies, with some studies highlighting conflicts, particularly concerning land and water resource pressures, food security, and carbon emissions from land-use changes [19, 20]. Additionally, technological uncertainties pose risks to sustainability [17, 21]. Another significant issue arising from this transition is the "food vs. fuel" dilemma [17, 22]. Current technology cannot meet the world's energy demands solely with biofuels, and rising population will increase food demand, exacerbating food security concerns [20, 23]. Therefore, balancing fuel production with food security remains a critical challenge. Furthermore, bioeconomy implementation leads to concerns related with the land use change. Biofuel production can drive direct and indirect land use changes that may lead to higher CO_2 emissions than those saved by biofuels [24, 25]. Additionally, developed countries often outsource energy crop production to developing countries, shifting the problem and burdening local populations [26]. Land conversion for biofuels disrupts local ecosystems and biodiversity, and invasive species can be introduced to enhance biofuel production [27]. Lastly, the introduction of new technologies may result in unforeseen negative externalities, threatening the sustainability of bioeconomy efforts [17, 21].

5.2 *Regulatory Framework*

Policies and regulations are essential to the bioeconomic development. The shift from traditional economic models to bio-based economies requires a solid regulatory framework. As Borgström and Mauerhofer [28] note, intervention is needed to address challenges such as innovation, resource efficiency, sustainability, and potential conflicts related to natural resource use and technology risks. Bioeconomy governance can be divided into two categories, namely: enabling and constraining governance [29, 30]. Enabling governance supports bioeconomy growth by funding research, creation of infrastructures, and encouraging collaboration among stakeholders. It also promotes the development of bio-based industries through incentives like tax breaks and quotas. Enabling governance boosts industry competitiveness by specializing the workforce, creating clusters, protecting intellectual property rights, and connecting scientific output to market needs. Public awareness campaigns on bioeconomy are also key [30]. On the other hand, constraining governance seeks to minimize negative impacts on society, the economy, and the environment. This includes regulations, taxation, standards, and certification [30]. It ensures balance between stakeholder goals and promotes sustainable resource use [28, 31]. It also addresses ethical concerns, particularly in biotechnology [32]. National authorities should combine enabling and constraining governance strategies to tailor policies for their specific needs in order to achieve a desirable outcome.

5.3 Monitoring

Monitoring the bioeconomy is essential for advancing bio-based economies. A robust monitoring framework helps assess the current state of bioeconomy in specific regions and evaluate progress toward related goals, while also identifying sustainability achievements [33]. However, global bioeconomic monitoring is often inadequate, hindering the implementation of bioeconomy strategies. The absence of a universally accepted definition of bioeconomy and its associated activities leads to inconsistent monitoring efforts [34–37]. Definitions, priorities, and strategic goals vary widely among nations, and the sectors considered part of bioeconomy differ significantly. For example, the EU includes agriculture in its bioeconomy framework, while the US does not, making international comparisons challenging [38]. Another issue is the difficulty in quantifying bioeconomy's contribution to the overall economy due to a lack of data on bio-based versus non-bio-based production [33]. Partially bio-based sectors further complicate this, as isolating the bio-based share is difficult [38]. Additionally, monitoring efforts often prioritizing economic indicators, such as GDP, leading to neglection of environmental and social dimensions due to data limitations and traditional analytical methods hegemony [12, 39]. Data gaps and poor-quality data further hinder analysis [40]. To address these challenges, further research is needed to develop a consistent monitoring framework and indicators that capture the full spectrum of bioeconomy, including its environmental and social impacts.

5.4 Demand Side

Today's economic systems are typically linear, with fossil resources driving most economic processes. Shifting to bio-based economies demands intervention across all aspects of social and economic life to modify these established patterns [30]. Even if the supply side transitions to bioeconomy, there must also be a simultaneous increase in demand for bio-based products [41]. Society plays a crucial role as a key stakeholder in this transition [42]. Changing consumer's behavior is essential, but consumption patterns rooted in fossil-based economies and the sociocultural factors influencing them are challenging to alter [7, 30, 43, 44]. As Hempel et al. [45] noted, the general public remains largely unaware of bioeconomy and its related concepts. Consequently, raising public awareness about the benefits of bioeconomy and bio-based products, along with addressing misconceptions about new technologies, is essential [30]. Increased awareness can also help mitigate opposition to new bioeconomic facilities, such as biorefineries, through the "Not in my backyard" phenomenon [42]. To successfully implement a bioeconomy, promoting sustainable consumption practices like waste management, recycling, and circularity, as well as using labelling and clear communication of scientific findings, will help drive demand for bio-based products [30, 41, 42].

6 Discussion

The analysis indicated the complexity of bioeconomy as a scientific field. Despite the increasing number of publications regarding the topic there are still things to be done in order to formulate a robust bioeconomic framework. The first obstacle is the creation of a globally accepted definition. As it is indicated in Fig. 3, a large number of scientific papers investigating issues concerning visions, perspectives, bioeconomy concepts, and definitions, indicating an effort to create a theoretical framework. The lack of a theoretical background creates problems concerning the implementation strategies and the monitoring efforts. Moreover, bioeconomy is perceived as a holistic paradigm of socioeconomic organization. As a result, it is an interdisciplinary topic that requires analysis from a large number of scientific fields. As it is indicated in Fig. 2, the analysis from different angles will increase the knowledge flows, helping in the successful implementation of bioeconomy strategies. Furthermore, as Figs. 2 and 3 indicating, recent trends of research appear to combining circular economy with bioeconomy leading to the creation of mixed approaches, e.g., the turn toward circularity in the European Union (EU) could be tracked in official reports and publications. Another remark is the further integration of environmental aspects in the analysis, such as environmental resource management and biodiversity concern issues, leading to the conjecture that environmental parameters becoming more influential in the political agenda. Finally, Figs. 2 and 3 are indicating a trend toward regional implementation of bioeconomy strategies which confirms the importance of local characteristics and resources in the bioeconomic development.

The results of the study could be applied to the formulation of robust and efficient management plans in bioeconomy related sectors. Specifically, the main points of the analysis are summarized in the need for collaboration and engagement of individuals with diversified scientific backgrounds due to the interdisciplinary face of bioeconomy. Furthermore, the identification of the regulatory framework and its characteristics is essential in order to implement bioeconomy related plans. Moreover, stakeholder identification and engagement as well as the establishment of proper metrics or Key Productivity Indicators (KPIs), not only related with the economic dimension but also with the social and environmental ones, are important in the contextualization and structuralization of the bioeconomy management plans.

7 Conclusions

The term "bioeconomy" appeared in the public speech in the beginning of twentieth century. Nowadays the term is used to describe an alternative economic reality, however, there is no globally accepted definition for the bioeconomy concept leading to different visions, priorities or goals. Nevertheless, the initial definitions changed during the years leading to three distinct bioeconomy visions: The biotechnology vision which aims to facilitate knowledge which derives from biological and natural sciences in order to promote growth; the bioresource vision which aims to change system's linearity by using biomass instead of fossil resources and targeting both in sustainability, as well as in growth; and the bioecology vision which takes into account natural limits and ecosystems preservation. The bioeconomy visions are affecting the formulation of bioeconomy strategies which sometimes combine traits of more than one approaches leading to hybrid paradigms. However, there are still things to be done for this alternative socio-economic paradigm to fulfil a real transformation of the current systems. The major issues remain the relationship of bioeconomy with sustainability, the characteristics of the regulatory framework which affect the implementation of bioeconomy strategies, the issues related with difficulties in monitoring, as well as the need for increasing information flows among citizens and stakeholders, about bio-based products and other aspects concerning both the supply and the demand side of the economy. Furthermore, the recent research efforts seem to emphasizing in the further inclusion of environmental parameters in the analysis of bioeconomy. Moreover, bioeconomy tends to adopt characteristics of circular economy leading to the appearance of mixed approaches such as circular bioeconomy. Finally, the implementation of bioeconomy strategies after 2021 appear to focus on regional level, especially in the EU.

Acknowledgements The present study has been benefited by the European Union through the funding of the Horizon Europe project "BIO2REG" (https://bio2reg.eu/): "*Enabling transition toward circular and systemic Bioeconomy model regions by a Regions-to-Regions approach*" (Grant agreement ID: 101135420). Views and opinions expressed are however those of the authors only and do not necessarily reflect those of the European Union or European Research Executive Agency (REA). Neither the European Union nor REA can be held responsible for them.

References

1. Proestou, M., Schulz, N., Feindt, P.H.: A global analysis of bioeconomy visions in governmental bioeconomy strategies. Ambio **53**(3), 376–388 (2024). https://doi.org/10.1007/s13280-023-01958-6
2. Gould, H., Kelleher, L., O'Neill, E.: Trends and policy in bioeconomy literature: a bibliometric review. EFB Bioeconomy Journal **3**, 100047 (2023). https://doi.org/10.1016/j.bioeco.2023.100047
3. Staffas, L., Gustavsson, M., McCormick, K.: Strategies and policies for the bioeconomy and bio-based economy: an analysis of official national approaches. Sustainability **5**(6), 2751–2769 (2013). https://doi.org/10.3390/su5062751
4. Vivien, F.-D., Nieddu, M., Befort, N., Debref, R., Giampietro, M.: The hijacking of the bioeconomy. Ecol. Econ. **159**, 189–197 (2019). https://doi.org/10.1016/j.ecolecon.2019.01.027
5. Georgescu-Roegen, N.: Inequality, limits and growth from a bioeconomic viewpoint. Rev. Soc. Econ. **35**(3), 361–375 (1977)
6. Bugge, M.M., Hansen, T., Klitkou, A.: What Is the bioeconomy? A Review of the Literature. Sustainability **8**(7), 691 (2016). https://doi.org/10.3390/su8070691
7. Birner, R.: Bioeconomy concepts. In: Lewandowski, I. (eds.) Springer International Publishing. Cham (2018). https://doi.org/10.1007/978-3-319-68152-8_3
8. VOSviewer Homepage. https://www.vosviewer.com//, last accessed 2024/11/18

9. Hausknost, D., Schriefl, E., Lauk, C., Kalt, G.: A transition to which bioeconomy? an exploration of diverging techno-political choices. Sustainability **9**(4), 669 (2017). https://doi.org/10.3390/su9040669
10. Johnson, F.X., Canales, N., Fielding, M., Gladkykh, G., Aung, M.T., Bailis, R., Ogeya, M., Olsson, O.: A comparative analysis of bioeconomy visions and pathways based on stakeholder dialogues in Colombia, Rwanda, Sweden, and Thailand. J. Environ. Planning Policy Manage. **24**(6), 680–700 (2022). https://doi.org/10.1080/1523908X.2022.2037412
11. Canales, N., González, J.G.: Policy dialogue on a bioeconomy for sustainable development in Colombia (2020). https://www.sei.org/publications/policy-dialogue-bioeconomy-sustainable-development-colombia/
12. Bracco, S., Calicioglu, O., Gomez San Juan, M., Flammini, A.: Assessing the contribution of bioeconomy to the total economy: a review of national frameworks. Sustainability **10**(6), Article 6 (2018). https://doi.org/10.3390/su10061698
13. Hoff, H., Johnson, F. X., Allen, B., Biber-Freudenberger, L., Förster, J.J.: Sustainable bio-resource pathways towards a fossil-free world: the European bioeconomy in a global development context, In: IEEP Think2030 conference, Brussels (2018)
14. Meyer, R.: Bioeconomy strategies: contexts, visions, guiding implementation principles and resulting debates. Sustainability **9**(6), 1031 (2017). https://doi.org/10.3390/su9061031
15. Barcelos, M.C.S., Lupki, F.B., Campolina, G.A., Nelson, D.L., Molina, G.: The colors of biotechnology: general overview and developments of white, green and blue areas. FEMS Microbiol. Lett. **365**(21), fny239 (2018). https://doi.org/10.1093/femsle/fny239
16. Gartland, K., Bruschi, F., Dundar, M., Gahan, P., Viola Magni, M., Akbarova, Y.: Progress towards the 'Golden Age' of biotechnology. Curr. Opin. Biotechnol. **24**, S6–S13 (2013). https://doi.org/10.1016/j.copbio.2013.05.011
17. Gawel, E., Pannicke, N., Hagemann, N.: A path transition towards a bioeconomy—the crucial role of sustainability. Sustainability **11**(11), Article 11 (2019). https://doi.org/10.3390/su11113005
18. Stegmann, P., Londo, M., Junginger, M.: The circular bioeconomy: Its elements and role in European bioeconomy clusters. Resour., Conserv. & Recycl.: X **6**, 100029 (2020). https://doi.org/10.1016/j.rcrx.2019.100029
19. Heimann, T.: Bioeconomy and SDGs: does the bioeconomy support the achievement of the SDGs? Earth's Future **7**(1), 43–57 (2019). https://doi.org/10.1029/2018EF001014
20. Landeweerd, L., Surette, M., van Driel, C.: From petrochemistry to biotech: a European perspective on the bio-based economy. Interface Focus. **1**(2), 189–195 (2011). https://doi.org/10.1098/rsfs.2010.0014
21. Zilberman, D., Kim, E., Kirschner, S., Kaplan, S., Reeves, J.: Technology and the future bioeconomy. Agric. Econ. **44**(s1), 95–102 (2013). https://doi.org/10.1111/agec.12054
22. Bos, R.T., Dam, J.E.G.V.: Sustainability, polysaccharide science, and bio-economy. Carbohyd. Polym. **93**(1), 3–8 (2013). https://doi.org/10.1016/j.carbpol.2012.01.087
23. Sheppard, A.W., Gillespie, I., Hirsch, M., Begley, C.: Biosecurity and sustainability within the growing global bioeconomy. Curr. Opin. Environ. Sustain. **3**(1), 4–10 (2011). https://doi.org/10.1016/j.cosust.2010.12.011
24. Pfau, S.F., Hagens, J.E., Dankbaar, B., Smits, A.J.M.: Visions of sustainability in bioeconomy research. Sustainability **6**(3), 1222–1249 (2014). https://doi.org/10.3390/su6031222
25. Gitz, V., Ciais, P.: Amplifying effects of land-use change on future atmospheric CO2 levels. Glob. Biogeochem. Cycles **17**(1) (2003). https://doi.org/10.1029/2002GB001963
26. Gibbs, H.K., Johnston, M., Foley, J.A., Holloway, T., Monfreda, C., Ramankutty, N., Zaks, D.: Carbon payback times for crop-based biofuel expansion in the tropics: the effects of changing yield and technology. Environ. Res. Lett. **3**(3) (2008). https://doi.org/10.1088/1748-9326/3/3/034001
27. Barney, J.N., DiTomaso, J.M.: Nonnative species and bioenergy: are we cultivating the next invader? Bioscience **58**(1), 64–70 (2008). https://doi.org/10.1641/B580111
28. Borgström, S., Mauerhofer, V.: Developing law for the bioeconomy. J. Energy & Nat. Resour. Law **34**(4), 373–406 (2016). https://doi.org/10.1080/02646811.2016.1200349

29. Lamprinakis, L.: Circular regulations (CR) for bioeconomy development. (2020). https://doi.org/10.20900/jsr20200021
30. Dietz, T., Börner, J., Förster, J.J., Von Braun, J.: Governance of the bioeconomy: a global comparative study of national bioeconomy strategies. Sustainability **10**(9), 3190 (2018). https://doi.org/10.3390/su10093190
31. Pender, A., Kelleher, L., O'Neill, E.: Regulation of the bioeconomy: Barriers, drivers and potential for innovation in the case of Ireland. Clean. Circ. Bioeconomy **7**, 100070 (2024). https://doi.org/10.1016/j.clcb.2023.100070
32. Gremmen, B. (2022). Bioeconomy and ethics. In: Lanzerath, D., Schurr, U., Pinsdorf, C., & Stake, M. (eds.). Springer International Publishing, Cham (2022). https://doi.org/10.1007/978-3-030-87402-5_15
33. Jander, W., Wydra, S., Wackerbauer, J., Grundmann, P., Piotrowski, S.: Monitoring bioeconomy transitions with economic-environmental and innovation indicators: addressing data gaps in the short term. Sustainability **12**(11), 4683 (2020). https://doi.org/10.3390/su12114683
34. Bracco, S., Tani, A., Çalıcıoğlu, Ö., Gomez San Juan, M., Bogdanski, A.: Indicators to monitor and evaluate the sustainability of bioeconomy: overview and a proposed way forward. Environment and Natural Resources Management. Working Paper (FAO) eng no. 77 (2019)
35. Dolge, K., Balode, L., Laktuka, K., Kirsanovs, V., Barisa, A., Kubule, A.: A comparative analysis of bioeconomy development in european union countries. Environ. Manage. **71**(2), 215–233 (2023). https://doi.org/10.1007/s00267-022-01751-3
36. Ferreira, V., Fabregat-Aibar, L., Pié, L., Terceño, A.: Research trends and hotspots in bioeconomy impact analysis: a study of economic, social and environmental impacts. Environ. Impact Assess. Rev. **96**, 106842 (2022). https://doi.org/10.1016/j.eiar.2022.106842
37. Vogelpohl, T., Töller, A.E.: Perspectives on the bioeconomy as an emerging policy field. J. Environ. Planning Policy Manage. **23**(2), 143–151 (2021). https://doi.org/10.1080/1523908X.2021.1901394
38. Gardossi, L., Philp, J., Fava, F., Winickoff, D., D'Aprile, L., Dell'Anno, B., Marvik, O.J., Lenzi, A.: Bioeconomy national strategies in the G20 and OECD countries: sharing experiences and comparing existing policies. EFB Bioeconomy Journal **3**, 100053 (2023). https://doi.org/10.1016/j.bioeco.2023.100053
39. Wesseler, J., Braun, J. von.: Measuring the bioeconomy: economics and policies. Annu. Rev. Resour. Econ. **9**(1), 275–298 (2017). https://doi.org/10.1146/annurev-resource-100516-053701
40. Mubareka, S., Giuntoli, J., Sanchez Lopez, J., Lasarte Lopez, J., M'barek, R., Ronzon, T., Renner, A., Avraamides, M.: Trends in the EU bioeconomy, Publications Office of the European Union (2023). https://doi.org/10.2760/835046
41. De Besi, M., McCormick, K.: Towards a bioeconomy in Europe: national regional and industrial strategies. Sustainability **7**(8), 10461–10478 (2015). https://doi.org/10.3390/su70810461
42. Kiresiewa, D.Z.: Improving the Public Acceptance of Bio-Based Products and Processes at Regional and Local Level Homepage, https://www.ecologic.eu/18039 (2020)
43. Aguilar, A., Twardowski, T., Wohlgemuth, R.: Bioeconomy for sustainable development. Biotechnol. J. **14**(8), 1800638 (2019). https://doi.org/10.1002/biot.201800638
44. Eversberg, D., Koch, P., Lehmann, R., Saltelli, A., Ramcilovic-Suominen, S., Kovacic, Z.: The more things change, the more they stay the same: promises of bioeconomy and the economy of promises. Sustain. Sci. **18**(2), 557–568 (2023). https://doi.org/10.1007/s11625-023-01321-4
45. Hempel, C., Will, S., Zander, K.: Societal perspectives on a bio-economy in Germany: an explorative study using Q methodology. Int. J. Food Syst. Dyn. **10**, 21–37 (2019). https://doi.org/10.18461/ijfsd.v10i1.02

Exploring the Impact of Corporate Social Responsibility on Employee Mental Health, Resilience, and Job Satisfaction: Evidence from a Survey in the NGO

Kristina Kucheruk and Konstantinos Evangelinos

Abstract Corporate Social Responsibility (CSR) has emerged as a critical factor in shaping organizational success, fostering stakeholder trust, and enhancing corporate reputation. While the beneficial impacts of CSR on organizational performance and public perception are well-documented, the effects of CSR initiatives on Employee Mental Health, job engagement, resilience, and overall satisfaction remain less explored. This study employs a detailed questionnaire to examine how CSR-related practices impact employees' mental resilience, job satisfaction, turnover intentions, and adaptability, with a focus on companies within the NGO's landscape. By collecting comprehensive data on employees' experiences with CSR initiatives—including environmental practices, social responsibility measures, leadership styles, and communication within the organization—this research seeks to identify key correlations and differentiations among variables related to mental health and job satisfaction. The findings aim to provide actionable insights for organizational leaders and HR professionals on integrating CSR into organizational policies to promote sustainable employee well-being and engagement. This study's outcomes are anticipated to support the development of CSR strategies that foster a resilient, engaged, and satisfied workforce.

Keywords Corporate Social Responsibility · Corporate sustainability · Employee Mental Health · Organizational change

JEL Classification I21 · I31 · I38 · J28 · J81Q01 · Q52 · Q56

K. Kucheruk · K. Evangelinos (✉)
Department of Environment, University of the Aegean, Mytilene, Lesvos, Greece
e-mail: kevag@aegean.gr

G. Halkos et al. (eds.), *Environmental Risk Analysis*, Springer Proceedings in Mathematics & Statistics 540, https://doi.org/10.1007/978-3-032-24155-9_11

1 Introduction

Corporate Social Responsibility (CSR) has emerged as a pivotal element in organizational strategies, transcending its traditional focus on external stakeholder engagement to influence internal dynamics, including employee well-being. Broadly defined, CSR encompasses voluntary corporate initiatives aimed at addressing societal, environmental, and economic concerns, aligning business practices with the broader good [1]. Organizations that actively invest in CSR not only enhance their public reputation but also cultivate trust and loyalty among employees, fostering a positive workplace culture [2].

While a substantial body of research examines CSR's external impacts, its influence on employees—particularly on mental health, resilience, and job satisfaction—remains underexplored. Most studies have concentrated on how CSR enhances organizational performance or public perception, with limited emphasis on the psychological and behavioral outcomes for employees [3]. Given the increasing prevalence of workplace stress and the demand for organizational support in fostering employee resilience, this gap warrants urgent attention. While some evidence suggests that CSR initiatives may improve job satisfaction and engagement, their specific effects on employees' mental health and adaptability in challenging circumstances are not well-documented [4].

The focus on the NGO sector in this study is particularly pertinent. NGOs often operate under resource constraints, mission-driven objectives, and strong value-based cultures, placing a unique emphasis on social and environmental contributions. Employees in NGOs may be more attuned to ethical values and collective purpose, potentially amplifying the perceived impact of CSR initiatives on well-being. At the same time, the high demands, emotional labor, and often unstable funding characteristic of NGOs can heighten stress levels, making resilience and job satisfaction critical for workforce sustainability. These sector-specific dynamics provide a compelling context in which to explore the role of CSR as a driver of psychological and organizational health.

This study aims to bridge the existing knowledge gap by investigating the relationships between CSR practices and key employee well-being metrics, including mental resilience, job satisfaction, and turnover intentions. By analyzing data from employees within NGO contexts, this research seeks to illuminate the nuanced ways in which CSR initiatives influence workforce dynamics. The insights gained are expected to provide actionable recommendations for organizational leaders and HR practitioners on leveraging CSR to foster sustainable employee engagement and mental health, ultimately contributing to a more resilient and satisfied workforce.

2 Literature Review

2.1 *Corporate Social Responsibility and Organizational Dynamics*

CSR encompasses a range of voluntary corporate activities designed to address ethical, environmental, and social issues while contributing to business sustainability [5]. CSR has traditionally focused on external stakeholders—such as customers, regulators, and communities—but growing evidence suggests it also profoundly influences internal organizational dynamics [2]. Numerous studies confirm the association between CSR practices and enhanced organizational performance, improved brand reputation, and stakeholder trust [6].

From a strategic perspective, CSR supports stakeholder theory, which argues that businesses must meet the expectations of various stakeholders to ensure long-term success [7]. This alignment between social responsibility and strategic intent fosters reputational capital and legitimizes business operations within society.

2.2 *CSR and Employee Outcomes: Engagement, Satisfaction, and Commitment*

Recent scholarship has shifted attention to the internal benefits of CSR, particularly those affecting employee attitudes and behaviors. Research by Brammer, Millington, and Rayton [8] found that employees who perceive their organizations as socially responsible demonstrate higher levels of organizational commitment. Similarly, Kim et al. [9] established that CSR enhances employees' identification with their organization, increasing both job satisfaction and engagement.

Employees tend to perceive CSR initiatives as expressions of shared values, which enhances their emotional attachment to the organization [10]. This emotional connection can translate into lower turnover intentions and stronger loyalty. Turker [11] supports this, showing that employees' perceptions of CSR practices directly influence their organizational citizenship behavior and job involvement.

2.3 *CSR and Psychological Well-Being: Mental Health, Resilience, and Stress*

Despite the mounting interest in CSR's organizational effects, its psychological implications for employees—especially in relation to mental health—remain underexplored. Bauman and Skitka [12] demonstrated that perceived ethical behavior on the part of an organization contributes to employee well-being, reducing psychological strain and increasing personal alignment with corporate goals. Employees working in socially responsible organizations tend to experience lower stress levels and higher resilience, as they feel their work contributes meaningfully to society [13].

De Roeck and Farooq [14] found that ethical and CSR-driven leadership contributes to psychological safety, a crucial component in fostering mental resilience and adaptive coping. Similarly, Hameed et al. [15] observed that CSR perceptions reduce emotional exhaustion and burnout in service employees.

2.4 The Role of Leadership in Amplifying CSR's Effects

Leadership is pivotal in translating CSR philosophy into internal outcomes. Transformational and ethical leadership styles have been shown to act as conduits between CSR initiatives and employee engagement [16]. When leaders model socially responsible behavior and emphasize moral values, employees are more likely to internalize CSR principles [17]. This leads to a positive psychological climate and promotes collective efficacy, both of which are associated with mental resilience and adaptability.

Moreover, empowering leadership encourages employee participation in CSR initiatives, which boosts engagement and fosters a sense of control—factors strongly associated with psychological well-being [18]. Empowerment also enhances perceptions of fairness and inclusivity, reducing cynicism and fostering commitment.

2.5 CSR, Organizational Culture, and the Importance of Communication

Organizational culture serves as a crucial mechanism through which CSR is enacted and experienced. A learning-oriented and inclusive culture, for example, reinforces CSR values and motivates employees to adopt prosocial behaviors [19]. Effective internal communication about CSR goals and outcomes enhances transparency and trust, enabling employees to feel valued and informed [20].

Communication also mediates the relationship between CSR and job satisfaction. When CSR efforts are clearly communicated and linked to organizational values, employees report higher job clarity, reduced ambiguity, and greater psychological alignment with their roles [21]. These elements directly contribute to greater resilience and lower turnover intentions.

2.6 Sectoral and Demographic Variations in CSR Perceptions

Research indicates that perceptions of CSR and its impact on employee outcomes can differ across sectors and demographic lines. Employees in NGOs and non-profit organizations often place greater emphasis on ethical values and social contribution, making CSR especially relevant in such contexts [22]. Similarly, demographic variables such as age, gender, education, and parental status influence how CSR is perceived and experienced [23].

For instance, employees with higher education levels may exhibit more critical engagement with CSR initiatives, emphasizing authenticity and structural integration, while younger employees might focus on environmental sustainability [24]. This suggests the need for CSR strategies that are tailored to workforce composition to maximize psychological and organizational benefits.

Despite growing interest in CSR's internal implications, a significant research gap persists in understanding its influence on employees' mental health and resilience. Existing studies predominantly focus on engagement and satisfaction, neglecting critical dimensions such as stress reduction, adaptability, and psychological well-being [13]. With increasing workplace complexities and mental health challenges, examining how CSR practices mitigate stress and foster resilience is essential. Addressing this gap will provide a holistic understanding of CSR's impact, enabling organizations to craft strategies that align employee well-being with broader sustainability goals.

3 Methodology

3.1 Research Questions and Hypotheses

To guide the empirical investigation, the study was framed by the following central research question:

RQ1: How do CSR-related organizational practices affect employee mental resilience, job satisfaction, and turnover intentions within the NGO sector?

From this core question, several sub-questions were derived to explore specific dimensions of CSR and their psychological impact on employees:

RQ2: Do environmental and social sustainability practices correlate with higher employee engagement and resilience?

RQ3: Does empowering leadership moderate the relationship between CSR initiatives and employee well-being?

RQ4: Do demographic factors (e.g., gender, educational level, parental status) influence how employees perceive and are affected by CSR practices?

Based on these research questions and supported by existing theoretical frameworks, the following hypotheses were formulated:

H1: Employees who perceive higher levels of CSR engagement within their organization report significantly greater job satisfaction and mental resilience.
H2 Empowering leadership positively moderates the relationship between CSR practices and employee outcomes such as engagement and adaptability.
H3: Social sustainability dimensions of CSR have a stronger positive correlation with job satisfaction and resilience than environmental dimensions.
H4: Demographic variables such as gender and parental status significantly influence perceptions of CSR and associated psychological outcomes.

These hypotheses were tested using Chi-Square tests, ANOVA, and Spearman correlation analyses, enabling a multidimensional understanding of the relationship between CSR practices and employee well-being across various organizational and individual characteristics.

3.2 Data Collection

This study employed a structured questionnaire to collect comprehensive data from employees working within the NGO sector. The questionnaire was designed to capture respondents' perceptions of CSR-related practices, including environmental sustainability policies, social responsibility measures, leadership styles, communication quality, and organizational culture. Questions were tailored to measure both the direct and indirect effects of these practices on employees' mental resilience, job satisfaction, and intentions to leave their organization. Participants were selected through purposive sampling to ensure representation from diverse organizational roles and backgrounds.

3.3 Sample

The study draws on data collected from 126 employees working within the NGO sector in Greece and other countries. Data collection took place between [03/2023] and [11/2023] through an online questionnaire distributed to targeted organizations. A purposive sampling approach ensured diversity in roles, functional areas, and cultural backgrounds.

The sample comprised 54% female and 46% male respondents, with ages ranging from 19 to 78 years (average 42). Regarding nationality, 24% were from Greece, 11% from Turkey, 9% from Finland, 7% from Austria, 7% from Belgium, 7% from Ukraine, 4% from Australia, 4% from Canada, 4% from Egypt, and smaller proportions from other countries. In terms of family status, 77% were married or in a relationship, and 61% had children. Disabilities or special needs were reported by 14% of participants.

Participants' positions included 53% employees, 12% supervisors/responsible staff, 6% directors or executives, and 29% volunteers. The main sectors of activity were community development (15%), disability-related work (14%), employment and labor (13%), children's services (12%), and refugee/asylum-seeker support (12%), with other areas also represented.

This demographic and sectoral diversity enhances the representativeness of the study and allows for a nuanced analysis of how CSR initiatives are perceived and experienced within diverse NGO contexts.

3.4 Variables Studied

The study focused on both independent and dependent variables:

- Independent Variables: CSR dimensions (e.g., environmental sustainability, social responsibility), leadership approaches (e.g., empowering leadership), communication practices, and employee participation.
- Dependent Variables: Employee mental resilience, job satisfaction, turnover intentions, job engagement, and adaptability to workplace challenges.

The data collected allowed for a multidimensional analysis of how variations in CSR practices and organizational dynamics influenced employee well-being outcomes.

3.5 Statistical Techniques

To identify significant associations and test the study's hypotheses, a range of statistical techniques were applied:

- Chi-Square Tests: Used to explore relationships between categorical variables, such as demographic factors and perceptions of CSR practices.
- Analysis of Variance (ANOVA): Applied to compare means among different groups, such as levels of job satisfaction across various leadership styles.
- Spearman Correlations: Employed to assess the strength and direction of associations between ranked variables, such as the relationship between CSR dimensions and mental resilience scores.

These methods provided a robust framework to examine the interplay between CSR initiatives and employee outcomes, ensuring that the findings were both statistically significant and practically relevant.

Table 1 Associations between demographic factors and CSR-related employee outcomes (Chi-Square test results)

Factor variable	Differentiated variable	Chi-Square statistic	p-value	Significance
Gender	Employee resilience	9.90	0.042	Significant
Gender	Intentions to turnover	13.85	0.031	Significant
Do you have children?	Social sustainability aspects	17.34	0.001	Highly significant
Do you have children?	Empowering leadership	11.72	0.039	Significant
Do you have children?	Employee resilience	12.33	0.015	Significant

4 Results

4.1 Chi-Square Test Results

The Chi-Square tests were conducted to examine the relationships between categorical demographic factors and differentiated CSR-related variables. The key findings are presented in Table (Table 1).

The results indicate significant associations between demographic factors (e.g., gender and parental status) and specific CSR-related variables. For instance, employees with children displayed stronger associations with social sustainability aspects and empowering leadership. Similarly, gender differences were evident in resilience and turnover intentions.

These findings suggest that demographic characteristics influence how employees perceive and are impacted by CSR initiatives, underscoring the need for tailored CSR strategies to address diverse workforce needs.

4.2 ANOVA Results

The ANOVA tests examined the effects of categorical demographic factors on continuous CSR-related variables. The significant results are summarized below (Table 2).

These results demonstrate that demographic factors, such as educational level and years of experience in the NGO sector, significantly influence perceptions of and engagement with CSR practices. For example, employees with higher educational attainment displayed greater sensitivity to social sustainability aspects, while years of experience shaped perceptions of leadership and communication practices.

Table 2 Effects of demographic variables on CSR dimensions and employee outcomes (ANOVA results)

Independent variable	Factor variable	p-value	Significance
Total environmental sustainability policies	If yes, what kind of special needs do you have	0.029	Significant
Total environmental sustainability policies	How many years are you working in the field of NGOs	0.001	Highly significant
Social sustainability aspects	Educational level	0.015	Significant
Cultivation of learning culture	How many years are you working in the field of NGOs	0.005	Significant
Empowering leadership	How many years are you working in the field of NGOs	0.008	Significant
Employee participation	How many years are you working in the field of NGOs	0.023	Significant
Communication	How many years are you working in the field of NGOs	0.002	Highly significant
Employee resilience	Educational level	0.006	Significant
Job engagement	Educational level	0.005	Significant
Job engagement	How many years are you working in the field of NGOs	0.036	Significant
Intentions to turnover	Educational level	0.003	Significant
Mental resilience	Educational level	0.032	Significant

The ANOVA findings highlight the importance of tailoring CSR initiatives to align with employees' educational and professional backgrounds, ensuring that programs resonate across diverse workforce demographics.

4.3 Spearman Correlation Results

The Spearman correlation tests examined the strength and direction of relationships between ranked variables, focusing on CSR practices and employee outcomes. The significant correlations are summarized below:

The results demonstrate strong positive correlations between various CSR dimensions and employee resilience, engagement, and satisfaction. Notably, "Cultivation of Learning Culture" exhibited robust correlations with "Empowering Leadership" and "Communication," underscoring their interdependence. Additionally, "Social Sustainability Aspects" strongly correlated with "Employee Resilience" and "Job Engagement," suggesting that these CSR dimensions play a pivotal role in fostering a motivated and resilient workforce (Table 3).

Table 3 Spearman correlation coefficients between CSR dimensions and employee psychological outcomes

Variable pair	Correlation coefficient (ρ)	p-value	Significance
Position in company versus areas of work interest	−0.260	0.003	Significant
Position in company versus environmental sustainability policies	0.206	0.021	Significant
Position in company versus social sustainability aspects	−0.200	0.026	Significant
Social sustainability aspects versus employee resilience	0.706	<0.001	Highly significant
Social sustainability aspects versus job engagement	0.577	<0.001	Highly significant
Cultivation of learning culture versus empowering leadership	0.753	<0.001	Highly significant
Communication versus employee resilience	0.500	<0.001	Highly significant
Employee resilience versus mental resilience	0.782	<0.001	Highly significant
Job engagement versus job satisfaction	0.496	<0.001	Highly significant

These insights highlight the critical role of CSR practices in shaping employee well-being and organizational dynamics, emphasizing the need for integrated strategies to maximize these benefits.

5 Discussion

The results provide evidence that Corporate Social Responsibility (CSR) practices are associated with higher employee well-being within the NGO sector. Addressing the core research question, CSR relates positively to mental resilience, job satisfaction, and engagement, and negatively to turnover intentions—positioning CSR as a strategic lever for psychologically healthy and committed workforces [2, 9]. The strong links between social sustainability and both resilience ($\rho = 0.706$) and job engagement ($\rho = 0.577$) reinforce this pattern, in line with work showing that CSR signals organizational care and fairness [10, 12, 19].

Empowering leadership appears to amplify these effects. Its ties to learning culture and communication align with theories that inclusive, autonomy-supportive leadership makes CSR more credible and actionable [16, 17]. Together, these dynamics help explain the observed associations with resilience and adaptability.

Importantly, these findings should be interpreted within the NGO context and the countries represented in our sample. Features commonly present in NGOs—mission-driven work and high demands on staff—may heighten the salience of social sustainability practices, which could partly account for the stronger associations observed relative to environmental dimensions. The mechanisms identified here may not operate identically in for-profit settings, where organizational goals and incentive structures differ; thus, caution is needed when generalizing beyond NGOs.

Demographic patterns (e.g., gender, parental status, and education) further indicate that CSR is not experienced uniformly across employees, underscoring the value of tailoring CSR and internal communication to workforce composition [23, 24].

6 Limitations and Future Research

This study is cross-sectional, relying on self-reported survey data; therefore, causal inferences cannot be drawn. The use of non-probability sampling and the specific national composition of the sample limit generalizability beyond NGOs and beyond the countries included. Future research should employ longitudinal designs and/or mixed-methods approaches and include cross-sector comparisons to test whether the observed relationships hold in different organizational contexts and over time.

7 Conclusion

This study has highlighted the significant influence of Corporate Social Responsibility (CSR) practices on key aspects of employee well-being, including mental resilience, job satisfaction, and turnover intentions. The results suggest that CSR initiatives, particularly those related to social sustainability, empowering leadership, communication, and employee participation, can create a work environment that promotes psychological strength, engagement, and organizational loyalty.

The findings also reveal that demographic and professional characteristics—such as educational background, parental status, and years of experience in the NGO sector—moderate the effects of CSR, indicating the importance of designing inclusive and targeted CSR strategies that address the diverse needs of employees. Organizations that adopt such tailored approaches may foster a more adaptive and committed workforce, better equipped to navigate the demands of modern workplaces.

Looking ahead, there are several directions for future research. Longitudinal studies are needed to examine the sustained impact of CSR on mental health and resilience over time. Additionally, qualitative approaches could offer deeper insights into employees' lived experiences and perceptions of CSR within organizational settings. Comparative research across sectors—such as private companies, public institutions, and non-profits—may also uncover sector-specific dynamics that shape how CSR is perceived and experienced by employees.

Another promising avenue for further exploration involves the role of cultural and geographical contexts. Understanding how CSR practices influence employee outcomes in different national or multicultural environments could enhance the global applicability of CSR strategies. Such insights would be particularly valuable for multinational organizations seeking to align social responsibility with employee well-being on a broader scale.

In sum, this study affirms the value of CSR not only as a mechanism for external accountability but as a powerful internal driver of organizational health. Integrating CSR into the core of organizational culture and leadership can serve as a catalyst for fostering resilience, satisfaction, and long-term employee engagement.

References

1. Carroll, A.B.: The pyramid of corporate social responsibility: toward the moral management of organizational stakeholders. Bus. Horiz. **34**, 39–48 (1991). https://doi.org/10.1016/0007-6813(91)90005-G
2. Glavas, A.: Corporate social responsibility and organizational psychology: an integrative review. Front. Psychol. **7** (2016). https://doi.org/10.3389/fpsyg.2016.00144
3. Aguinis, H., Glavas, A.: What we know and don't know about corporate social responsibility. J. Manag. **38**, 932–968 (2012). https://doi.org/10.1177/0149206311436079
4. Dhanesh, G.S.: CSR as organization-employee relationship management strategy. Manag. Commun. Q. **28**, 130–149 (2014). https://doi.org/10.1177/0893318913517238
5. Carroll, A.B., Shabana, K.M.: The business case for corporate social responsibility: a review of concepts, research and practice. Int. J. Manag. Rev. **12**, 85–105 (2010). https://doi.org/10.1111/j.1468-2370.2009.00275.x
6. Orlitzky, M., Schmidt, F.L., Rynes, S.L.: Corporate social and financial performance: a meta-analysis. Organ. Stud. **24**, 403–441 (2003). https://doi.org/10.1177/0170840603024003910
7. Freeman, R.E.: Strategic Management. Cambridge University Press (2010). https://doi.org/10.1017/CBO9781139192675
8. Brammer, S., Millington, A., Rayton, B.: The contribution of corporate social responsibility to organizational commitment. Int. J. Hum. Resour. Manag. **18**, 1701–1719 (2007). https://doi.org/10.1080/09585190701570866
9. Kim, H.-R., Lee, M., Lee, H.-T., Kim, N.-M.: Corporate social responsibility and employee-company identification. J. Bus. Ethics **95**, 557–569 (2010). https://doi.org/10.1007/s10551-010-0440-2
10. Rupp, D.E., Ganapathi, J., Aguilera, R.V., Williams, C.A.: Employee reactions to corporate social responsibility: an organizational justice framework. J. Organ. Behav. **27**, 537–543 (2006). https://doi.org/10.1002/job.380
11. Turker, D.: How corporate social responsibility influences organizational commitment. J. Bus. Ethics **89**, 189–204 (2009). https://doi.org/10.1007/s10551-008-9993-8
12. Bauman, C.W., Skitka, L.J.: Corporate social responsibility as a source of employee satisfaction. Res. Organ. Behav. **32**, 63–86 (2012). https://doi.org/10.1016/j.riob.2012.11.002
13. Glavas, A., Piderit, S.K.: How does doing good matter? J. Corp. Citizsh. **2009**, 51–70 (2009). https://doi.org/10.9774/GLEAF.4700.2009.wi.00007
14. De Roeck, K., Farooq, O.: Corporate social responsibility and ethical leadership: investigating their interactive effect on employees' socially responsible behaviors. J. Bus. Ethics **151**, 923–939 (2018). https://doi.org/10.1007/s10551-017-3656-6
15. Hameed, I., Riaz, Z., Arain, G.A., Farooq, O.: How do internal and external CSR affect employees' organizational identification? A perspective from the group engagement model. Front. Psychol. **7** (2016). https://doi.org/10.3389/fpsyg.2016.00788

16. Avolio, B.J., Walumbwa, F.O., Weber, T.J.: Leadership: current theories, research, and future directions. Annu. Rev. Psychol. **60**, 421–449 (2009). https://doi.org/10.1146/annurev.psych.60.110707.163621
17. Liu, W., Zhu, R., Yang, Y.: I warn you because I like you: voice behavior, employee identifications, and transformational leadership. Leadersh. Q. **21**, 189–202 (2010). https://doi.org/10.1016/j.leaqua.2009.10.014
18. Amabile, T.M., Schatzel, E.A., Moneta, G.B., Kramer, S.J.: Leader behaviors and the work environment for creativity: perceived leader support. Leadersh. Q. **15**, 5–32 (2004). https://doi.org/10.1016/j.leaqua.2003.12.003
19. Valentine, S., Godkin, L., Fleischman, G.M., Kidwell, R.: Corporate ethical values, group creativity, job satisfaction and turnover intention: the impact of work context on work response. J. Bus. Ethics **98**, 353–372 (2011). https://doi.org/10.1007/s10551-010-0554-6
20. Du, S., Bhattacharya, C.B., Sen, S.: Maximizing business returns to corporate social responsibility (CSR): the role of CSR communication. Int. J. Manag. Rev. **12**, 8–19 (2010). https://doi.org/10.1111/j.1468-2370.2009.00276.x
21. Men, L.R., Stacks, D.W.: The impact of leadership style and employee empowerment on perceived organizational reputation. J. Commun. Manag. **17**, 171–192 (2013). https://doi.org/10.1108/13632541311318765
22. Bice, S.: Corporate social responsibility as institution: a social mechanisms framework. J. Bus. Ethics **143**, 17–34 (2017). https://doi.org/10.1007/s10551-015-2791-1
23. Lee, E.M., Park, S.-Y., Lee, H.J.: Employee perception of CSR activities: its antecedents and consequences. J. Bus. Res. **66**, 1716–1724 (2013). https://doi.org/10.1016/j.jbusres.2012.11.008
24. Farooq, O., Payaud, M., Merunka, D., Valette-Florence, P.: The impact of corporate social responsibility on organizational commitment: exploring multiple mediation mechanisms. J. Bus. Ethics **125**, 563–580 (2014). https://doi.org/10.1007/s10551-013-1928-3

Overcoming Environmental Risks in European SMEs: A Policy Review of EMS Integration Within ESG

Panagiotis-Stavros C. Aslanidis, Jaime Moll de Alba, and George Halkos

Abstract Environmental risks pose a significant threat to small-medium enterprises (SMEs), impacting operational continuity, regulatory compliance, and long-term competitiveness in an increasingly sustainability-oriented and data-driven market. The aim of this review is to present a comprehensive framework of Environmental Management Systems (EMS), specifically ISO 14001 and EMAS, and their linkages to Corporate Social Responsibility (CSR) and Environmental, Social, and Governance (ESG) frameworks. Drawing on 47 peer-reviewed publications from Scopus, Web of Science, and Google Scholar, the review also explores the emerging role of the European Sustainability Reporting Standards (ESRS) in enhancing environmental performance, particularly for SMEs. The novelty of this study is rooted in the integration of traditional EMS with evolving ESG and ESRS standards. Moreover, the study pursues two key objectives: (i) to map the interconnections among a holistic ESG framework and (ii) to assess their collective potential in improving SME resilience to environmental risks. The findings suggest that adopting a synergistic approach not only supports compliance but also fosters long-term commitment to the achievement of sustainable development goals (SDGs). Ultimately, this review offers actionable insights and policy implications for SMEs on integrating EMS under the scope of ESG, aiming to raise resilience against environmental risks and align with future regulatory and market expectations under the scope of sustainability.

Keywords CSR · ESG · CSRD · SDGs · SMEs · European green deal

---

P.-S. C. Aslanidis (✉) · G. Halkos
Laboratory of Operations Research, Department of Economics, University of Thessaly, Volos, Greece
e-mail: paslanidis@uth.gr

G. Halkos
e-mail: halkos@uth.gr

J. Moll de Alba
United Nations Industrial Development Organization (UNIDO), Vienna International Centre, Vienna, Austria
e-mail: J.Moll-de-Alba@unido.org

G. Halkos et al. (eds.), *Environmental Risk Analysis*, Springer Proceedings in Mathematics & Statistics 540, https://doi.org/10.1007/978-3-032-24155-9_12

Abbreviations

Carbon border adjustment mechanism (CBAM)	Circular economy action plan (CEAP)	Corporate social responsibility (CSR)
Corporate sustainability disclosure standards (CSDS)	Corporate sustainability due diligence directive (CS3D)	Corporate sustainability reporting directive (CSRD)
Eco-Management and audit scheme (EMAS)	Environmental management systems (EMS)	Environmental, social, and corporate governance (ESG)
European financial reporting advisory group (EFRAG)	European green deal (EGD)	European sustainability reporting standards (ESRS)
Global reporting initiative (GRI)	International financial reporting standards (IFRS) foundation	International organization for standardization (ISO)
International sustainability standards board (ISSB)	Life cycle assessment (LCA)	Organisation for economic co-operation and development (OECD)
Plan-Do-Check-Act (PDCA)	Small and medium-sized enterprises (SMEs)	Sustainable development goals (SDGs)
Task force of climate-related financial disclosures (TCFD)	United nations (UN)	Waste framework directive (WFD)

1 Introduction

In recent decades, the escalating environmental risks, ranging from climate change and biodiversity loss to resource depletion and pollution, have intensified pressure on both governments and the private sector to adopt more sustainable practices against the occurrence of several crises, or as often called multi-crisis [1, 2]. Businesses, particularly those operating in resource-intensive or high-emission sectors, are increasingly held accountable for their environmental impact [3, 4]. Characteristically, beyond compliance with environmental regulations, there is a growing expectation for companies to proactively mitigate environmental risks, adopt cleaner technologies, and contribute to the transition towards a low-carbon, sustainable economy [5–7]. In this sense, the present review seeks to shed light on factors that can minimize environmental or other risks that many companies face in Europe, particularly smaller businesses.

To begin with, environmental management systems (EMS) have emerged as essential frameworks for guiding organizations in systematically managing their environmental responsibilities [8]. Among the most widely recognized EMS standards are ISO 14001 and the EU's Eco-Management and Audit Scheme (EMAS), both of which help organizations identify, monitor, and reduce environmental impacts while promoting continuous improvement on environmental, social, and corporate governance (ESG) issues [9, 10]. The EMS focuses on important factors such as structured processes, documentation, and stakeholder engagement, but it has been proven that EMS implementation not only enhances regulatory compliance but also drives operational efficiencies, risk management, and reputational benefits [11–13].

In parallel with EMS adoption, there has been a rapid evolution of corporate sustainability frameworks such as Corporate Social Responsibility (CSR), or, nowadays, the ESG criteria. To exemplify, CSR emphasizes a company's voluntary commitment to ethical, social, and environmental values, while ESG provides a more structured and investor-focused approach to measuring non-financial performance [14, 15]. Increasingly, businesses are expected to integrate ESG factors into their core strategies and reporting practices, reflecting a broader shift from traditional compliance models towards value-based and stakeholder-centric governance [16, 17]. Nevertheless, companies are facing various challenges on ESG issues, especially the small and medium-sized enterprises (SMEs).

SMEs constitute the backbone of most economies and play a significant role in global supply chains, more specifically, SMEs in Europe represent 99.8% of all enterprises according to the European Commission. In 2024, SMEs faced a minor reduction in real value added (-0.2%) even though they experienced an increase in employment (1.1%); furthermore, micro-SMEs seem to have potential factors that have strengthened their resilience and growth in comparison to the rest of the companies [18]. Despite their economic importance, SMEs often face disproportionate challenges in adopting comprehensive sustainability frameworks due to limited financial, technical, and human resources [19, 20]. For example, the Organisation for Economic Co-operation and Development (OECD) highlighted the pivotal role of governments that support SMEs with short- and long-term financial measures, emphasizing their role in the green transition; despite progress, SMEs still face barriers to sustainable finance, requiring continued policy focus [21].

However, SMEs are also particularly vulnerable to environmental risks such as supply chain disruptions, resource scarcity, and climate-related pressures [22, 23]. These challenges have led policymakers to propose standards such as the European Sustainability Reporting Standards (ESRS) that can boost companies' ESG performance by standardizing sustainability reporting, hence these policy-related efforts pave the way for the integration with EMS for improved compliance, transparency, and performance tracking [24–26]. To summarize, understanding how EMS, CSR, and ESG frameworks can be tailored to support SME resilience is both timely and critical for fostering inclusive, sustainable development.

The novelty of this study lies in integrating traditional EMS with evolving ESG and ESRS standards to address environmental risks more effectively. The aim of the study is to showcase the importance of ensuring data-driven compliance with environmental (or other) standards, to support risk-based thinking with quantifiable evidence, and to enhance quality control, sampling, and performance monitoring in companies (especially SMEs). Moreover, the study pursues two key objectives: (i) to map the interconnections among a holistic ESG framework and (ii) to assess their collective potential in improving SME resilience to environmental risks. The present review structure is the following: Section 2 introduces the literature review on how EMS implementation can be strengthened by—international or European—standards for the diffusion of ESG frameworks in SMEs in Europe. Section 3 presents the methodology, whereas Section 4 includes the results and discussion. Section 5 summarizes the key points, highlighting the theoretical and practical implications, mentioning the limitations of the study, and concluding with future research directions.

2 Literature Review

The following sub-sections outline how ESG implementation is supported by standards and EMS to highlight the alignment between European and global frameworks to ensure consistency in sustainability reporting. Additionally, they explain how ISO 14001 and EMAS function as EMSs, with EMAS offering more rigorous transparency and public accountability even though there are important obstacles in its adoption. Overall, the text emphasizes that these standards enhance environmental performance, improve reporting quality, and support stakeholder engagement, with EMS frameworks serving as practical tools to translate sustainability goals into measurable actions.

2.1 Environmental Compliance in ESG Framework

The ESRS encompass a broad range of ESG issues, including climate change, biodiversity, and human rights. The ESRS are mandatory reporting standards created by the European Financial Reporting Advisory Group (EFRAG) under the Corporate Sustainability Reporting Directive (CSRD), which took effect in January 2023 [27]. This framework promotes clear, comparable sustainability reporting aligned with the EU's climate goals, such as the European Green Deal (EGD), and supports stakeholders in understanding environmental risks and impacts [28, 29]. The ESRS align with global frameworks such as the Global Reporting Initiative (GRI) and the International Sustainability Standards Board (ISSB), the Task Force of Climate-related Financial Disclosures (TCFD), and the United Nations (UN) Sustainable Development Goals (SDGs) to ensure consistency, reduce redundant reporting, and support a unified sustainability approach [26].

The ESRS adopt GRI standards to promote clear, comparable disclosures and follows ISSB principles to meet global investor and market needs, because the GRI standards can support organizations to address different aspects of ESG [30]. Moreover, during 2021–2022, the International Financial Reporting Standards (IFRS) Foundation initiated the ISSB in order to provide sustainability standards that also address stakeholders' needs, to boost corporate sustainability reporting and diffuse information more adequately regarding corporate sustainability reporting to all stakeholders [31]. Furthermore, the ESRS covers enough environmental issues as in TCFD, mainly of the environmental risks (i.e., physical risks) and other transitional (e.g., legal or market) risks that companies face [32].

From 2015, there are several initiatives that try to combine ESG and SDGs, where the SDGs can place the policy framework around the ESG targets, for example, corporate sustainability commitments can be pivotal for actively achieving SDGs [33–35]. However, it should be mentioned that there are differences between these frameworks, namely SDGs have a specific time constraint till the year 2030, while ESG framework does not have a specific timeline [36].

2.2 Beyond Compliance Through the Implementation of Environmental Management Systems

The International Organization for Standardization (ISO) provides a structured framework for organizations to manage environmental responsibilities effectively, through different standards such as the ISO 14001 based on the Plan-Do-Check-Act (PDCA) cycle [8]. The ISO 14001 encourages a systematic approach to environmental management with the following key principles such as legal compliance, risk-based thinking, continual improvement, life cycle thinking, and stakeholder engagement [37]. Overall, these elements help organizations align their operations with sustainability goals while addressing both regulatory and environmental risks.

There are several benefits for firms implementing ISO 14001, as it offers a range of environmental and financial benefits. Initially, organizations typically see improved environmental performance through reduced emissions, waste, and resource consumption, which also translates into cost savings and greater operational efficiency [7]. Moreover, ISO 14001 enhances regulatory compliance and reduces the risk of legal penalties, while boosting reputation and stakeholder trust. Although certification is voluntary, it is widely recognized and often expected across supply chains as a marker of credibility and strong risk management [16]. ISO 14001:2015 was the second most applicable standard (the most common was ISO 9001) with 300,410 total valid certificates in 526,046 sites based on the ISO 2023 survey that included 39 sectors by which the sectors of (i) construction, (ii) basic metals and fabricated metal products, and (iii) wholesale and retail trade (motor vehicles, motorcycles, and personal & household goods) covered the top-three places with 7.79%, 7.13%, and 5.90% of the sample respectively [38].

Additionally, the EMAS builds upon ISO 14001 by adding requirements for transparency, public reporting, and third-party verification. EMAS mandates an initial environmental review and evaluation of six core indicators: energy, materials, water, waste, biodiversity, and emissions. The annual environmental statement, validated by independent verifiers, ensures transparency and accountability [20]. Essentially, EMAS also prioritizes stakeholder communication and employee involvement, encouraging continuous improvement through participatory environmental initiatives [8]. As per April 2025, EMAS has been implemented by 4,114 organizations in 15,815 sites, most of whose organizations are in Italy (1155), Germany (1,109), and Spain (853) [39]. The main leading sectors that apply EMAS are (i) the waste sectors, public administration, and specialized construction activities, and, moreover, in terms of company's size, the micro and small business are covering the 10.6% and 29.5% which is indeed an important statistic that should be mentioned.

Both EMS (i.e., ISO 14001 and EMAS) support sustainability reporting by providing structured methods for identifying environmental risks, setting objectives, and tracking performance. They ensure that reported data is accurate, consistent, and compliant with stakeholder expectations; for example, the continual improvement process embedded in EMS drives progress towards environmental goals, while documentation and reporting requirements enhance the clarity and credibility of sustainability disclosures [9, 12, 40, 41]. Despite EMAS's more rigorous demands, both systems are complementary in fostering environmental responsibility and long-term resilience.

2.3 *Aligning Standards: Linkages Between EMSs and ESG Frameworks*

The core characteristics of standards and EMSs relate to the scope, type, purpose, and applicability as illustrated in Table 1. The global scopes of GRI and ISO standards have been pivotal for the structure of the ESRS and EMAS respectively. The EMAS and ISO 14001 are certifiable EMS, whereas the ESRS and GRI are reporting frameworks with the former being mandatory while the latter voluntary. The purpose of both EMSs (i.e., ISO 14001 and EMAS) is to guide companies on improving their environmental performance. Moreover, GRI helps companies to disclose non-financial information such as ESG elements, and ESRS also refer to sustainability disclosures based on the CSRD directive. Furthermore, regarding the applicability, the ISO is for all types of organizations, and EMAS is also largely applicable to different types of organizations. Additionally, the GRI standards are applicable and voluntary to all organizations, whereas the ESRS due to its legal elements is mandatory to EU-based companies and listed SMEs.

In Table 2, the main linkages between the above frameworks are explained based on the ESRS framework. Firstly, the environmental ESRS elements include five core categories: (i) climate change, (ii) pollution, (iii) water and marine resources, (iv) biodiversity and ecosystems, and (v) resource use and circular economy. In this sense, the GRI standards have more specific standards than the ESRS for all the five categories; the GRI standards are presented in more detail in Appendix A in Table 4. The EMS also successfully cover these five categories; to exemplify, the EMAS has specific directions for companies' impacts such as emissions, water, biodiversity, and energy or material resources, and waste flows. Moreover, the main environmental-related ISO standard is ISO 14001 but is also complemented by ISO 14046 (on water footprint), ISO 14067 (on carbon footprints), ISO 14040 and ISO 14044 (on life cycle assessment, LCA), ISO 50001 (on energy management), and ISO 59004 (on circular economy), with more information in Appendix A in Table 4. To conclude, companies

Table 1 Comparison of the characteristics of selected the ISO, GRI, and ESRS frameworks. *Source* Authors' creation

Feature	ESRS	GRI	EMAS	ISO
Scope	EU—wide	Global	EU—wide	Global
Type	Mandatory reporting standards	Voluntary reporting framework	Standard (certifiable EMS)	Standard (certifiable EMS)
Purpose	Define mandatory sustainability disclosures under CSRD	Help companies report ESG impacts to stakeholders	Guide companies in managing and improving environmental performance	Guide companies in managing and improving environmental performance
Applicability	EU-based large companies and listed SMEs (mandatory)	All organizations (voluntary)	EMAS is open to any organization, public or private, regardless of size, type, or sector	All types of organizations (certifiable)

should further address environmental risks effectively, and frameworks like ESRS, GRI, ISO 14001, and EMAS provide structured guidance across key areas such as climate, pollution, water, biodiversity, and circular economy. Therefore, these standards ensure comprehensive, credible, and actionable environmental management, supporting sustainability goals and regulatory compliance.

However, companies do not target only environmental issues but also social and governance issues. Again, based on the ESRS, there are four core social-related issues on (i) company's workforce, (ii) workers, (iii) affected communities, and (iv) consumers. The GRI standards also further enhance transparency and accountability, empowering society with reliable information on organizations' ESG impacts. In the EMAS, there are specific elements for own workforce and workers in the value chain at all levels of companies' structure, whereas regarding the impacts towards affected communities and consumers the EMAS necessitate that companies invest on openness, transparency, and periodic provision of environmental information. In the ISO standards, also all the ESRS categories can be underpinned by social-centric standards, specifically with ISO 10002 (quality management), ISO 20400 (sustainable procurement), ISO 22000 (safe food), ISO 26000 (social responsibility), ISO 30414 (human resource management), ISO 37000 (governance), and ISO 45001 (health and safety).

Finally, the ESRS has only one category for corporate governance. The GRI standards for corporate governance focus on four issues: (i) economic performance, (ii) market presence, (iii) anti-corruption, and (iv) taxation. In addition, the EMAS provides a robust governance-centric framework in which there is evaluation of the company's performance against its objectives. Additionally, the ISO standards also cover, as mentioned before, social aspects such as issues of social responsibility, governance of organizations through different principles and practices to fulfil ESG

Table 2 Linkages between the selected frameworks based on ESRS categories. *Source* Authors' creation

Scope	ESRS	GRI	EMAS	ISO standards
Climate change	ESRS—E1	GRI 201–202, 302, and 305	Emissions	ISO 14001 (indirectly), 14067
Pollution	ESRS—E2	GRI 305, 306	Emissions	ISO 14001
Water and marine resources	ESRS—E3	GRI 303	Water	ISO 14001, 14046
Biodiversity and ecosystems	ESRS—E4	GRI 101, 304	Biodiversity	ISO 14001 (indirectly)
Resource use and circular economy	ESRS—E5	GRI 301, 302, 305, 306	Energy efficiency, Material efficiency, and Waste	ISO 14001, 14040, 14044, 50001, 59004
Own workforce	ESRS—S1	GRI 401–407	At all levels	ISO 45001, 30414, and 26000
Workers in the value chain	ESRS—S2	GRI 407–409, 414	At all levels	ISO 20400 and 26000
Affected communities	ESRS—S3	GRI 411, 413, 416	OTEI*	ISO 26000 and 37000:
Consumers and end-users	ESRS—S4	GRI 416–418	OTEI*	ISO 10002, 22000, and 26000
Business conduct	ESRS—G1	GRI 201, 202, 205, 207	Evaluation of organization's performance against its objectives	ISO 26000, 37000, 37301

* OTEI: openness, transparency, and periodic provision of environmental information.

responsibilities, and compliance management systems that guide organizations on ensuring legal and ethical adherence.

3 Methodology

A review paper serves as a vital academic tool that systematically gathers, synthesizes, and critically evaluates existing literature on a given topic. Its purpose is not only to map the current state of knowledge but also to identify research gaps, inconsistencies, and emerging trends. Through this comprehensive analysis, a well-conducted review can formulate nuanced, evidence-based policy recommendations that are both credible and relevant. These recommendations are typically supported by a robust and

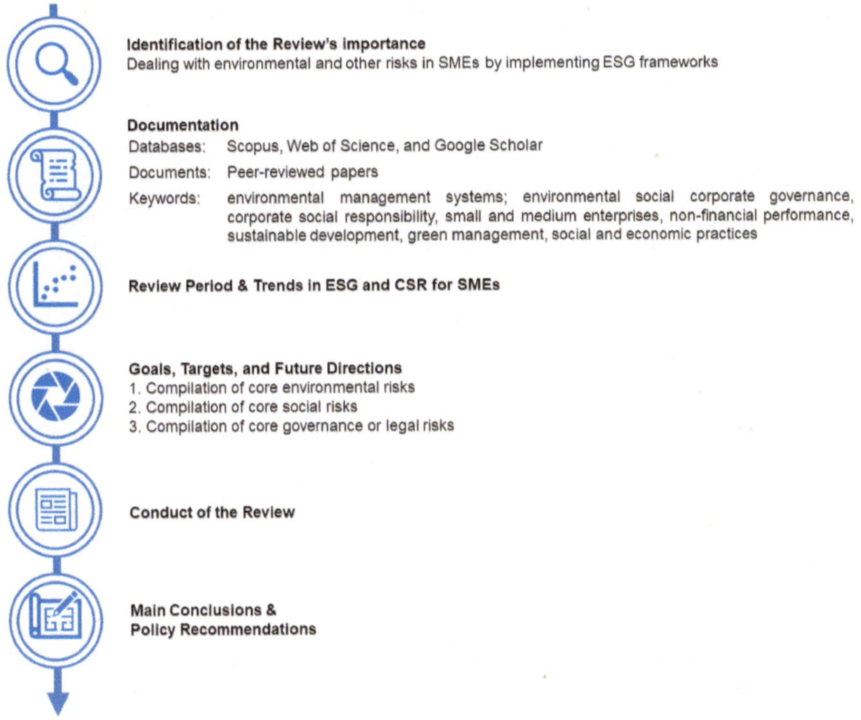

Fig. 1 The methodological flow of the review

multifaceted argumentation structure that aligns with established eligibility criteria and responds to a diverse set of research objectives (see Fig. 1).

In the context of the present study, the review seeks to advance the interdisciplinary nature of environmental standards that can be combined with sustainability reporting schemes under the scope of ESG. Specifically, the paper emphasizes the ways in which SMEs can be resilient towards environmental or other risks by adopting EMS, and further report their impacts and identify market opportunities. Overall, proper sustainability-driven decision-making can boost not only the economic, but also the environmental performance of SMEs.

This review presents a thorough analysis of key academic publications identified through an extensive search across the Scopus, Web of Science, and Google Scholar databases. These databases were selected for their widespread recognition and reliability in literature reviews, particularly within fields addressing EMS, CSR, and ESG dimensions and their comprehensive coverage which ensures access to high-quality and relevant scholarly work.

Furthermore, in order to establish a balanced and insightful perspective, the review incorporates a synthesis of both foundational and contemporary studies, capturing the historical evolution as well as current trends in the academic discourse. All the selected publications (n=47, period 2001–2024) underwent a rigorous and independent cross-verification process by the authors to ensure accuracy, consistency, and objectivity in the findings. Additionally, an advanced screening protocol was applied to the search results, which involved excluding certain records based on specific criteria: (i) papers listed as articles in press, (ii) non-English language publications, (iii) duplicate entries, and (iv) those failing to meet the broader inclusion requirements. This meticulous approach ensures that the final pool of literature reflects both quality and relevance to the study's objectives.

4 Results and Discussion

The present review monitors 47 relevant publications that can be linked to the EMS literature and covers also aspects of ESG and CSR strands, highlighting the need to offer companies (especially SMEs) practical tools to deal with environmental or other risks. From the 47 studies, 65.96% (31 studies) are linked to SMEs; moreover, 48.94% (23 studies) refer to CSR and 42.55% (20 studies) are connected with ISO 14001 implementation, followed by EMAS implementation with 17.02% (eight studies), and finally 4.26% (two studies) on ESRS application. According to the studies' relevance, as illustrated at Fig. 2, in the last twenty years, and especially after 2009 and 2016, there are rising trends both in cumulative citations and in the volume of publications.

The present review gathered studies from 16 journals (see Table 3), from which the first three journals are the *Journal of Cleaner Production*, *Corporate Social Responsibility and Environmental Management*, and *Business Strategy and the Environment*. Initially, in the Journal of Cleaner Production, overall, 16 studies were extracted with

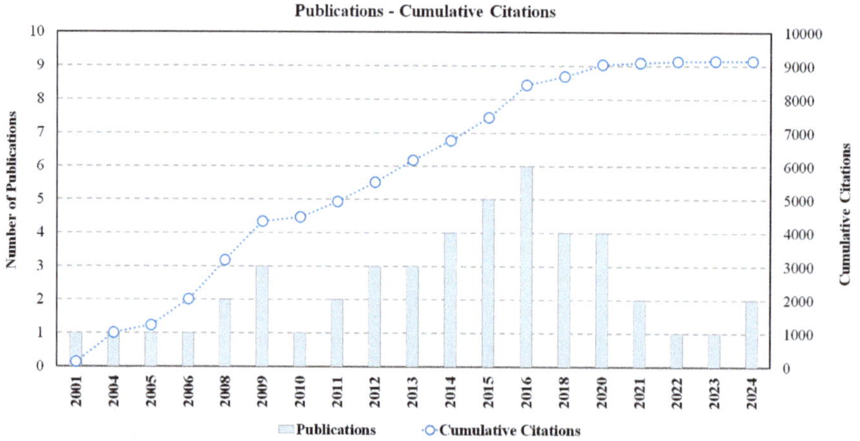

Fig. 2 The summary of the present review of publications and their cumulative citations in the period 2001–2024

Table 3 The journals and the relevant studies that have been included in the present review

Journal	No. of studies
Journal of cleaner production	16
Corporate social responsibility and environmental management	9
Business strategy and the environment	6
Journal of environmental management	3
Journal of business ethics	2
Bioresource technology	1
Business & society	1
Environment and planning C: politics and space	1
International small business journal: researching entrepreneurship	1
Journal of environmental planning and management	1
Journal of global responsibility	1
Journal of management control	1
Journal of small business management	1
Organization and environment	1
Quality and quantity	1
Symphonya emerging issues in management	1

the majority focusing on ISO 14001 implementation (14 out of 16 studies) and most of which also observe the performance of SMEs (eight out of 16 studies). Additionally, nine studies were found in the journal of Corporate Social Responsibility and Environmental Management, focusing on SMEs' performance and CSR/ESG adoption strategies with eight and seven studies respectively. Last, six studies were found in the journal of Business Strategy and the Environment with five focusing on CSR/ESG issues and four of them focusing also on issues related to SMEs. Moreover, Fig. 3 presents a citation map of the 47 articles that have been included in the present review; the 47 papers have been arranged in this way to show their date of publication (on the horizontal axis) and their interconnection (on the vertical axis).

4.1 The Role of EMS Frameworks in Enhancing SME Resilience

Before implementing EMS, it has been alerted in the literature that socio-political factors can influence initial adoption decisions, but real performance improvements stem from legitimacy-driven, symbolic reasoning [42]; for example, a cluster analysis revealed that legal compliance and environmental aspects are key priorities for

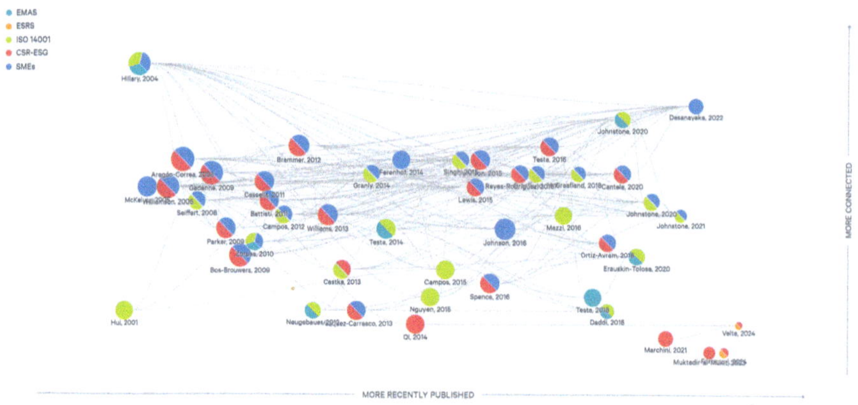

Fig. 3 Studies that focus on the integration of EMS in ESG framework. *Source* Figure created by the authors from litmaps.com

companies, reflecting strong concern for meeting regulations and conserving environmental resources [37]. These challenges led to the need for adopting EMS as a way to deal with important factors that can affect a company's performance.

Influence of ISO 14001

Regarding the ISO 14001 standard, the benefits of EMS implementation involve a strengthened policy setting, proper planning on environmental and legal aspects; however, EMS require structured operation though training (e.g., employees) and documentation and verification through monitoring, a company's compliance evaluation, and audits (internal or external) [43, 44]. External and internal factors can affect a company's decision to adopt EMS. An interesting example was that cooperative systems engineering approach helped SMEs overcome challenges in implementing EMS, effectively addressing environmental impacts at global, regional, and local levels, including climate change, pollution, and resource depletion [7]. On the other hand, participation in external networks mediates the positive impact of ISO 14001 certification on SMEs' ecological performance [45].

Eco-driven aspects can strongly influence a company's performance. For example, although only a minority of SMEs set environmental targets, adopting even simple formal CSR processes can significantly improve environmental outcomes; thus, this challenges the view that formal tools are unsuitable and highlights the need for tailored support and greater awareness [46]. Similarly, regarding waste management performance, ISO 14001 certification was a significant factor of waste minimization that boosted performance to 25%, primarily through resource reduction and waste separation [47].

However, there are several issues raised due to stakeholders' pressure, potential obstacles, and employees' involvement. In Vietnam, certified plants seem to outperform non-certified ones in management and operations, with ISO 14001 showing potential to enhance environmental performance despite early implementation challenges in rapidly growing industries [48]. In Norway, the comparison of ISO 14001 with Eco-Lighthouse showed that market benefits and cost reduction drive Eco-Lighthouse adoption, while customer pressure and improved routines favour ISO 14001, which faces greater challenges in employee buy-in, competence, and time [49]. Having these issues in mind, a critical conclusion of ISO 14001 implementation was that even though firms gain reputational benefits from the certification due to secondary stakeholder pressure, there is a pressure that does not lead to genuine internalization of the standard [16].

Regarding the employees' involvement and EMS adoption there are also necessary steps to be taken. It has been highlighted that most firms view green manufacturing and EMS as driving forces of competitiveness, especially by identifying the crucial factor of teamwork and morale in employee involvement [50]. Similarly, the combination of formal controls with fostering employee autonomy and values, as both organizational systems and personal beliefs of owner-managers and employees, shape sustainable practices [51].

The ISO 14001 and EMAS

There is a need to clarify how environmental performance is defined and how its dimensions interact, and to investigate the link between EMS (either ISO or EMAS) implementation and environmental performance in SMEs [10, 40]. It is common in the literature to mention that the adoption of environmental standards like EMAS leads to varied behaviours among firms, with the degree of internalizing requirements crucial for achieving genuine performance improvements [40], showing the need to clarify how environmental performance is defined and its dimensions interact, especially on SMEs' environmental performance [12].

Characteristically, ISO 14001 is mainly adopted due to external pressure or mandates, while EMAS is driven by internal values and expected benefits; currently competing, the two standards may complement each other in the future, with adoption motivations ranging from reputational concerns to anticipated environmental and financial gains [52]. In more detail, the EMAS has been characterized as "the right system to have"; however, some companies want to abandon it, but they fear potential reputational damage. On the contrary, ISO 14001 seems to be more applicable, but companies implement it because it is mandatory due to external stakeholders' pressure.

In many SMEs, the main obstacles of adopting EMS focus on the internal and external barriers. On the one hand, the internal barriers include lack of understanding and perception problems, issues on the company culture, and obstacles due to lack of proper resources; on the other hand, some external barriers are linked to economic incapacities, institutional pressures, lack of support or guidance, and also potential misunderstanding between the company and certifiers of the EMS [20], although the implementation of EMS is essential for SMEs because it can offer benefits like improved environmental performance, compliance, risk reduction, cost efficiency, and greater employee environmental awareness [8].

To recapitulate, EMS can be helpful to deal with environmental risks as well as to address other potential institutional or operational barriers. An interesting conclusion is that EMS, like ISO 14001 or EMAS, offer short-term environmental gains, but lasting impact depends on strong internal commitment beyond mere certification [9]. However, companies and especially SMEs should consider some drawbacks; for example, the decline of EMAS has been attributed to limited financial and human resources, low market and stakeholder recognition, unclear added value, and concerns over sensitive data [13].

4.2 Unlocking the Potential of ESG Frameworks in SMEs

As mentioned before, key themes in CSR for SMEs include factors such as unique and flexible management approaches as well as antecedents, barriers, and potential impacts shaping environmental risk reduction practices [53]. In these terms, the academic literature on CSR is categorized into ethical values and relationships, business context and performance, formal integration processes, and relevant political issues [15]; moreover, four categories reflect SMEs' efforts to (i) reduce emissions, (ii) manage waste responsibly, (iii) use sustainable materials, and (iv) formalize environmental commitment through EMS and eco-labels [6]. Essentially, such CSR practices have paved the way for integrating sustainability into innovation, which ultimately can lead SMEs to create novel products and collaborate more with stakeholders [54].

Barriers and Obstacles in ESG for SMEs

SMEs face various internal and external barriers when trying to adopt EMS and apply CSR practices so policymakers should have such problems in mind when aiming to deal with rising environmental risks and financial instability. Some issues have been analysed previously in the literature at some point, but such aspects have a greater impact when referring to SMEs. Especially, key barriers to implementing EMS include the costs associated with documentation and audits, the complexity of legislative requirements, and limited involvement and training of both management and employees [55].

Furthermore, barriers to adopting CSR practices include financial constraints, time limitations, and concerns about losing competitive advantage, at the same time, external pressures arise from customer expectations, community interests, regulatory demands, and competitor behaviour [56]. Although legislation helps SMEs to raise environmental awareness, many firms continue to operate informally and remain unaware of the potential cost savings associated with environmental practices [19]; and, even more, market-driven pressures often discourage SMEs from implementing environmentally motivated policies [57]. Therefore, such barriers and obstacles can deviate SMEs from implementing successfully their CSR efforts.

How SMEs Benefit from Addressing Environmental Issues

SMEs should target personal values and their desire to make a meaningful contribution in order to achieve effective environmental policy [23], meaning that SMEs' environmental engagement is driven by three core parameters: (i) shared vision, (ii) strategic proactivity, and (iii) effective stakeholder management [58]. Another important parameter for successful environmental policy is "environmental proactivity" that can be driven by external pressures and entrepreneurs' attitudes, with real environmental performance linked to genuine internal investment and commitment, highlighting the need for supportive policies that go beyond symbolic compliance [59].

Environmental responsibility is not just good ethics, but it can also be smart business as in the following two examples. The first example is a study that found how environmental performance can positively impact financial performance, especially when supported by resource slack and industry conditions, urging policymakers to enforce strict standards and offer incentives to promote sustainable innovation [60]. Another example from Denmark showed that when SMEs implement environmental initiatives driven mainly by strategic intent, they can have enhanced competitiveness due to cost savings [61]. To conclude, most SMEs engage in environmental actions, but smaller businesses perceive fewer benefits, leading to varied levels of engagement [22]. In conclusion, while many SMEs recognize the value of environmental initiatives, the perceived benefits, and thus the level of engagement, vary widely.

Benefits in Social and Governance Issues for SMEs

In the ESG literature, the three core pillars, environmental, social, and governance, are deeply interconnected, with no strict boundaries between them. As a result, improvements in environmental performance can positively influence social and governance outcomes, and vice versa. A publication showed SMEs' owners' experiences can affect the company's environmental responsibility, emphasizing company's diversity and prioritizing environmental performance over financial gains [62], but the social factors also play an important role. In addition, SMEs' social responsibility bridges mainstream CSR research and management studies, advocating for a holistic, care-based, feminist-informed framework to address the traditional neglect of small firms in CSR discourse [14].

Under this perspective, two studies show that board gender diversity and independence can enhance environmental performance, notably by reducing GHG emissions, while women directors contribute through their commitment and sustainability expertise, supporting upper echelons and resource dependence policies [17, 63]. However, while larger boards may boost environmental investments, they can also negatively affect overall environmental outcomes [63]. To conclude, ESG factors are interlinked, with environmental performance influenced by social elements like diversity, leadership values, and inclusive CSR approaches, especially on the social and governance parameters, where the board gender diversity can further enhance environmental outcomes.

A Changing Political and Policy World Landscape and Its Impact on ESG and Sustainability Regulations

Recent changes in the political and policy environment throughout the world coupled with limited economic growth may have led to consider introducing changes to facilitate the implementation of existing ESG and sustainability frameworks. In Europe, the Omnibus Package presented by the European Commission in February 2025, currently under negotiation, seeks to streamline funding access and administrative and reporting processes, enabling SMEs to focus their efforts and more easily implement ESG standards and contribute to the SDGs.

The Omnibus Package includes proposals to simplify the legislation and implementation of related regulations such as the CSRD, the Corporate Sustainability Due Diligence Directive (CS3D), the EU Taxonomy Regulation, and the Carbon Border Adjustment Mechanism (CBAM) [64]. The ongoing revision of the existing EU framework may be partly seen as a response to the quest for faster and sustained economic growth in Europe underpinned by enhanced competitiveness. The 2024 report of the future of European competitiveness put forward proposals to strengthen governance in Europe, notably through regulatory simplification with a specific reference to the EU's sustainability reporting and due diligence framework as "a major source of regulatory burden, magnified by a lack of guidance" [65, 66]. The ongoing discussion in Europe may have also contributed to the decision of the Federal Council of Switzerland to put on hold the Ordinance on Climate Disclosures regulating reporting requirements by companies [67].

Recent policy changes in the USA have resulted in a decline in the interest in ESG compliance and reporting as signalled by the decision of the Securities and Exchange Commission to put on hold the implementation of disclosure requirements on climate risks and greenhouse gas emissions [68].

The above contrasts with the efforts undertaken in China to enhance corporate ESG and sustainability disclosure. The country introduced in 2024 the first ever national Corporate Sustainability Disclosure Standards (CSDS), Basic Standards, based on those of the ISSB, to enhance ESG disclosure by companies [69]. Also, the three stock markets of Beijing [70], Shanghai [71], and Shenzhen [72] published guidance on ESG corporate reporting.

It would be of interest to monitor future developments throughout the world in terms of ESG and sustainability and identify good practices that help strike a balance that ensures that ESG and sustainability regulatory simplification is not done at the expense of environmental protection and the move towards a net-zero economy.

The Future of SMEs in Europe

Unfortunately, some cases show that SMEs report very low levels of actual implementation of sustainability management tools [73]; the most common environmental, social, governance, and legal risks that SMEs face when implementing ESG frameworks are summarized in Figure 4. Several factors affect this trend, more specifically, EMS implementation in SMEs which is influenced by age, education, legislation, awareness, and stakeholder pressure, particularly from customers and employees,

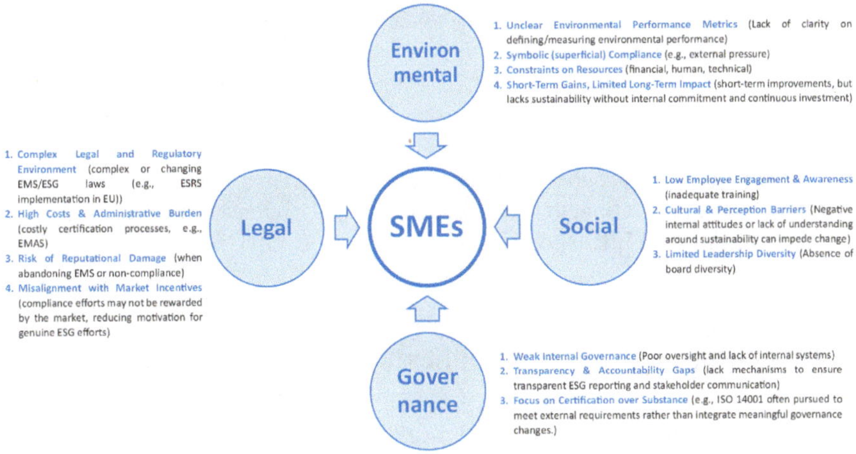

Fig. 4 Environmental, social, governance, and legal risks that SMEs face

yet remains low due to resource constraints, industry perceptions, and competing priorities [74].

Policymakers, academics, and practitioners in ESG should inspect the aforementioned parameters to blueprint the proper policies and strategies. For example, research on EMS implementation in SMEs highlights challenges like knowledge gaps, training needs, initial costs, and emphasizes medium-term benefits including sustainability, cost savings, reputation, and competitive advantage [75]. Moreover, a study identified several internal barriers like resource shortages, low employee engagement, and cultural issues, alongside external challenges such as limited government support and high certification costs, as key obstacles to EMS adoption [76].

An interesting yet ambiguous initiative is the adoption of ESRS in Europe, and especially how these standards would be applied to the SME context. The ESRS framework integrates ESG factors into corporate strategy, increasing operational complexity and requiring cross-departmental coordination, enhanced stakeholder engagement, greater director accountability, and stronger governance to boost transparency, trust, and sustainable performance [25]. The ESRS-E5 standard supports the shift to circular business models by using resource flow data and aligns with key EU policies like the Circular Economy Action Plan (CEAP) and Waste Framework Directive (WFD) [77, 78], aiming to enhance sustainability through resource efficiency, waste reduction, and environmental protection [79].

Additionally, ESG factors such as board gender diversity positively influence environmental outcomes, especially in reducing emissions and improving resource efficiency and circular economy performance [24]. Therefore, the adoption of ESRS in Europe presents both opportunity and complexity, particularly for SMEs, by embedding ESG into corporate strategy and governance to drive transparency, accountability, and sustainable performance.

To summarize, SMEs display varying levels of environmental commitment, requiring tailored intervention strategies for each category of ESG framework [5]. Thus, in order to boost SMEs' adoption of sustainability tools by SMEs, governments should first use soft policies like education and partnerships before imposing strict regulations [80]. In short, SMEs face significant barriers to implementing sustainability tools, highlighting the need for tailored, supportive policies, starting with education and collaboration, to bridge knowledge gaps, reduce resource constraints, and foster meaningful ESG adoption.

5 Conclusions and Policy Implications

This review presents an integrated framework linking EMS, such as ISO 14001 and EMAS, with CSR, ESG, and the emerging ESRS, focusing on SMEs. The review collects and analyses 47 peer-reviewed studies, which highlight how combining traditional EMS with evolving ESG and ESRS standards can enhance SME resilience to environmental risks, support regulatory compliance, and align with SDGs. The study maps these interconnections and offers actionable insights and policy guidance for SMEs navigating the sustainability transition.

5.1 Theoretical Implications

The adoption of ESG frameworks and related standards (e.g., ESRS, GRI) represents a significant evolution in corporate theory, merging traditional economic rationalism with stakeholder and institutional theory. ESG frameworks emphasize that firms are no longer evaluated solely by financial metrics, but also on their broader social and environmental impact. This shift challenges the conventional shareholder-centric model and supports a more inclusive, long-term view of corporate responsibility and value creation.

Theoretically, this also aligns with CSR theory, where organizations must align with societal values to maintain social acceptance to operate. The institutionalization of ESG standards, such as the ESRS, demonstrates how regulatory bodies are formalizing previously voluntary disclosures, thus reshaping the theoretical boundaries of corporate governance, transparency, and accountability. These developments demand that scholars reevaluate performance measurement systems, decision-making frameworks, and governance structures within the context of sustainable development (e.g., SDGs).

5.2 Practical Implications

In practice, the application of ESG frameworks presents both a roadmap and a challenge for organizations, particularly SMEs, by requiring structured approaches to managing ESG risks and opportunities. Implementing ESG standards, such as the ESRS, faces substantial changes in data collection, reporting infrastructure, stakeholder engagement, and internal governance.

For SMEs, this means overcoming resource and capability constraints through targeted support, capacity building, and tailored guidance. Essentially, firms should integrate ESG considerations into strategic planning, supply chain management, and employee training, while directors and managers are expected to demonstrate an increased sense of ecological driven leadership, accountability, and foresight.

Additionally, ESG compliance is becoming a determinant of market access, investment readiness, and consumer trust, thereby linking sustainability performance directly with the company's competitiveness. Therefore, practical adoption of ESG standards is no longer a reputational exercise, but a core element of strategic risk management and long-term value creation in a rapidly evolving regulatory and stakeholder landscape.

5.3 Limitations, Future Directions, and Concluding Remarks

One of the most critical limitations of the present review is the limited availability of empirical data, especially concerning SMEs, which are often underrepresented in the literature, especially when addressing the environmental risks. Moreover, the heterogeneity in how organizations adopt and apply environmental or other standards across different sectors and regions further complicates generalization. Additionally, the evolving nature of frameworks like the ESRS introduces uncertainty, as regulatory interpretations and compliance practices are still developing. The overlap and conceptual ambiguity between ESG reporting standards and EMS also pose challenges in drawing clear analytical boundaries. Finally, inconsistent performance metrics, publication bias towards positive outcomes, and the lack of longitudinal studies limit the ability to assess the long-term, real-world effectiveness of these frameworks in promoting genuine sustainability and governance improvements.

To recapitulate, SMEs face persistent challenges in adopting sustainability management tools, particularly EMS, due to limited resources, knowledge gaps, cultural resistance, and high implementation costs. Despite increasing regulatory demands and stakeholder pressure, the actual uptake remains low, for example, the ESRS present a valuable yet complex framework for embedding ESG principles into SME strategies, enhancing transparency, accountability, and governance. Moreover, there is the opportunity to assimilate international standards (e.g., GRI) to further strengthen the ESRS framework.

In this sense, future policy directions should focus on applicability issues, because owners or managers in small businesses do not have enough time or resources to comply with complex legislation. Hence, policymakers in order to effectively align SMEs with the SDGs, must prioritize soft policy tools, such as targeted education, training, financial incentives, and collaborative partnerships to overcome adoption or implementation barriers. Aligning SDGs with ESG criteria offers organizations a structured path to sustainability. This integration strengthens governance, supports environmental preservation, enhances social impact, and promotes ethical economic growth. Together, SDGs and ESG frameworks help businesses operate transparently, responsibly, and sustainably. More specifically, policymakers should target especially the SDGs related to climate action (SDG 13), responsible consumption (SDG12), inclusive growth (SDG8), and institutions (SDG16). In a nutshell, by building internal capacity and creating an enabling environment, SMEs can meaningfully integrate ESG standards, driving long-term resilience to environmental risks, boost innovation, and achieve measurable progress towards the SDGs.

Statement Equal author contribution.

Compliance with Ethical Standards

Conflict of Interest The authors declare no conflict of interest.

Research Involving Human Participants and/or Animals Not applicable.

Informed Consent Not applicable.

Disclaimer The views expressed herein are those of the author(s) and do not necessarily reflect the views of the United Nations Industrial Development Organization (as provided for in Administrative Circular UNIDO/DA/PS/AC.69 of 17 December 1990). This document has been produced without formal United Nations editing. The designations employed and the presentation of the material in this document do not imply the expression of any opinion whatsoever on the part of the Secretariat of the United Nations Industrial Development Organization (UNIDO) concerning the legal status of any country, territory, city, or area or its authorities, or concerning the delimitation of its frontiers or boundaries, or its economic system or degree of development. Designations, such as ‘developed’, ‘industrialized’, and ‘developing’, are intended for statistical convenience and do not necessarily express a judgement about the stage reached by a particular country or area in the development process. Mentioning firm names or commercial products does not constitute an endorsement by UNIDO. The opinions, statistical data, and estimates contained herein are the responsibility of the author(s) and should not necessarily be considered as reflecting the views of bearing the endorsement of UNIDO.

Appendix A

See Table 5.

Table 4 Brief description of GRI standards in the present study. *Source* [30]

GRI	Scope	Description of GRI standards on reporting...
101	Biodiversity	Biodiversity-related impacts, aligned with global biodiversity goals (e.g., Targets in the Kunming-Montreal Global Biodiversity Framework), covering impact locations, biodiversity loss drivers, and ecosystem service changes
201	Economic performance	Economic value distribution, climate-related financial risks and opportunities, retirement plan obligations, and government financial assistance received
202	Market presence	Wage equality by gender is relative to local minimum wage and the percentage of senior management hired from the local community
205	Anti-corruption	Corruption risk assessments, anti-corruption training and communication, and reported incidents of corruption along with actions taken
207	Tax	Tax management approach, including governance, risk control, stakeholder engagement, and country-by-country tax reporting
301	Materials	Total materials used, the proportion of recycled inputs, and the number of reclaimed products and packaging materials
302	Energy	Inclusion of energy consumption inside and outside the organization, energy intensity, and efforts to reduce energy use both overall and in products and services
303	Water and effluents	Water management practices, including interactions with shared water resources, management of water discharge impacts, and data on water withdrawal, discharge, and consumption
304	Biodiversity	Operational sites near protected or high-biodiversity areas, significant impacts on biodiversity by having in mind IUCN Red List, efforts to protect or restore habitats, and species of conservation concern affected by operations
305	Emissions	Emissions management, including direct (Scope 1) and indirect (Scopes 2 and 3) GHG emissions, emission intensity, reduction efforts, use of offsets, and emissions of ozone-depleting substances and other significant air pollutants
306	Effluent and waste	Waste management practices (waste generation), impacts, and the amounts of waste diverted from disposal versus waste sent to disposal
401	Employment	New hires and employee turnover, benefits for full-time versus temporary or part-time employees, and parental leave policies
402	Labor/management eelations	Minimum notice periods organizations must provide to employees regarding operational changes

(continued)

Table 4 (continued)

GRI	Scope	Description of GRI standards on reporting…
403	Occupational health and safety	Occupational health and safety management systems, including risk assessment, worker participation and training, health promotion, coverage of workers, and incidents of work-related injuries and illnesses
404	Training and education	Average employee training hours, programs for skill development and career transitions, and the percentage of employees receiving performance reviews
405	Diversity and equal opportunity	Diversity within governance bodies and employees, as well as the gender pay ratio between women and men
406	Non-discrimination	Reporting on the total number of discrimination incidents during the reporting period, remediation actions taken, and outcomes. These incidents should include any form of discrimination (e.g., racial, social, or religious) or other relevant forms affecting internal and external stakeholders
407	Freedom of association and collective bargaining	Operations and suppliers where the right to freedom of association and collective bargaining might be at risk
408	Child labor	Operations and suppliers that are at significant risk of child labour incidents
409	Forced or compulsory labor	Identification of operations and suppliers at significant risk of forced or compulsory labour incidents
411	Rights of indigenous peoples	Disclosure on any incidents of violations involving the rights of Indigenous peoples, as recognized by the relevant UN Declaration and ILO Convention 169
413	Local communities	Operations involving local community engagement and development programmes
414	Supplier social assessment	New suppliers screened using social criteria and any negative social impacts identified in the supply chain, along with actions taken to address them
416	Customer health and safety	Reporting on how product and service categories are assessed for health and safety impacts, and any incidents of non-compliance related to those impacts
417	Marketing and labelling	Requirements for product and service information and labelling, along with any incidents of non-compliance related to labelling and marketing communications
418	Customer privacy	Substantiated complaints related to breaches of customer privacy and any losses of customer data

Appendix B

See Table 6.

Table 5 Brief description of ISO standards in the present study. *Source* [81]

ISO	Scope	Description
10002	Quality management, customer satisfaction	This standard focuses on establishing an effective complaints-handling process to enhance customer satisfaction through open feedback, strong top management support, and a user-friendly system
14001	EMS	Provides a framework for organizations to develop and improve an EMS, aiming to reduce their environmental impact, comply with legal requirements, and achieve environmental goals through ongoing improvement
14046	Water Footprint	It provides guidelines for water footprint assessment based on LCA, focusing on water-related impacts. It includes reporting but excludes external communication like labels or declarations
14040, 14044	Life cycle assessment	These standards provide the principles, framework, requirements, and guidelines for conducting life cycle assessment (LCA) and life cycle inventory (LCI) and life cycle impact assessment (LCIA) studies
14067	Greenhouse gases: Carbon footprint of products	This standard outlines principles, requirements, and guidelines for quantifying and reporting the carbon footprint of a product based on LCA, focusing solely on climate change as the impact category, excluding carbon offsetting, communication of results, and any social, economic, or other environmental impacts
14075	Social Life Cycle Assessment	This specific standard addresses the SDGs and there is provision of goal and scope definition, inventory analysis, impact assessment, interpretation, and reporting of the S-LCA of a product
20400	Sustainable procurement	This standard provides guidance to organizations, independent of their activity or size, on integrating sustainability within procurement, as described in ISO 26000. It is intended for stakeholders involved in, or impacted by, procurement decisions and processes
22000	Food safety management	This standard helps organizations identify and manage food safety hazards, applicable to all producers; it enhances trust and safety across the global food supply chain
26000	Social responsibility	This standard offers guidance for organizations that value respect for society and the environment as key to success. Beyond being ethical, it helps assess an organization's commitment to sustainability and overall performance
30414	Human resource management	This standard provides guidelines for transparent human capital reporting to support workforce sustainability across all organizations. It covers key areas like diversity, leadership, health, and skills, with tailored recommendations for SMEs

(continued)

Table 5 (continued)

ISO	Scope	Description
37000	Governance of organizations	This standard provides universal guidance on governance principles and practices to help organizations' governing bodies fulfil their responsibilities and achieve their purpose
37301	Compliance management systems	Guides organizations in establishing and improving effective compliance management systems. It helps ensure legal and ethical adherence, reduces risks, fosters integrity, and enhances governance and stakeholder trust
45001	Occupational health and safety management systems	Provides a framework for organizations to manage occupational health and safety risks systematically. It emphasizes leadership, worker participation, hazard identification, legal compliance, and continual improvement using the PDCA approach
50001	Energy management	This standard helps organizations in any sector improve energy efficiency by establishing an energy management system
59004	Circular Economy	Guidance for all types of organizations to transition to a circular economy. It outlines key principles and actions to support sustainable resource use and reduce environmental impact

Table 6 The aim of the studies in the present review

Aim of the study	Country & sample	Categories	Refs.
Examination of the environmental strategies spectrum adopted by SMEs, ranging from basic compliance to proactive environmental leadership	Spain, 108 SMEs in the automotive repair sector	CSR-ESG, SMEs	[58]
Observation of the barriers, drivers, and opportunities for EMS adoption in SMEs, focusing on voluntary standards like EMAS and ISO 14001 for improving environmental performance	European Union, 33 studies	EMAS, ISO 14001, SMEs	[20]
Assess of the gap between positive environmental attitudes of SME owner/managers and their limited implementation of eco-friendly practices	Australia, 166 questionnaires	CSR-ESG, SMEs	[19]

(continued)

Table 6 (continued)

Aim of the study	Country & sample	Categories	Refs.
Identification of the key factors that influence the implementation of sustainable innovation in SMEs, particularly within the rubber and plastics industry, with a focus on eco-efficiency and cost reduction	Netherlands, 26 companies (most SMEs within the rubber and plastics industry (RPI))	CSR-ESG, SMEs	[54]
To investigate how business performance and regulatory pressures influence SME environmental behaviour, and to compare engagement and perceptions across SMEs	UK, 31 manufacturing SMEs	CSR-ESG, SMEs	[57]
Analysis of how SMEs adopt environmental practices and how perceptions of environmental pressures and benefits differ between SMEs	UK, 102 SMEs in the extractive, metals and engineering, chemicals, printing and paper and timber sectors	CSR-ESG, SMEs	[22]
Exploration on how small businesses in the East of England engage with pro-environmental actions, particularly in response to climate change	UK, 9 in-depth interviews of SME managers	CSR-ESG, SMEs	[23]
To examine the motivations and justifications for why SMEs should adopt sustainability management tools	N/A, 112 studies	SMEs	[73]
The need for a comprehensive mix of interventions to effectively enhance SME engagement and environmental performance	N/A, 47 studies	CSR-ESG, SMEs	[5]
Investigation on how awareness and attitudes towards environmental issues influence the actual adoption of environmental practices in SMEs	New Zealand, 148 questionnaires	CSR-ESG, SMEs	[6]

(continued)

Table 6 (continued)

Aim of the study	Country & sample	Categories	Refs.
Evaluation of the ongoing debate about the effectiveness of ISO 14001 and EMAS in delivering measurable environmental performance improvements in SMEs	Italy, 229 energy intensive plants	EMAS, ISO 14001	[9]
Identification on the managerial and organizational traits that can influence the likelihood of SMEs adopting sustainability management tools	Germany, 176 SME managers	CSR-ESG, SMEs	[80]
To investigate the internal, external, and moderating factors that affect EMS implementation in a select group of SMEs	Australia, 1,000 owner-managers of SMEs	SMEs	[74]
Reinterpretation of CSR theories by integrating the ethic of care and feminist insights, making them more applicable to the context of small businesses	N/A	CSR-ESG, SMEs	[14]
Critique on the applicability of large-firm benchmarks to SMEs and advocate for more appropriate measures of sustainable business practices tailored to small enterprises	New Zealand, 50 small-business owners	CSR-ESG, SMEs	[62]
Analysis of the current theoretical landscape of CSR in SMEs by identifying its key antecedents and consequences, providing a foundation for future empirical research	N/A, 28 articles	CSR-ESG, SMEs	[53]
Observation on how external pressures, internal factors, and entrepreneurs' environmental attitudes influence proactive environmental strategies in small and micro firms, and their effect on investment and performance	Italy, 355 questionnaires	CSR-ESG, SMEs	[59]

(continued)

Table 6 (continued)

Aim of the study	Country & sample	Categories	Refs.
Assessment of the current state of EMS and green manufacturing practices among businesses in through survey analysis	Hong Kong, 11 medium-sized companies	ISO 14001	[50]
Exploration of the unresolved relationship between environmental and financial performances, focusing on potential moderating factors influencing this link	China, 39 industries	CSR-ESG	[60]
Monitoring of environmental performance indicators for ongoing improvement and management in ISO 14001 certified companies	Brazil, 62 industrial companies (mainly large)	ISO 14001	[37]
Explanation on whether pressure from secondary stakeholders can encourage firms to internalize third-party certifications rather than adopt them symbolically	Australia and New Zealand, 328 firms	CSR-ESG, ISO 14001	[16]
To explore why SMEs adopt EMS, emphasizing its role in sustainably monitoring and managing waste production	N/A	EMAS, ISO 14001, SMEs	[8]
Examination of SME owner(s)'/managers' attitudes towards environmental issues, regulations, voluntary standards, and inter-firm collaboration	New Zealand, 740 owner-managers	CSR-ESG, SMEs	[55]
To develop and test a decision-making model showing how SME owners/managers respond to sustainability pressures, balancing perceived benefits and barriers	Italy, 348 questionnaires	CSR-ESG, SMEs	[56]

(continued)

Table 6 (continued)

Aim of the study	Country & sample	Categories	Refs.
Identification of EMS' components, such as certification, fault analysis, responsibility, and impact mitigation that can be integrated into SME production processes	N/A, 27 articles	SMEs	[75]
Evaluation of the impact from ISO-compliant EMS on environmental performance in Vietnam's cement industry by comparing pre/post-certification and certified vs. non-certified plants	Vietnam, 15 ISO 14001 certified and non-certified cement plants	ISO 14001	[48]
To trace the evolution of motivators, environmental initiatives, and their perceived effects on SMEs' competitive advantage	Denmark, manufacturing SMEs	CSR-ESG, SMEs	[61]
Observing the critical importance of accurate environmental impact evaluation methodologies within ISO 14001 EMS for global market access	Brazil, 20 simple case studies	ISO 14001, SMEs	[7]
Examination of the ongoing debate around the effectiveness and managerial motivations behind adopting ISO 14001 versus EMAS in SMEs	In all EU member states, 224 questionnaires (mainly to SMEs)	EMAS	[40]
Analysis of the growing conceptual and terminological fragmentation surrounding CSR in SMEs through a systematic literature review	N/A, 118 studies	CSR-ESG, SMEs	[15]
Comparison between the perspectives of small and large companies in Brazil on which ISO 14001 EMS requirements best align with small business needs	Brazil, 83 questionnaires for small and 24 for large businesses	ISO 14001, SMEs	[43]

(continued)

Table 6 (continued)

Aim of the study	Country & sample	Categories	Refs.
Monitoring of the challenges and benefits experienced by Italian ISO 14001-certified organizations over time through national survey data	Italy, ISO 14001 certified organizations	ISO 14001	[44]
Observation on how adopting voluntary environmental certifications influences corporate environmental performance	N/A, 53 studies	EMAS, ISO 14001	[10]
Comparison between EMS implementation experiences of manufacturing SMEs using the Eco-Lighthouse system versus ISO 14001	Norway, 9 manufacturing SMEs that are ISO 14001 or Eco-Lighthouse certified	ISO 14001, SMEs	[49]
A study on the role of waste minimization as a critical environmental performance indicator under ISO 14001 certification in Indian SMEs	India, 63 firms	ISO 14001, SMEs	[47]
Exploration on why firms adopt both EMAS and ISO 14001 and whether the two standards act as complements or substitutes	Germany, 21 interviews with industrial and institutional representatives	EMAS, ISO 14001	[41]
Overview of the EMS adoption in SMEs through a management accounting and control perspective, focusing on drivers, implementation, and outcomes	N/A, 23 articles	EMAS, ISO 14001	[12]
Testing whether simple formal management tools like target-setting improve environmental impacts in SMEs	12 European countries, 5,205 SMEs	ISO 14001, SMEs	[46]
Observation of how contextual factors influence ISO 14001 adoption and its effect on environmental performance in SMEs	N/A, 23 semi-structured interviews	ISO 14001, SMEs	[42]
To highlight overlooked mediators that may explain ISO 14001 indirect positive impact on ecological outcomes	12 European countries, 3,633 SMEs	ISO 14001, SMEs	[45]

(continued)

Table 6 (continued)

Aim of the study	Country & sample	Categories	Refs.
To identify internal and external triggers and barriers influencing SME adoption of environmental management practices using institutional, resource-based, and contingency theory frameworks	Global, 54 articles (mainly European (n=32), and also from Asia (n=13), Oceania (n=5), and Africa (n=1))	SMEs	[76]
Investigation of the declining EMAS participation by analysing trends in other certification schemes and exploring why organizations abandon or avoid EMAS in favour of ISO 14001	Germany, Italy, Estonia, Bulgaria, Spain, 17 interviews (7 from SMEs)	EMAS, ISO 14001	[13]
Assessment of how mandatory gender quotas on corporate boards impact environmental performance indicators among Italian listed companies	Italy, 40 firms	CSR-ESG	[17]
The aim is to find how SMEs apply sustainability control systems through ISO 14001-certified EMS	Northern Europe, 18 SMEs, and seven auditors	ISO 14001, SMEs	[51]
Examination of how corporate governance and environmental behaviour interact under the UK's mandatory carbon reporting regime, focusing on board composition and carbon emission outcomes	UK-listed firms, 382 firms	CSR-ESG	[63]
Analysis of the progression of non-financial reporting in Europe in light of regulatory shifts such as the CSRD and ESRS adoption	N/A	CSR-ESG, ESRS	[25]
Evaluation of how firm- and country-level factors influence the value and effectiveness of non-carbon-related environmental goals under the EU Taxonomy and ESRS frameworks	Global, 80 studies	CSR-ESG, ESRS	[24]

References

1. Halkos, G.E., Aslanidis, P.-S.C.: Sustainable energy development in an era of geopolitical multi-crisis. Applying productivity indices within institutional framework. Resour. Policy. **85**, 103854 (2023). https://doi.org/10.1016/j.resourpol.2023.103854
2. Tooze, A.: Welcome to the world of the polycrisis, https://www.ft.com/content/498398e7-11b1-494b-9cd3-6d669dc3de33, last accessed 2023/03/10
3. Halkos, G.E., Moll de Alba, J., Aslanidis, P.-S.C.: Examining energy and carbon intensity of the european fashion industry: a hybrid DEA analysis for circular economy. Circ. Econ. Sustain. (2024). https://doi.org/10.1007/s43615-024-00488-x
4. Halkos, G.E., Moll de Alba, J., Aslanidis, P.-S.C., Bampatsou, C.: Monitoring eco-efficiency and its convergence towards sustainability in the european rubber and plastics industry through circular economy transition. Sustainability. **17**, 1272 (2025). https://doi.org/10.3390/su17031272
5. Parker, C.M., Redmond, J., Simpson, M.: A review of interventions to encourage SMEs to make environmental improvements. Environ. Plan. C Gov. Policy. **27**, 279–301 (2009). https://doi.org/10.1068/c0859b
6. Cassells, S., Lewis, K.: SMEs and environmental responsibility: do actions reflect attitudes? Corp. Soc. Responsib. Environ. Manag. **18**, 186–199 (2011). https://doi.org/10.1002/csr.269
7. Seiffert, M.E.B.: Environmental impact evaluation using a cooperative model for implementing EMS (ISO 14001) in small and medium-sized enterprises. J. Clean. Prod. **16**, 1447–1461 (2008). https://doi.org/10.1016/j.jclepro.2007.10.001
8. Zorpas, A.: Environmental management systems as sustainable tools in the way of life for the SMEs and VSMEs. Bioresour. Technol. **101**, 1544–1557 (2010). https://doi.org/10.1016/j.biortech.2009.10.022
9. Testa, F., Rizzi, F., Daddi, T., Gusmerotti, N.M., Frey, M., Iraldo, F.: EMAS and ISO 14001: the differences in effectively improving environmental performance. J. Clean. Prod. **68**, 165–173 (2014). https://doi.org/10.1016/j.jclepro.2013.12.061
10. Erauskin-Tolosa, A., Zubeltzu-Jaka, E., Heras-Saizarbitoria, I., Boiral, O.: ISO 14001, EMAS and environmental performance: A meta-analysis. Bus. Strateg. Environ. **29**, 1145–1159 (2020). https://doi.org/10.1002/bse.2422
11. Morse-Jones, S., Bateman, I.J., Kontoleon, A., Ferrini, S., Burgess, N.D., Turner, R.K.: Stated preferences for tropical wildlife conservation amongst distant beneficiaries: Charisma, endemism, scope and substitution effects. Ecol. Econ. **78**, 9–18 (2012). https://doi.org/10.1016/j.ecolecon.2011.11.002
12. Johnstone, L.: A systematic analysis of environmental management systems in SMEs: Possible research directions from a management accounting and control stance. J. Clean. Prod. **244**, 118802 (2020). https://doi.org/10.1016/j.jclepro.2019.118802
13. Daddi, T., De Giacomo, M.R., Frey, M., Iraldo, F.: Analysing the causes of environmental management and audit scheme (EMAS) decrease in Europe. J. Environ. Plan. Manag. **61**, 2358–2377 (2018). https://doi.org/10.1080/09640568.2017.1395316
14. Spence, L.J.: Small Business Social Responsibility. Bus. Soc. **55**, 23–55 (2016). https://doi.org/10.1177/0007650314523256
15. Ortiz-Avram, D., Domnanovich, J., Kronenberg, C., Scholz, M.: Exploring the integration of corporate social responsibility into the strategies of small- and medium-sized enterprises: A systematic literature review. J. Clean. Prod. **201**, 254–271 (2018). https://doi.org/10.1016/j.jclepro.2018.08.011
16. Castka, P., Prajogo, D.: The effect of pressure from secondary stakeholders on the internalization of ISO 14001. J. Clean. Prod. **47**, 245–252 (2013). https://doi.org/10.1016/j.jclepro.2012.12.034
17. Marchini, P.L., Tibiletti, V., Mazza, T., Gabrielli, G.: Gender quotas and the environment: Environmental performance and enforcement. Corp. Soc. Responsib. Environ. Manag. **29**, 256–272 (2022). https://doi.org/10.1002/csr.2200

18. Schulze Brock, P., Katsinis, A., Lagüera González, J., Di Bella, L., Odenthal, L., Hell, M., Lozar, B., Secades Casino, B.: Annual Report on European SMEs 2024/2025. SME performance review. Publ. Off. Eur, Union (2025)
19. Gadenne, D.L., Kennedy, J., McKeiver, C.: An Empirical Study of Environmental Awareness and Practices in SMEs. J. Bus. Ethics **84**, 45–63 (2009). https://doi.org/10.1007/s10551-008-9672-9
20. Hillary, R.: Environmental management systems and the smaller enterprise. J. Clean. Prod. **12**, 561–569 (2004). https://doi.org/10.1016/j.jclepro.2003.08.006
21. OECD: Financing SMEs and Entrepreneurs 2024: An OECD Scoreboard. OECD (2024). https://doi.org/10.1787/fa521246-en
22. Brammer, S., Hoejmose, S., Marchant, K.: Environmental management in SMEs in the UK: practices, pressures and perceived benefits. Bus. Strateg. Environ. **21**, 423–434 (2012). https://doi.org/10.1002/bse.717
23. Williams, S., Schaefer, A.: Small and medium-sized enterprises and sustainability: managers' values and engagement with environmental and climate change issues. Bus. Strateg. Environ. **22**, 173–186 (2013). https://doi.org/10.1002/bse.1740
24. Velte, P.: A literature review concerning the non-carbon-related environmental goals of the EU Taxonomy Regulation and the European Sustainability Reporting Standards (ESRS). J. Glob. Responsib. (2024). https://doi.org/10.1108/JGR-02-2024-0032
25. Fornasari, T., Traversi, M.: The impact of the CSRD and the ESRS on non-financial disclosure. Symphonya. Emerg. Issues Manag. 117–133 (2024). https://doi.org/10.4468/2024.1.07fornasari.traversi
26. Halkos, G.E., Aslanidis, P.-S.C.: integrating environmental economics into ESG reporting in europe: the environmental topic-specific cross-sector european sustainability reporting standards (ESRS). Munich Pers. RePEc Arch. MPRA Pap. No. 124991. (2025)
27. European Parliament: Directive (EU) 2022/2464 on corporate sustainability reporting, https://eur-lex.europa.eu/legal-content/EN/TXT/PDF/?uri=CELEX:32022L2464, last accessed 2023/09/15
28. Halkos, G.E., Aslanidis, P.-S.C.: New circular economy perspectives on measuring sustainable waste management productivity. Econ. Anal. Policy. **77**, 764–779 (2023). https://doi.org/10.1016/j.eap.2023.01.001
29. Von-der-Leyen, U.: A Union that strives for more: My agenda for Europe, https://commission.europa.eu/system/files/2020-04/political-guidelines-next-commission_en_0.pdf, last accessed 2022/11/19
30. GRI: Consolidated Set of the GRI Standards, https://www.globalreporting.org/how-to-use-the-gri-standards/gri-standards-english-language/, last accessed 2025/06/10
31. IFRS: International Sustainability Standards Board, https://www.ifrs.org/groups/international-sustainability-standards-board/, last accessed 2025/06/10
32. TCFD: Recommendations of the Task Force on Climate-related Financial Disclosures, https://assets.bbhub.io/company/sites/60/2021/10/FINAL-2017-TCFD-Report.pdf, last accessed 2023/09/15
33. UN: The 17 goals., https://sdgs.un.org/goals, last accessed 2023/11/10
34. Franco, C. De, Nicolle, J., Tran, L.-A.: Sustainable Investing: ESG versus SDG. J. Impact ESG Invest. jesg.2021.1.019 (2021). https://doi.org/10.3905/jesg.2021.1.019
35. Kumar Soni, T.: Demystifying the relationship between ESG and SDG performance: Study of emerging economies. Invest. Manag. Financ. Innov. **20**, 1–12 (2023). https://doi.org/10.21511/imfi.20(3).2023.01
36. Halkos, G.E., Aslanidis, P.-S.C.: Reviewing environmental aspects under the scope of ESG. Munich Pers. RePEc Arch. MPRA Pap. No. 120298. (2024)
37. Campos, L.M.S., de Melo Heizen, D.A., Verdinelli, M.A., Cauchick Miguel, P.A.: Environmental performance indicators: a study on ISO 14001 certified companies. J. Clean. Prod. **99**, 286–296 (2015). https://doi.org/10.1016/j.jclepro.2015.03.019

38. ISO: ISO Survey of certifications to management system standards - Full results, https://www.iso.org/committee/54998.html?t=KomURwikWDLiuB1P1c7SjLMLEAgXOA7emZHKGWyn8f3KQUTU3m287NxnpA3DIuxm&view=documents#section-isodocuments-top, last accessed 2025/07/01
39. EMAS: EMAS: Facts and Figures, https://green-forum.ec.europa.eu/emas/about-emas/emas-statistics-and-graphs_en, last accessed 2025/07/01
40. Testa, F., Iraldo, F., Daddi, T.: The effectiveness of EMAS as a management tool: a key role for the internalization of environmental practices. Organ. Environ. **31**, 48–69 (2018). https://doi.org/10.1177/1086026616687609
41. Neugebauer, F.: EMAS and ISO 14001 in the German industry—complements or substitutes? J. Clean. Prod. **37**, 249–256 (2012). https://doi.org/10.1016/j.jclepro.2012.07.021
42. Johnstone, L., Hallberg, P.: ISO 14001 adoption and environmental performance in small to medium sized enterprises. J. Environ. Manage. **266**, 110592 (2020). https://doi.org/10.1016/j.jenvman.2020.110592
43. Campos, L.M.S.: Environmental management systems (EMS) for small companies: a study in Southern Brazil. J. Clean. Prod. **32**, 141–148 (2012). https://doi.org/10.1016/j.jclepro.2012.03.029
44. Mazzi, A., Toniolo, S., Mason, M., Aguiari, F., Scipioni, A.: What are the benefits and difficulties in adopting an environmental management system? The opinion of Italian organizations. J. Clean. Prod. **139**, 873–885 (2016). https://doi.org/10.1016/j.jclepro.2016.08.053
45. Graafland, J.J.: Ecological impacts of the ISO14001 certification of small and medium sized enterprises in Europe and the mediating role of networks. J. Clean. Prod. **174**, 273–282 (2018). https://doi.org/10.1016/j.jclepro.2017.10.322
46. Graafland, J., Smid, H.: Environmental Impacts of SMEs and the Effects of Formal Management Tools: Evidence from EU's Largest Survey. Corp. Soc. Responsib. Environ. Manag. **23**, 297–307 (2016). https://doi.org/10.1002/csr.1376
47. Singh, M., Brueckner, M., Padhy, P.K.: Environmental management system ISO 14001: effective waste minimisation in small and medium enterprises in India. J. Clean. Prod. **102**, 285–301 (2015). https://doi.org/10.1016/j.jclepro.2015.04.028
48. Nguyen, Q.A., Hens, L.: Environmental performance of the cement industry in Vietnam: the influence of ISO 14001 certification. J. Clean. Prod. **96**, 362–378 (2015). https://doi.org/10.1016/j.jclepro.2013.09.032
49. Granly, B.M., Welo, T.: EMS and sustainability: experiences with ISO 14001 and Eco-Lighthouse in Norwegian metal processing SMEs. J. Clean. Prod. **64**, 194–204 (2014). https://doi.org/10.1016/j.jclepro.2013.08.007
50. Hui, I., Chan, A.H., Pun, K.: A study of the environmental management system implementation practices. J. Clean. Prod. **9**, 269–276 (2001). https://doi.org/10.1016/S0959-6526(00)00061-5
51. Johnstone, L.: Facilitating sustainability control in SMEs through the implementation of an environmental management system. J. Manag. Control. **32**, 559–605 (2021). https://doi.org/10.1007/s00187-021-00329-0
52. Villalobos, C., Chávez, C., Uribe, A.: Energy poverty measures and the identification of the energy poor: A comparison between the utilitarian and capability-based approaches in Chile. Energy Policy. **152**, (2021). https://doi.org/10.1016/j.enpol.2021.112146
53. Vázquez-Carrasco, R., López-Pérez, M.E.: Small & medium-sized enterprises and corporate social responsibility: a systematic review of the literature. Qual. Quant. **47**, 3205–3218 (2013). https://doi.org/10.1007/s11135-012-9713-4
54. Bos-Brouwers, H.E.J.: Corporate sustainability and innovation in SMEs: Evidence of themes and activities in practice. Bus. Strateg. Environ. **19**, 417–435 (2010). https://doi.org/10.1002/bse.652
55. Lewis, K.V., Cassells, S., Roxas, H.: SMEs and the potential for a collaborative path to environmental responsibility. Bus. Strateg. Environ. **24**, 750–764 (2015). https://doi.org/10.1002/bse.1843
56. Cantele, S., Zardini, A.: What drives small and medium enterprises towards sustainability? Role of interactions between pressures, barriers, and benefits. Corp. Soc. Responsib. Environ. Manag. **27**, 126–136 (2020). https://doi.org/10.1002/csr.1778

57. Williamson, D., Lynch-Wood, G., Ramsay, J.: Drivers of environmental behaviour in manufacturing SMEs and the implications for CSR. J. Bus. Ethics **67**, 317–330 (2006). https://doi.org/10.1007/s10551-006-9187-1
58. Aragón-Correa, J.A., Hurtado-Torres, N., Sharma, S., García-Morales, V.J.: Environmental strategy and performance in small firms: A resource-based perspective. J. Environ. Manage. **86**, 88–103 (2008). https://doi.org/10.1016/j.jenvman.2006.11.022
59. Testa, F., Gusmerottia, N.M., Corsini, F., Passetti, E., Iraldo, F.: Factors affecting environmental management by small and micro firms: the importance of entrepreneurs' attitudes and environmental investment. Corp. Soc. Responsib. Environ. Manag. **23**, 373–385 (2016). https://doi.org/10.1002/csr.1382
60. Qi, G.Y., Zeng, S.X., Shi, J.J., Meng, X.H., Lin, H., Yang, Q.X.: Revisiting the relationship between environmental and financial performance in Chinese industry. J. Environ. Manage. **145**, 349–356 (2014). https://doi.org/10.1016/j.jenvman.2014.07.010
61. Reyes-Rodríguez, J.F., Ulhøi, J.P., Madsen, H.: Corporate environmental sustainability in danish SMEs: a longitudinal study of motivators, initiatives, and strategic effects. Corp. Soc. Responsib. Environ. Manag. **23**, 193–212 (2016). https://doi.org/10.1002/csr.1359
62. Battisti, M., Perry, M.: Walking the talk? Environmental responsibility from the perspective of small-business owners. Corp. Soc. Responsib. Environ. Manag. **18**, 172–185 (2011). https://doi.org/10.1002/csr.266
63. Muktadir-Al-Mukit, D., Bhaiyat, F.H.: Impact of corporate governance diversity on carbon emission under environmental policy via the mandatory nonfinancial reporting regulation. Bus. Strateg. Environ. **33**, 1397–1417 (2024). https://doi.org/10.1002/bse.3555
64. European Commission: Questions and answers on simplification omnibus I and II, https://ec.europa.eu/commission/presscorner/detail/en/qanda_25_615, last accessed 2025/07/01
65. European Union: The future of European competitiveness. Part A: A competitiveness strategy for Europe, https://commission.europa.eu/document/download/97e481fd-2dc3-412d-be4c-f152a8232961_en?filename=The future of European competitiveness _ A competitiveness strategy for Europe.pdf, last accessed 2025/07/01
66. European Union: The future of European competitiveness. Part B: In-depth analysis and recommendations, https://commission.europa.eu/document/download/ec1409c1-d4b4-4882-8bdd-3519f86bbb92_en, last accessed 2025/07/01
67. The Swiss Confederation: Federal Council decides on next steps with regard to companies' climate disclosures, https://www.news.admin.ch/en/newnsb/o-9C-JRLbzdn2YfnJFzh7, last accessed 2025/07/01
68. US Securities and Exchange Commission: Press Release: SEC Votes to End Defense of Climate Disclosure Rules, https://www.sec.gov/newsroom/press-releases/2025-58, last accessed 2025/07/01
69. Ministry of Finance of the People's Republic of China: Corporate Sustainability Disclosure Standards—Basic Standards, trial version (In Chinese), https://kjs.mof.gov.cn/zhengcefabu/202412/t20241216_3949745.htm, last accessed 2025/07/01
70. Beijing Stock Exchange: Beijing Stock Exchange Listed Companies Continuous Supervision Guideline No. 11—Sustainability Report (Trial) (In Chinese), https://www.bse.cn/cxjg_list/200021393.html, last accessed 2025/07/01
71. Shanghai Stock Exchange: Shanghai Stock Exchange Listed Companies Self-Regulatory Supervision Guideline No. 14 — Sustainability Report (Trial) (In Chinese), http://www.sse.com.cn/lawandrules/sselawsrules/stocks/mainipo/c/c_20240412_5737862.shtml, last accessed 2025/07/01
72. Exchange, S.S.: Shenzhen stock exchange listed companies self-regulatory supervision guideline No. 17—sustainability report (Trial) (In Chinese), https://www.szse.cn/lawrules/rule/stock/supervision/currency/t20240412_606839.html, last accessed 2025/07/01
73. Johnson, M.P., Schaltegger, S.: Two decades of sustainability management tools for SMEs: how far have we come? J. Small Bus. Manag. **54**, 481–505 (2016). https://doi.org/10.1111/jsbm.12154

74. McKeiver, C., Gadenne, D.: Environmental management systems in small and medium businesses. Int. Small Bus. J. Res. Entrep. **23**, 513–537 (2005). https://doi.org/10.1177/0266242605055910
75. Ferenhof, H.A., Vignochi, L., Selig, P.M., Lezana, Á.G.R., Campos, L.M.S.: Environmental management systems in small and medium-sized enterprises: an analysis and systematic review. J. Clean. Prod. **74**, 44–53 (2014). https://doi.org/10.1016/j.jclepro.2014.03.027
76. Dasanayaka, C.H., Gunarathne, N., Murphy, D.F., Nagirikandalage, P.: Triggers for and barriers to the adoption of environmental management practices by small and medium-sized enterprises: A critical review. Corp. Soc. Responsib. Environ. Manag. **29**, 749–764 (2022). https://doi.org/10.1002/csr.2244
77. EC: Circular Economy Action Plan, https://ec.europa.eu/environment/circular-economy/pdf/new_circular_economy_action_plan.pdf, last accessed 2022/11/19
78. EC: Directive 2008/98/EC of the European Parliament and of the Council of 19 November 2008 on waste and repealing certain Directives, https://eur-lex.europa.eu/legal-content/EN/TXT/PDF/?uri=CELEX:32008L0098, last accessed 2022/12/30
79. Halkos, G.E., Aslanidis, P.-S.C.: Reviewing the integrated institutional waste-related framework for circular economy in the European Union. Waste Manag. Bull. **2**, 28–35 (2024). https://doi.org/10.1016/j.wmb.2024.06.002
80. Johnson, M.P.: Sustainability management and small and medium-sized enterprises: managers' awareness and implementation of innovative tools. Corp. Soc. Responsib. Environ. Manag. **22**, 271–285 (2015). https://doi.org/10.1002/csr.1343
81. ISO: ISO Standards, https://www.iso.org/standards.html, last accessed 2025/06/10

A Note on Climate Change Accounting and Reporting in Greece: Insights from the Adoption of the TCFD Recommendations Among Large-Caps

Polyxeni-Panagiota Nana and Antonis Skouloudis

Abstract As the urgency of climate action steadily continues to mount, the pivotal importance of empirical research in informing and shaping effective climate reporting practices is hard to overstate. In this research note, we attempt to shed light on the climate-related accounting and reporting practices of the 25 Greek corporations comprising the large-cap index in the Athens Stock Exchange. To achieve this, a quantitative content analysis of publicly available climate-related disclosures was conducted using the recommendations of the Task Force on Climate-related Financial Disclosures as the assessment criteria for benchmarking. This allowed us to evaluate the reported information on climate risks and opportunities in terms of governance mechanisms in place, strategic directions, risk management processes, and the adoption of relevant performance metrics and target-setting. Findings reveal that climate change reporting is far from uniform among large domestic firms with considerable disparities across thematic areas. Companies tend to provide comparatively more information on corporate governance and risk assessment in relation to climate change implications while disclosures on strategic orientation and long-range planning as well as performance tracking and target-setting leave much to be desired. It also suggests the persistent need for more effective regulatory interventions that will stimulate comprehensive corporate climate change disclosure both in favour of transparency and in reducing information asymmetry, so as not to reduce such efforts to tick-boxing exercises or greenwash. As domestic firms navigate through these challenges, the demand for detailed and credible disclosures will likely continue to grow given the urgent need to incorporate the ESRS reporting standards and, thus, meet the requirements of the CSRD Directive.

Keywords Climate change disclosures · Corporate accounting and reporting · TCFD recommendations · Quantitative content analysis · Large-caps · Greece

P.-P. Nana · A. Skouloudis (✉)
Department of Environment, University of the Aegean, Lesvos, Greece
e-mail: skouloudis@aegean.gr

G. Halkos et al. (eds.), *Environmental Risk Analysis*, Springer Proceedings in Mathematics & Statistics 540, https://doi.org/10.1007/978-3-032-24155-9_13

1 Introduction—Background

Corporate climate change disclosure (CCCD) has become an essential aspect of sustainability reporting, and nowadays companies are increasingly expected to publicly disclose climate-specific risks and opportunities (CRO) in order to better inform both financial and nonfinancial stakeholders about expected impacts and exposure. Such expectation is primarily based on the realization that the ever-growing climate crisis poses financial risks which are far from negligible and can impact a business entity's long-term viability and market performance [1, 2]. Research indicates that effective climate change disclosure can enhance a firm's reputation, attract new investments, and strengthen financial performance [2]. Likewise, such accounting and reporting techniques have been shown to stimulate better decision-making by investors and promote stakeholder accountability and transparency [3, 4]. It is both internal and external parameters that have been identified as determinants of CCCD, with governance mechanisms-structures, board diversity, shareholder activism, and management characteristics indicated as critical internal factors affecting the quantity and comprehensiveness of relevant disclosures [5–9]. Regulatory pressures, (foreign) institutional investors/ownership control, and stakeholder claims-expectations are aspects of the external environment which have been found to affect the level of engagement in CCCD [1, 10–12] and to increase transparency around climate risks [13, 14].

The Recommendations suggested by the Task Force on Climate-related Financial Disclosures (TCFD) have transformed the CCCD landscape worldwide assisting organizations that seek to solidify their accountability in terms of climate-related risks and opportunities. Created in 2015 by the international Financial Stability Board (FSB) with the aim of developing consistent climate-related financial risk disclosures, through its set of recommendations, the TCFD offered a structured approach for comparable and consistent information on CRO facing a company as well as financial impacts these might exert on its operations and existing business model [15, 16]. This initiative of corporate guidelines is structured around four core elements, designed to provide a comprehensive approach to climate-related financial disclosures. The four thematic areas pertain to disclosures on (i) the organization's governance structures and mechanisms around CRO, (ii) the actual and potential CRO impacts on the entity's operations, strategy, and financial planning where such information is material, (iii) how the reporting organization identifies, assesses, and manages material climate-related risks, and (iv) the respective metrics and targets employed to assess and manage the CRO deemed to be critical [16]. Each of these disclosure elements comprises specific CCCD topics (see Table 1) that collectively aim to bridge the gap between the financial impacts of climate change and corporate reporting practice, provide to all interested parties a clearer understanding of CRO management within the enterprise, and allow investors, lenders, and/or insurers in particular to make more informed decisions by understanding the integration status of climate-related considerations into overall risk management processes, resilience efforts, and long-term planning.

Table 1 An outline of the TCFD recommendations and supporting recommended disclosures

Core elements *(Brief definition)*	Recommended disclosures *(Description of…)*
Governance *The entity's governance around CRO management*	The Board's oversight of CRO
	The management's role in assessing and managing CRO
Strategy *The actual and potential impacts of CRO on the entity's businesses, strategy, and financial planning*	The climate-related risks and opportunities the organization has identified over the short, medium, and long terms
	The impact of CRO on the organization's businesses, strategy, and financial planning
	The resilience of the organization's strategy, taking into consideration different climate-related scenarios, including a 2 °C or lower scenario
Risk Management *The processes used by the organization to identify, assess, and manage climate-related risks*	The organization's processes for identifying and assessing climate-related risks
	The organization's processes for managing climate-related risks
	How processes for identifying, assessing, and managing climate-related risks are integrated into the organization's overall risk management
Metrics and Targets *The metrics and targets used to assess and manage relevant CRO*	The metrics used by the organization to assess CRO in line with its strategy and risk management process
	Scope 1, Scope 2, and, if appropriate, Scope 3 greenhouse gas (GHG) emissions, and the related risks
	The targets used by the organization to manage CRO and performance against targets

Source Adapted from TCFD [16]

Aligning the corporate climate action agenda with broader sustainability goals, such as the UN SDGs [17], the adoption of these CCCD guidelines continues to rise with the most recent TCFD Status Report to indicate that in 2023 it was 58% from a sample of 1,365 of large enterprises that reported information in line with at least five of the recommended disclosures, a 40% increase since 2020 [15]. Nevertheless, alignment and integration levels remain inconsistent across countries and sectors [18] while the TCFD penetration levels are largely unidentified in many regions and national settings, highlighting the need for both further policy incentives and research to address these gaps. Marco et al. [19] denote that, in Europe, the United Kingdom and France lead in implementing these recommendations, with adoption rates exceeding 50% among large corporations. In contrast, countries in Eastern Europe exhibit lower adoption rates, often below 30%, suggesting moderate levels of regulatory pressure and governance standards. Demaria and Rigot [20] and Amar et al. [21] examine the reporting practices of 40 of the largest corporations listed in the French stock exchange using the TCFD recommendations as the benchmark for climate-related financial transparency. Using a scoring system approach and a composite index, they assess the reports published between 2015 and 2018; their study confirms a moderate improvement in terms of comprehensiveness over time along with marked disparities across industries and management areas. Focusing on the CCCDs of the STOXX Europe 600 banks during the 2017–2020 period, Friedrich et al. [22] also pinpoint a gradual increase in the quality of climate reporting as well as significant deficiencies mainly in the inclusion of future projections or scenario analyses and the insufficient adoption of TCFD recommendations by banks. De Bernardi et al. [23] investigate the adherence of the largest Italian nonfinancial enterprises to the TCFD guidelines and pinpoint that sample firms are approximately covering 50% of the recommended disclosures; required information on risk management as well as metrics and targets are sufficiently reported (with average coverage of 80% and 65% of respective recommendations), while the description of governance mechanisms and corporate strategy to address climate-related impacts leaves much to be desired and can be characterized as particularly weak. In a similar vein, Bingler et al. [24] indicate that TCFD-supporters based in North America and Europe tend to disclose more information than those from Asian countries; North American companies provide comparatively more information on strategy but across all regions such disclosures along with data on metrics and targets are underrepresented, a shortcoming with detrimental effects of accountability and transparency. Relevant insights can be also drawn from the structured review of Ngo et al. [25], who stress that while several countries (such as New Zealand, South Korea, the United States, Russia, Brazil, and Japan) have announced or enacted regulations for climate-oriented disclosure mandates, data quality and availability constraints still prevent a comprehensive appraisal of climate-related risks in financial assets. The authors, hence, call for evidence on TCFD compliance from more national settings in order to substantiate whether reporting entities 'walk the climate talk' or merely engage in window-dressing reporting—also echoed in Eccles and Krzus [26], and Marco et al. [19].

This is where this study seeks to contribute by focusing on the large-cap enterprises listed in the Athens Stock Exchange, Greece. Greece is one of the few EU member states which saw a significant decrease in their total GHG emissions during the last decade (achieving a 28% reduction [27]). National policy measures towards climate change mitigation with direct effect on the Greek business sector have been the mandatory implementation of energy management systems, the creation of energy and environmental management centres in business parks, the promotion of voluntary agreements in industry and energy services in order to stimulate energy conservation along with the provision of subsidies to industrial and service sectors towards energy efficiency and renewable energy investments [28]. By attempting to shed light on CCCD from a country of the Balkan Peninsula, we attempt to shift the research lens from the leading national settings and economies (where prior research has placed more emphasis) towards national business systems where CCCD is still in nascent stages. Motivated by the growing body of empirical studies that examine the role of large enterprises towards net zero, our analysis offers insights on the status of CCCD among large Greek corporations and paves the way for further and comparative research in the southeast sub-region of Europe. In this context, the findings presented here offer insights (a) on how domestic corporations respond to the pressing issue of climate risk management and (b) on the quality and comprehensiveness of climate-related disclosures.

The remainder of this paper is structured as follows. Section 2 presents the material and methods, Sect. 3 presents the empirical findings and, finally, Sect. 4 concludes with explanatory remarks of the Greek case and relevant implications from this assessment.

2 Material and Methods

The sample used in this study consists of the 25 companies comprising the large-cap index in the Athens Stock Exchange. In order to explore the publicly available CCCDs, a web-based search was performed during the first quarter of 2024, locating the corporate websites of the sample firms and retrieving all the related information (most recent TCFD reports, sustainability/ESG reports, environmental statements along with relevant webpage/online information, etc.) (Table 2). In line with the structure and rationale of the TCFD recommendations and previous studies [20], CCCD is assessed according to a quantitative content analysis by applying numerical grading scheme (Table 3). In this scoring system, an approach well-established in the sustainability reporting literature, zero corresponds to nondisclosure of any sufficient and clear information. The value of 1 stands for disclosing very brief-vague information on the respective TCFD recommendation-criterion, 2 is assigned to disclosures that present with a degree of detail the company's approach on the relevant criterion albeit gaps-omissions are evident, and 3 in cases where the company provides extensive-comprehensive disclosures that are in full compliance with the given TCFD criterion. Riffe et al. [29] denote that 'quantitative content analysis is the systematic and replicable examination of symbols of communication, which have been assigned numeric values according to valid measurement rules and the analysis of relationships involving those values using statistical methods, to describe the communication, draw inferences about its meaning, or infer from the communication to its context, both of production and consumption' (p. 25). In line with Demaria and Rigot [20], we distinguish 38 unweighted scoring criteria ('topics', see Appendix A) into the 4 criteria blocks (i.e. the TCFD disclosure areas-themes: governance, strategy, risk management and metrics-targets) to allow for the classification of the different types of disclosed information.

Summarizing the available reports in Fig. 3, most companies published sustainability reports the previous year and only seven companies have compiled stand-alone TCFD reports. The 'other' category includes strictly online information through the website of the companies, social responsibility reports, Carbon Disclosure Project (CDP) reports, etc. (Fig. 1).

Table 2 The material gathered for the 25 large-caps of the sample and assessed in terms of CCCD

Company	Documents—Reports	Reference period
Piraeus Bank	TCFD Report	2023
Mitilineos	TCFD Report Sustainability Report	2022 2022
TITAN	Sustainability Report Integrated Annual Report Website section 'Sustainability'	2022 2023
Coca-Cola 3E	Sustainability Report CDP Report (Carbon Disclosure Project) Website section 'A more sustainable future'	2021 2023
Viohalco	TCFD Report	2023
Cenergy	TCFD Report Consolidated Annual Report	2023 2023
Elval Halcor	TCFD Report	2022
OTE	Sustainability Report	2022
Aegean Airlines	Corporate Social Responsibility Report Financial Report	2022 2023
Terna Energy	Sustainability Report Climate Change Disclosure	2022 2021
Terna	Sustainable Development Report	2022
Sarantis	Sustainability Report	2021–2022
Ellaktor	Sustainable Development Report Financial Report	2022 2023
Lamda	Sustainable Development Report	2022
Alpha Bank	TCFD Report	2022
National Bank of Greece	ESG Report	2022
Helleniq Energy	Sustainability Report	2022
Eurobank	TCFD Report Climate Report	2023 2022
Quest	Sustainable Development Report Financial Report Sust. Development Committee Report Website section 'Sustainability'	2021/2022 2022 2021
Autohellas	Sustainability Report	2022
Eydap	Sustainability Report	2022
OPAP	ESG Report Annual Report	2021 2022
Motor Oil	Sustainability Report	2022
Dei/Ppc	Sustainability Report Financial Report	2022 2022
Jumbo	-	-

Table 3 Basic rating qualification scale

Score	Rating qualification	Explanation/requirements
0	No coverage/ nondisclosure	The company does not disclose any sufficient information relevant to respective TCFD topic
1	Minimal disclosures	The company provides brief-vague relevant information; fails to sufficiently present its approach to the specific TCFD topic
2	Detailed disclosures; gaps-omissions are evident	The provided information is adequate and clear but despite the degree of detail there are still gaps-omissions in coverage of the TCFD topic
3	Extensive-full disclosure	The TCFD topic is covered in a comprehensive manner; reported information are in compliance with the recommended TCFD disclosure requirements

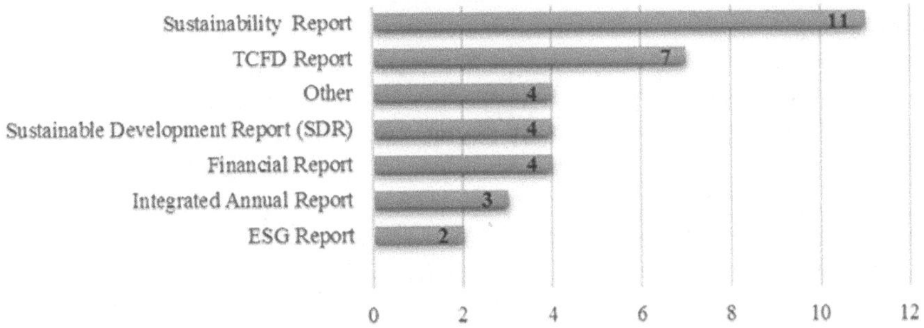

Fig. 1 Types of corporate reports included in the study

Based on the defined TCFD topics, a composite index was devised as follows:

$$CCCD_{(i)} = \sum\nolimits_{j=0,1,2,3}^{38} t_j$$

where t_j equals zero for nondisclosure, 1 if the company i discloses brief and/or insufficient information on the jth TCFD topic, 2 if it provides disclosures with a certain degree of detail but gaps-omissions are evident, and 3 when extensive coverage and/or comprehensive disclosure is provided that is in compliance with the respective TCFD criterion. The assessment was performed during the first half of 2024, independently by the two researchers. In the few cases-criteria where differences in disclosure evaluation were identified, these were commonly discussed by the coders and modified accordingly in order to minimize inter-coding errors and the need for examining such reliability issues. Results, presented in the following section, are expressed in percentages for both the overall disclosure index and the four TCFD core elements.

3 Findings

Overall findings reveal that companies have to a certain degree adopted key points the TCFD recommendations in structuring and preparing their climate disclosures. The content analysis scores range from zero points to 88% coverage of the TCFD disclosure requirements while the average score achieved by the large-cap companies of the sample reaching 49% of the maximum points. Figure 2 reveals that more than half of the 25 Greek large-cap companies provide adequate to significant climate-related information through their communication channels. In this regard, by relying on the mean score of the sample, we identify three distinct subgroups of reporting entities: the 'pace-setters' that scored well above the average (Mitilineos, Cenergy, Coca-Cola 3E, Piraeus Bank, Motor Oil, Eurobank, Dei/PPC, Alpha Bank, and National Bank of Greece) and lead CCCD in the domestic business sector, the 'fence-sitters' that scored close to the average (Viohalco, Ellaktor, Titan, Quest, Elval Halcor, OTE, EYDAP, Lamda, and Helleniq Energy) disclosing information with striking gaps and/or omissions, and a third group of 'laggards' that produced disclosures which left much to be desired and thus scored well-below the sample's average (Terna, OPAP, Auto Hellas, Sarantis, Aegean Airlines, and Jumbo). The top three companies in terms of disclosure comprehensiveness were as follows: Mitilineos with an overall score of 88%, Cenergy with a score of 82%, and Coca-Cola 3E with a total score of 75%. The companies with the lowest scores were Sarantis with 19%, Aegean with 11%, and lastly Jumbo which has not published any relevant report or any provided other online CCCDs through its communication channels. Out of the four thematic areas that the TCFD recommendation examines (governance, strategy, risk management, and metrics and targets), the reporters tend to provide relatively more information on aspects pertaining to risk management and governance mechanisms to address CRO (Appendix B) (Fig. 2).

Delving deeper in the comprehensives of provided information from the sample group, Table 4 presents in descending order the overall score (%) achieved in each one of the disclosure criteria. This allows a better understanding of the specific topics/ TCFD recommendations that companies tend to place emphasis on and analyse more thoroughly within their reports, as well as those that lack in key respects. The disclosure criteria are grouped in three sub-categories (using different shades of grey in the table to distinguish the groups): those with adequate to extensive coverage (above 65%) where no criteria pertaining to corporate strategy are included, those with moderate coverage (between the 35%–65% margin) where most TCFD requirements fall into, and those with low-to-minimal coverage (below 35%), a group dominated by disclosure recommendations on corporate strategy and performance metrics-targets.

Figure 2 presents the ranking of the sample firms based on the thematic area of 'Governance' of the TCFD recommendations. With an average disclosure score of 58%, more than 50% of the companies follow the respective recommendations in this aspect of climate-related disclosures. The top three companies are Mitilineos, Motor Oil, and Coca-Cola 3E with 96%, 96%, and 92%, respectively, which are

Fig. 2 Overall ranking of companies based on total CCCD index scores

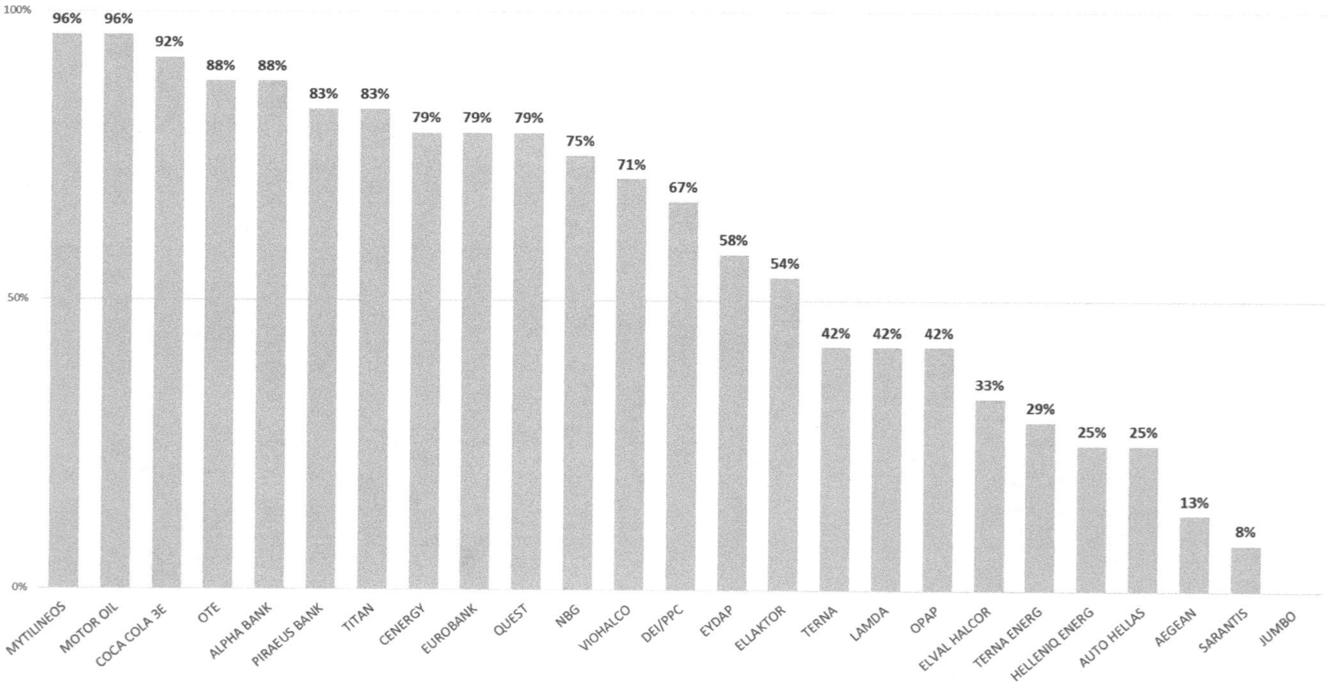

Fig. 3 Assessment scores in the 'Governance' thematic area

Table 4 Assessment scores per TCFD item

n/n	Recommended disclosures—assessment criteria	Score (%)
Q2/Risk	What is the materiality (in relation to climate risks)?	80
Q5/Metr	Carbon footprint, GHG emissions from Scopes 1 and 2	80
Q1/Gov	Is the Board informed about climate issues?	77
Q1/Metr	Indicators related to energy consumption, water, waste management, etc	77
Q1/Risk	Are the processes presented (related to climate risk management)?	73
Q2/Gov	How often (is the Board informed about climate issues)?	68
Q4/Metr	Evolution of indicators (for environmental footprint, managing business risks, and carbon pricing)	68
Q4/Gov	How does the Board assess progress (over the defined strategy)?	65
Q6/Metr	Scope 3 GHG emissions	64
Q8/Metr	Evolution of Scope 1, 2, and 3 performance indicators	63
Q3/Gov	Are climate topics taken into account in the evaluation and orientation of the strategy?	61
Q3/Risk	Are current and potential regulations taken into account (when assessing CRO)?	61
Q6/Gov	What are the responsibilities of managers related to climate change?	60
Q4/Risk	Existence of materiality study?	59
Q6/Str	Impact of risks and opportunities associated with climate change on the company's business and strategy	57
Q7/Gov	Climate information feedback process	56
Q8/Str	How are they integrated (the risks and opportunities associated with climate change)?	53
Q7/Risk	How the processes for identifying and assessing CRO are integrated into the overall risk management	53
Q8/Gov	How does the management manage climate issues?	52
Q7/Str	Identification of impacts (of risks and opportunities associated with climate change)?	52
Q6/Risk	What are the priorities identified?	52
Q5/Risk	Management of climate-related risks	45
Q9/Metr	Are objectives disclosed for Scope 1 and 2 indicators?	45
Q1/Str	Accuracy of the periods?	37
Q2/Str	Detail and costing of CRO—short term	37
Q3/Str	Detail and costing of CRO—medium term	37
Q4/Str	Detail and costing of CRO—long term	37

(continued)

Table 4 (continued)

n/n	Recommended disclosures—assessment criteria	Score (%)
Q12/Str	Horizon of these analyses (related to different scenarios of climate change implications)?	37
Q13/Str	Conclusions of the scenario analyses	36
Q10/Str	Study of the resilience of the business model?	35
Q2/Metr	Indicator similar to an indicator for managing business risks	32
Q11/Str	Analysis of different scenarios (related to climate change implications)?	29
Q7/Metr	Details of significant elements of Scope 3 GHG emissions	29
Q9/Str	Description of Capex and Opex related to CRO?	25
Q5/Gov	Variable remuneration linked to climate indicators?	23
Q10/Metr	Are objectives disclosed for Scope 3 indicators?	21
Q5/Str	Distribution of CRO at the sector and geographical levels	16
Q3/Metr	Assessment of an internal carbon price	3

the highest scores allocated among all four TCFD thematic areas. The companies with the lowest scores are Aegean Airlines (13%), Sarantis (8%), and Jumbo. This category has the second-highest average score out of the four thematic and the second most frequently covered disclosure with a score of 77% referring to clarifications on whether and how the Board is informed about climate issues. Most questions-criteria referring to this thematic area are to a certain degree covered with clarity by the majority of the companies, as relevant disclosure requirements pertaining to the first two groups identified in Table 3 above.

With a group average of 40%, the thematic area 'Strategy' reveals significant omissions and gaps that leave much to be desired as it is the category of CCCD with the lowest reporting comprehensiveness out of the four categories of the TCFD recommendations (Fig. 3). The highest score was achieved by Viohalco (90%), followed by Piraeus Bank, Mitilineos, and Cenergy (87%). Crucially, it is six companies that did not touch upon any of the relevant TCFD-recommended disclosures (i.e. Terna Energy, Ellaktror, Quest, Auto Hellas, EYDAP, and Jumbo) while an additional three (Aegean Airlines, OTE, and Terna) reported minimal information on strategic orientation to control and manage CRO. The most commonly addressed topic in this thematic area refers to the impact(s) of CRO on the company's business and strategy, with a score of 57% (see Table 3), while the majority of relevant disclosure requirements pertain to the two lower groups of covered criteria in terms of comprehensiveness identified in Table 3.

The highest average score among the four thematic areas is observed for the risk management disclosures (61%) as presented in Fig. 4. This domain of TCFD recommendations received the second highest overall reporting score, 95%, by DEI/PPC and Eurobank. Cenergy, Alpha Bank, and the National Bank of Greece follow with a score of 90% while the companies with the lowest score in terms of risk-related disclosures are the Aegean Airlines, Viohalco, and Jumbo (lower than 10%). This subgroup of CCCD items includes one of the most frequently covered TCFD recommendations referring to outlining the materiality assessment applied for climate risk (and opportunities) identification, with most reporting entities (approximately 80% of the sample group) including meaningful relevant information in key respects. This is also evidenced in Table 3 as most risk-related CCCD items-criteria pertain to the two upper groups of segmentation in terms of coverage.

In contrast, the average disclosure score observed in the Metrics & Targets thematic area is particularly low (48%), being the second lowest compared to the other three aspects of the TCFD recommendation (Fig. 5). Despite such low reporting performance, this category has one of the most widely covered disclosure criteria of the TCFD framework referring to the carbon footprint of the business entities in terms of Scope 1 and Scope 2 GHG emissions. The companies with the highest scores are Mitilineos, Coca-Cola 3E, and Motor Oil with 83%, 80%, and 77% respectively. On the other hand, the lowest scores were assigned to Aegean Airlines (20%), Auto Hellas (13%), and Jumbo (zero points). The absolute majority of the companies fails to report an 'assessment of an internal carbon price', a CCCD topic with an average disclosure score of a mere 3% as only two companies attempted to communicate relevant quantitative and/or descriptive information in their reports (i.e. Mitilineos and Coca-Cola 3E). Most disclosure criteria referring to this category of quantitative-performance data and target-setting information are on the lowest group of the TCFD criteria in terms of coverage, from 32% and below, with most companies either failing to provide a well-articulated presentation of performance data or presenting targets and evidence-based progress over time in a vague-fragmented manner (Fig. 6).

4 Discussion—Concluding Remarks

Our content analysis findings are in line with evidence from other national terrains of Europe and beyond, indicating considerable gaps and inconsistencies in the TCFD-required information [19–26, 30]. This reflects that alignment with such international CCCD initiatives and the relevant EU directives remains thin on the ground and can be considered being still in nascent stages for most large-caps in Greece. It also suggests the persistent need for more effective regulatory interventions that will stimulate comprehensive CCCD both in favour of transparency and in addressing information asymmetry in order to not reduce such efforts to tick-boxing exercises or greenwash. For instance, mandatory reporting, phased approaches, and/or capacity-building initiatives in aspects such as the internal carbon pricing assessments or in

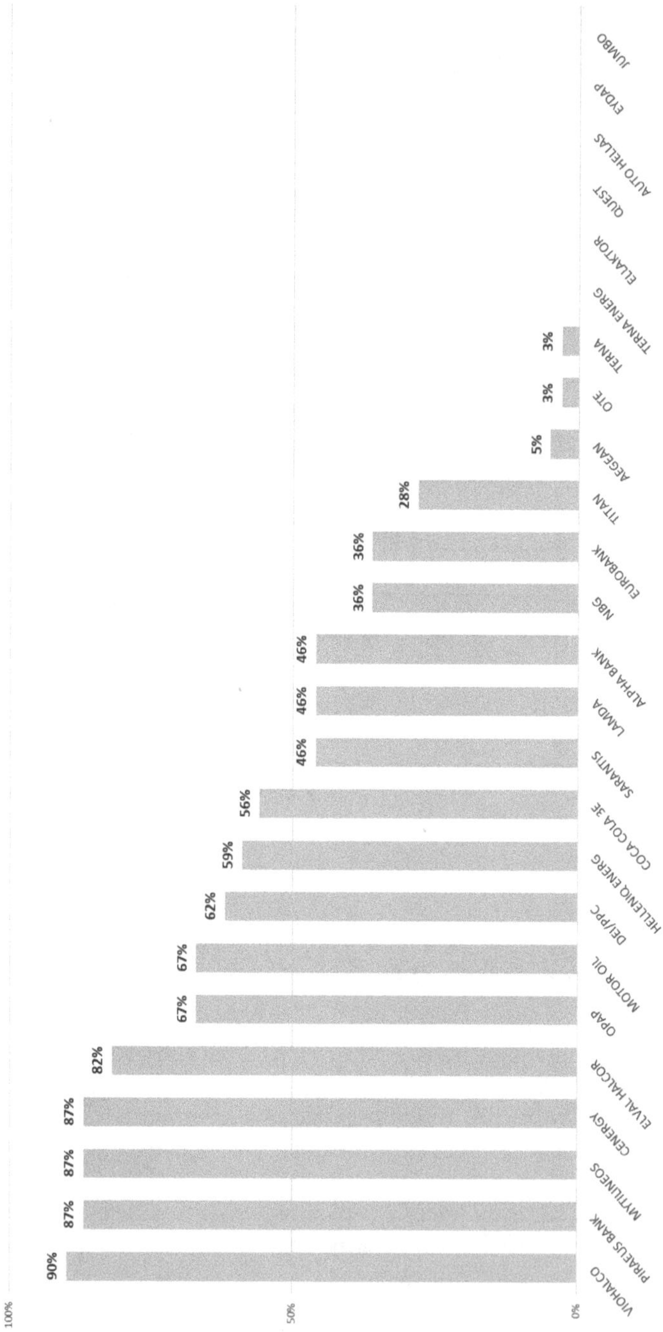

Fig. 4 Assessment scores in the ‘Strategy’ thematic area

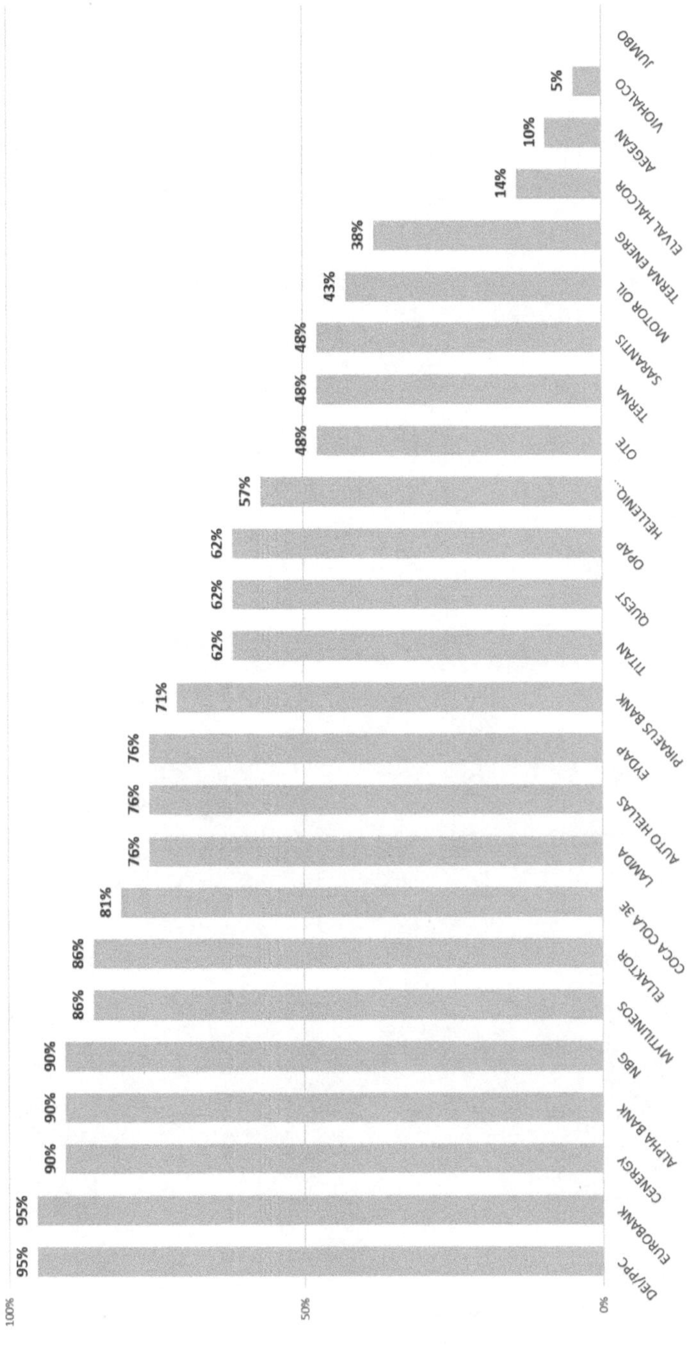

Fig. 5 Assessment scores in the 'Risk Management' thematic area

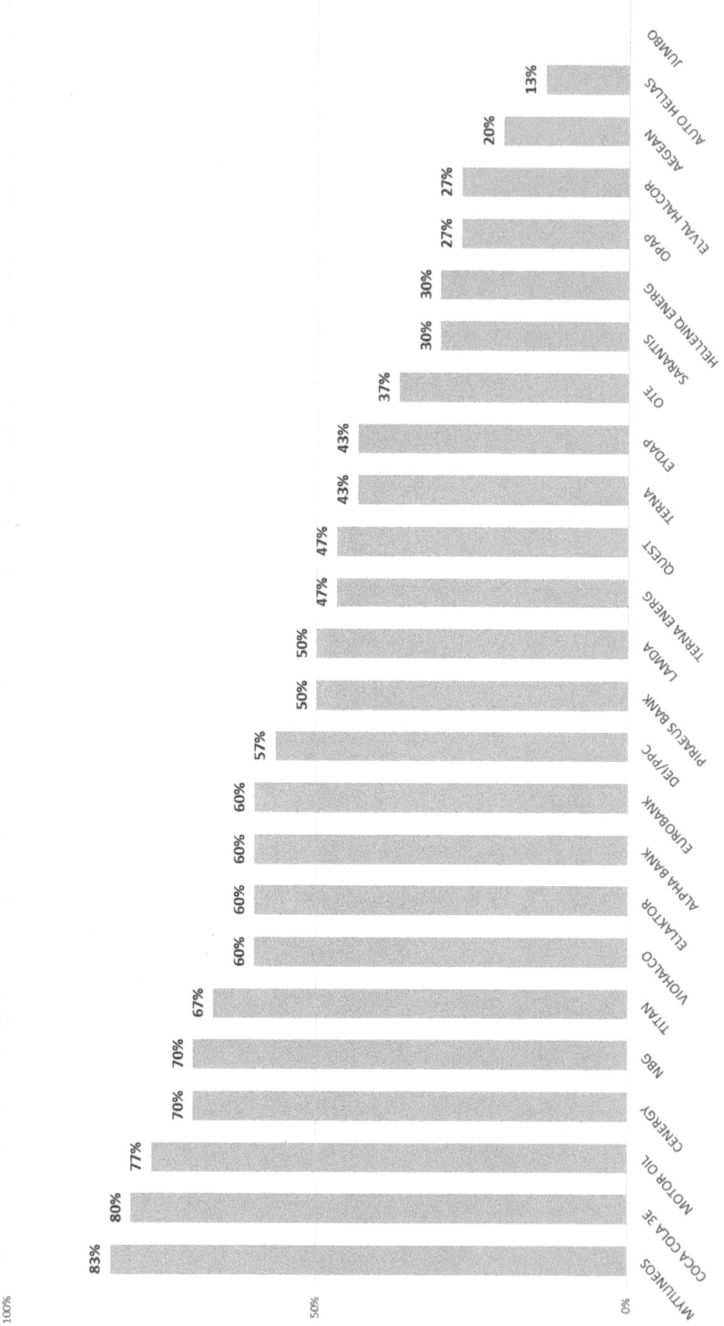

Fig. 6 Assessment scores in the 'Metrics & Targets' thematic area

disclosing robust objectives for Scope 3 emissions would yield material improvement in these areas.

Stakeholders, including providers of capital, investors, and consumers, are increasingly scrutinizing the substance and depth of corporate claims regarding climate mitigation and adaptation, which necessitates a shift towards more rigorous accounting and reporting processes as well as the adoption of robust standards for CRO disclosure [31, 32]. Since the debt crisis, awareness of environmental issues has been heightened, prompting both domestic corporations, social actors, and the public to reconsider their roles in addressing climate change [33]. This shift in perceptions has led to increased calls for nonfinancial accountability, as key members of the national business system demand that companies not only disclose their climate risks but also outline their strategic commitments for mitigating these risks [34]. Such challenge lies in ensuring that disclosures are not only comprehensive but also reflective of impactful interventions in corporate practices and CCCD commitments, confirming the 'talk-walk' hypothesis of corporate reporting that, ultimately, translates into patterns of tangible climate-related performance [35]. As domestic firms navigate these challenges, the demand for detailed and credible disclosures will likely continue to grow given the urgent need to incorporate the ESRS reporting standards and, thus, meet the requirements of the CSRD Directive.

Future research should focus on studies that track changes in CCCD practices over time, particularly in response to evolving regulatory frameworks and stakeholder expectations to better capture the nuances of corporate climate risk management. The increasing use of data analytics and machine learning in corporate reporting could revolutionize how firms disclose climate-related information, potentially leading to more accurate and timely disclosures, reflecting another particularly fruitful avenue for scientific inquiry [36]. Likewise, understanding how climate risk disclosures impact Greek institutional as well as small investor behaviour, decision-making, and corporate valuation will provide valuable insights for both academics and practitioners. In this respect, research could explore the effectiveness of domestic or regional regulatory frameworks in promoting high-quality climate-related disclosures through longitudinal or action research studies and further investigate implications of CCCD (and other pressing aspects of nonfinancial disclosure) to corporate governance and strategic decision-making, particularly in national business sectors largely overlooked in the literature [37, 38]. Lastly, qualitative studies exploring the motivations behind corporate disclosures could provide deeper and most likely actionable insights into the decision-making processes of Greek firms regarding climate change accounting and reporting.

Addressing the complexities of robust and impactful CCCD does require the concerted efforts from governmental bodies, business associations, the media, and educational institutions to foster a culture of accountability in climate-related reporting. Given that the urgency of climate action steadily continues to mount, the pivotal importance of empirical research in informing and shaping effective climate reporting practices is hard to overstate.

Disclosure of Interests The authors have no competing interests to declare that are relevant to the content of this article.

Appendix A: The (unweighted) Scoring Criteria Used in the Assessment

Note: Gov: Governance; Str: Strategy; Risk: Risk Management; Metr: Metrics & Targets.

n/n	Recommended disclosures—assessment criteria
Q1/Gov	Is the Board informed about climate issues?
Q2/Gov	How often (is the Board informed about climate issues)?
Q3/Gov	Are climate topics taken into account in the evaluation and orientation of the strategy?
Q4/Gov	How does the Board assess progress (over the defined strategy)?
Q5/Gov	Variable remuneration linked to climate indicators?
Q6/Gov	What are the responsibilities of managers related to climate change?
Q7/Gov	Climate information feedback process
Q8/Gov	How does the management manage climate issues?
Q1/Str	Accuracy of the periods?
Q2/Str	Detail and costing of CRO—short term
Q3/Str	Detail and costing of CRO—medium term
Q4/Str	Detail and costing of CRO—long term
Q5/Str	Distribution of CRO at the sector and geographical levels
Q6/Str	Impact of risks and opportunities associated with climate change on the company's business and strategy
Q7/Str	Identification of impacts (of risks and opportunities associated with climate change)?
Q8/Str	How are they integrated (the risks and opportunities associated with climate change)?
Q9/Str	Description of Capex and Opex related to CRO?
Q10/Str	Study of the resilience of the business model?
Q11/Str	Analysis of different scenarios (related to climate change implications)?
Q12/Str	Horizon of these analyses (related to different scenarios of climate change implications)?
Q13/Str	Conclusions of the scenario analyses
Q1/ Risk	Are the processes presented (related to climate risk management)?
Q2/ Risk	What is the materiality (in relation to climate risks)?
Q3/ Risk	Are current and potential regulations taken into account (when assessing CRO)?

(continued)

(continued)

n/n	Recommended disclosures—assessment criteria
Q4/ Risk	Existence of materiality study?
Q5/ Risk	Management of climate-related risks
Q6/ Risk	What are the priorities identified?
Q7/ Risk	How the processes for identifying and assessing CRO are integrated into the overall risk management
Q1/ Metr	Indicators related to energy consumption, water, waste management, etc
Q2/ Metr	Indicator similar to an indicator for managing business risks
Q3/ Metr	Assessment of an internal carbon price
Q4/ Metr	Evolution of indicators (for environmental footprint, managing business risks, and carbon pricing)
Q5/ Metr	Carbon footprint, GHG emissions from Scopes 1 and 2
Q6/ Metr	Scope 3 GHG emissions
Q7/ Metr	Details of significant elements of Scope 3 GHG emissions
Q8/ Metr	Evolution of Scope 1, 2, and 3 performance indicators
Q9/ Metr	Are objectives disclosed for Scope 1 and 2 indicators?
Q10/ Metr	Are objectives disclosed for Scope 3 indicators?

Appendix B: Content Analysis Results for the Four TCFD Thematic Areas

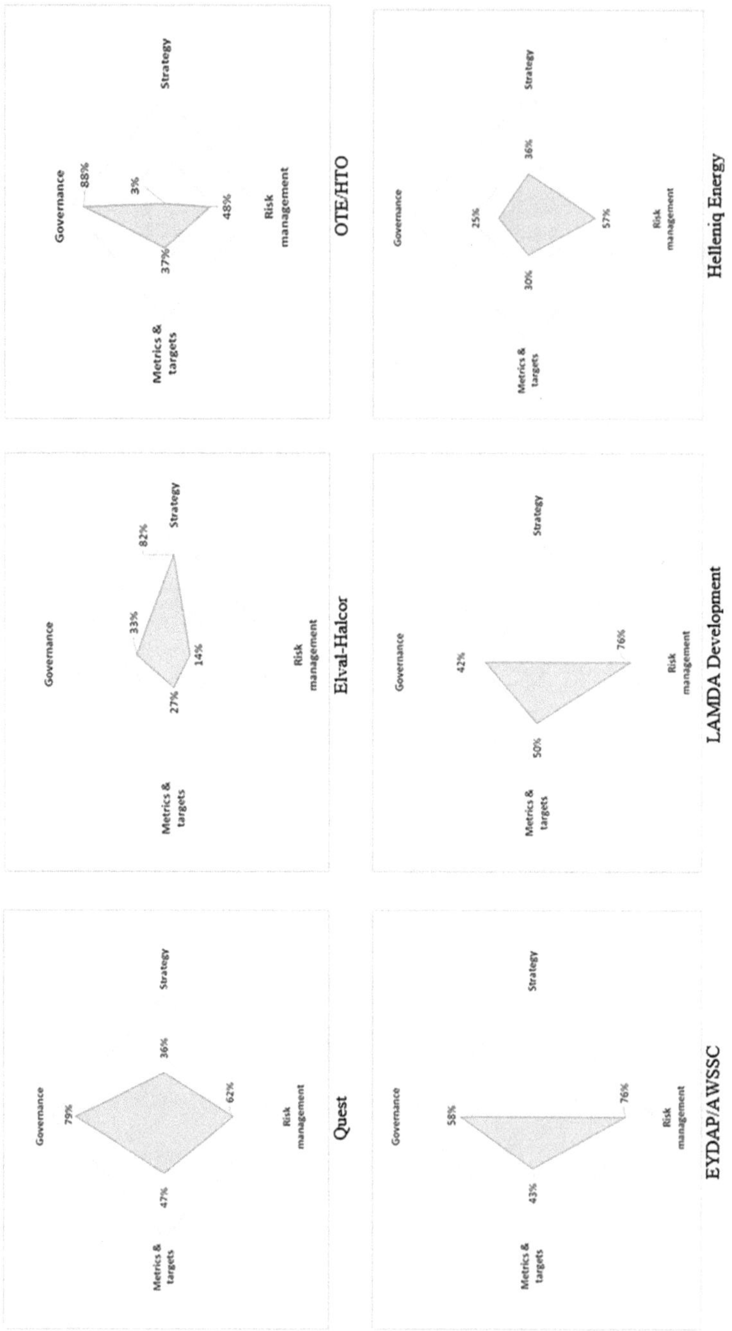
OTE/HTO
Governance
Strategy
Risk management
Metrics & targets
88%
3%
48%
37%
Helleniq Energy
Governance
Strategy
Risk management
Metrics & targets
25%
36%
57%
30%
Elval-Halcor
Governance
Strategy
Risk management
Metrics & targets
33%
82%
14%
27%
LAMDA Development
Governance
Strategy
Risk management
Metrics & targets
42%
76%
50%
Quest
Governance
Strategy
Risk management
Metrics & targets
79%
36%
62%
47%
EYDAP/AWSSC
Governance
Strategy
Risk management
Metrics & targets
58%
76%
43%

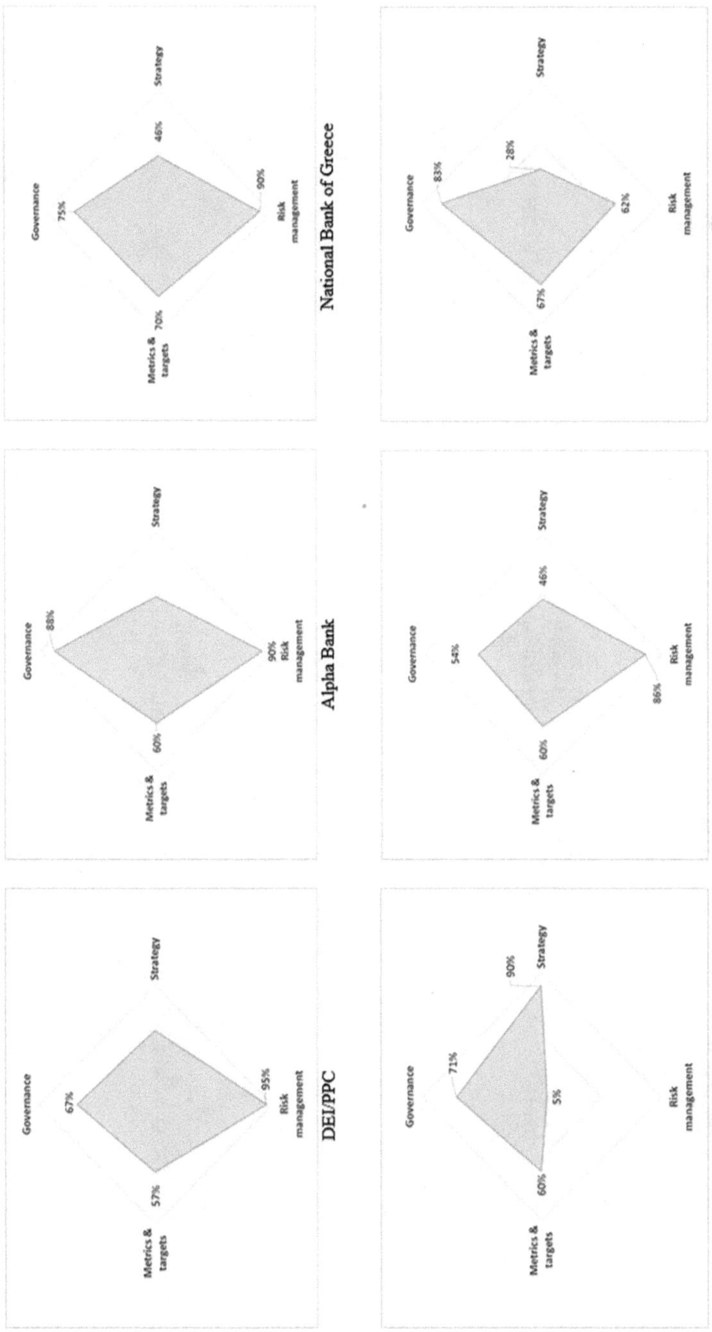
Governance
75%
Strategy
46%
Risk management
90%
Metrics & targets
70%
National Bank of Greece
Governance
83%
Strategy
28%
Risk management
62%
Metrics & targets
67%
Titan
Governance
88%
Strategy
Risk management
90%
Metrics & targets
60%
Alpha Bank
Governance
54%
Strategy
46%
Risk management
86%
Metrics & targets
60%
Ellaktor
Governance
67%
Strategy
Risk management
95%
Metrics & targets
57%
DEI/PPC
Governance
71%
Strategy
90%
Risk management
5%
Metrics & targets
60%
Viohalko

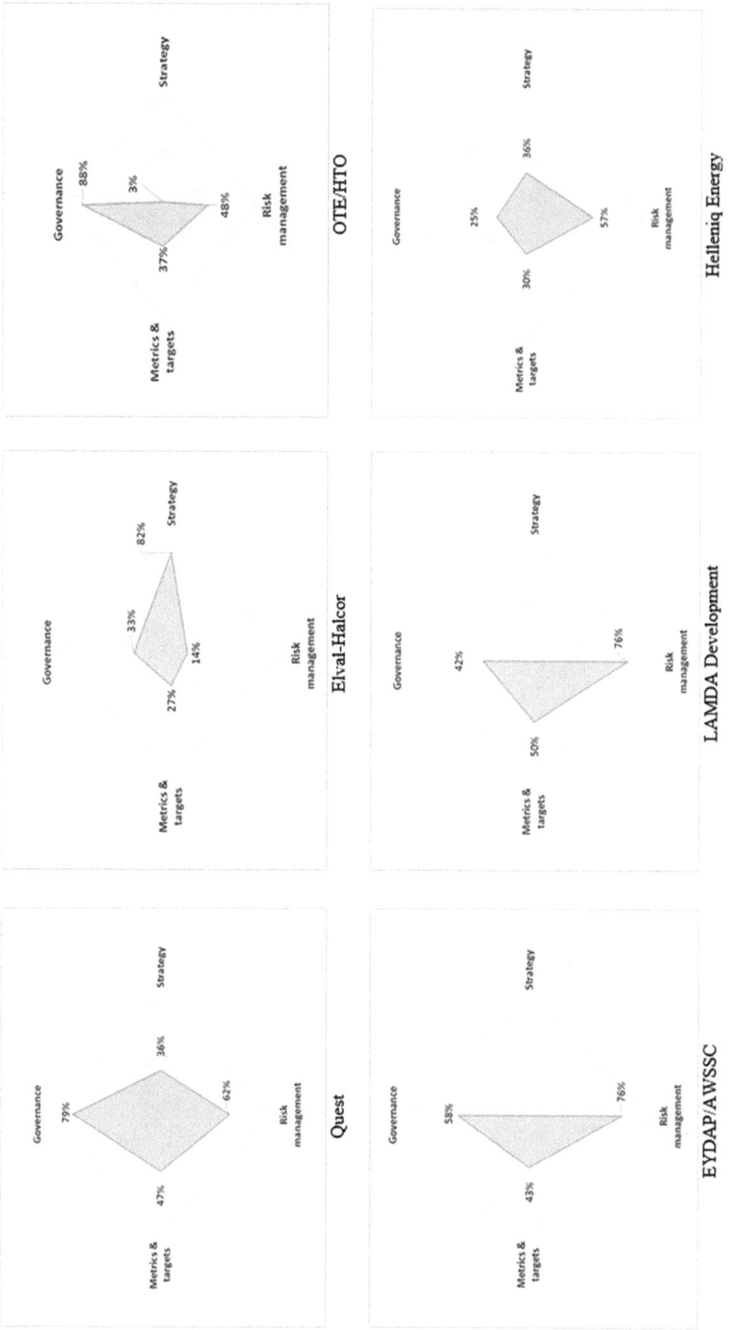
Governance
Strategy
Risk management
Metrics & targets
88%
3%
48%
37%
OTE/HTO
Governance
Strategy
Risk management
Metrics & targets
36%
25%
57%
30%
Helleniq Energy
Governance
Strategy
Risk management
Metrics & targets
82%
33%
14%
27%
Elval-Halcor
Governance
Strategy
Risk management
Metrics & targets
42%
76%
50%
LAMDA Development
Governance
Strategy
Risk management
Metrics & targets
79%
36%
62%
47%
Quest
Governance
Strategy
Risk management
Metrics & targets
58%
76%
43%
EYDAP/AWSSC

References

1. Bose, S., Lim, E., Minnick, K., Shams, S.: Do foreign institutional investors influence corporate climate change disclosure quality? International evidence. Corp. Gov.Ance Int. Rev. **32**(2), 322–347 (2023)
2. Maji, S., Kalita, N.: Climate change financial disclosure and firm performance: empirical evidence from Indian energy sector based on TCFD recommendations. Soc. Bus. Rev. **17**(4), 594–612 (2022)
3. Dahl, T., Fløttum, K.: Climate change as a corporate strategy issue: a discourse analysis of three climate reports from the energy sector. Corp. Commun.: Int. J. **24**(3), 499–514 (2019)
4. Eleftheriadis, I.M., Anagnostopoulou, E.G.: Relationship between corporate climate change disclosures and firm factors. Bus. Strateg. Environ. **24**(8), 780–789 (2015)
5. Elliott, C.: Rethinking complementarity: the co-evolution of public and private governance in corporate climate disclosure. Regul. and Gov.Ance **18**(3), 802–819 (2023)
6. Ben-Amar, W., McIlkenny, P.: Board effectiveness and the voluntary disclosure of climate change information. Bus. Strateg. Environ. **24**(8), 704–719 (2014)
7. Ben-Amar, W., Chang, M., McIlkenny, P.: Board gender diversity and corporate response to sustainability initiatives: evidence from the carbon disclosure project. J. Bus. Ethics **142**(2), 369–383 (2015)
8. Waard, D., Marra, T., Kranenburg, S., Oorschot, M.: Transparent carbon disclosures: depth in carbon-reporting of Dutch listed and non-listed companies. Maandblad Voor Accountancy en Bedrijfseconomie **94**(7/8), 275–284 (2020)
9. Galeone, G., Onorato, G., Shini, M., Dell'Atti, V.: Climate-related financial disclosure in integrated reporting: what is the impact on the business model? The case of Poste Italiane. Account. Res. J. **36**(1), 21–36 (2023)
10. Hossain, M., Farooque, O.: The emission trading system, risk management committee and voluntary corporate response to climate change - a CDP study. Int. J. Account. Inf. Manag. **27**(2), 262–283 (2019)
11. Zhao, J., Wang, X., Yang, D.C.: Climate change risk disclosure and accounting choice: evidence from U.S. oil and gas companies. Int. J. Bus. and Econ., **8**(2), 89–106 (2023)
12. Chen, J., Khoo, E., Peng, Z.: Climate change disclosure and the information environment in the initial public offering market. Account. Financ. **63**(S1), 907–952 (2023)
13. Ben-Amar, W., Herrera, D.C., Martinez, I.: Do climate risk disclosures matter to financial analysts? J. Bus. Financ. Acc. **51**(7–8), 2153–2180 (2024)
14. Eleftheriadis, I., Anagnostopoulou, E.: Measuring the level of corporate commitment regarding climate change strategies. Int. J. Clim. Chang. Strat. Manag. **9**(5), 626–644 (2017)
15. Task Force on Climate-related Financial Disclosures.: Status Report 2023. Task Force on Climate-related Financial Disclosures, Financial Stability Board (2023)
16. Taskforce on Climate-Related Financial Disclosures.: Status Report2021. Task Force on Climate-related Financial Disclosures, Financial Stability Board (2021)
17. Flammer, C., Toffel, M., Viswanathan, K.: Shareholder activism and firms' voluntary disclosure of climate change risks. Strateg. Manag. J. **42**(10), 1850–1879 (2021)
18. Bairagi, R., Ghosh, P.: Adoption and determinants of task force on climate-related financial disclosures (TCFD) reporting frameworks by Australian stock exchange (ASX) listed companies. Indian J. Appl. Econ. Bus. **5**(1), 1–18 (2022)
19. Marco, R., Dong, T., Malatincová, R., Reuter, M., Strömsten, T.: Symbol or substance? Scrutinizing the 'risk transparency premise' in marketized sustainable finance: the case of TCFD reporting. Bus. Strateg. Environ. **32**(6), 3027–3052 (2022)
20. Demaria, S., Rigot, S.: Corporate environmental reporting: Are French firms compliant with the task force on climate financial disclosures' recommendations? Bus. Strateg. Environ. **30**(1), 721–738 (2021)
21. Amar, J., Demaria, S., Rigot, S.: Enhancing financial transparency to mitigate climate change: Toward a climate risks and opportunities reporting index. Environ. Model. Assess. **27**(3), 425–439 (2022)

22. Friedrich, T.J., Velte, P., Wulf, I.: Corporate climate reporting of European banks: Are these institutions compliant with climate issues? Bus. Strateg. Environ. **32**(6), 2817–2834 (2023)
23. De Bernardi, P., Venuti, F., Bertello, A.: The relevance of climate change related risks on corporate financial and non-financial disclosure in Italian listed companies. Futur. Risk Manag., Vol. I: Perspect. Law, Healthc. Environ. 77–107 (2019)
24. Bingler, J.A., Kraus, M., Leippold, M., Webersinke, N.: Cheap talk and cherry-picking: What ClimateBert has to say on corporate climate risk disclosures. Financ. Res. Lett. **47**, 102776 (2022)
25. Ngo, T., Le, T., Ullah, S., Trinh, H.: Climate risk disclosures and global sustainability initiatives: a conceptual analysis and agenda for future research. Bus. Strateg. Environ. **32**(6), 3705–3720 (2022)
26. Eccles, R., Krzus, M.: Implementing the task force on climate-related financial disclosures recommendations: an assessment of corporate readiness. Schmalenbach Business Review **71**(2), 287–293 (2018)
27. EDGAR.: GHG emissions of all world countries: JRC/IEA 2024 Report. JRC-EDGAR-JRC138862, Luxembourg (2024). https://data.europa.eu/doi/https://doi.org/10.2760/4002897
28. Landis, F., Schenker, O., Tovar Reaños, M., Vonnahme, C., Zitzelsberger, S.: An overview on current climate policies in the European Union and its member states. Centre for European Economic Research, ZEW Mannheim (2012)
29. Riffe, D., Lacy, S., Watson, B. R., Lovejoy, J.: Analyzing media messages: Using quantitative content analysis in research. Routledge (2023)
30. Lombardi, R., Schimperna, F., Paoloni, P., Galeotti, M.: The climate-related information in the changing EU directive on non-financial reporting and disclosure: first evidence by Italian large companies. J. Appl. Acc. Res. **23**(1), 250–273 (2021)
31. Dutta, P., Dutta, A.: Impact of external assurance on corporate climate change disclosures: new evidence from Finland. J. Appl. Acc. Res. **22**(2), 252–285 (2020)
32. Kılıç, M., Kuzey, C.: Determinants of climate change disclosures in the Turkish banking industry. The Internat. J. Bank Mark. **37**(3), 901–926 (2019)
33. Papoulis, D., Kaika, D., Bampatsou, C., Zervas, E.: Public perception of climate change in a period of economic crisis. Climate **3**(3), 715–726 (2015)
34. Spyridi, D., Vlachokostas, C., Michailidou, A., Sioutas, C., Moussiopoulos, N: Strategic planning for climate change mitigation and adaptation: the case of Greece. Internat. J. Climate Chang. Strat. Manag. **7**(3), 272–289 (2015)
35. Coen, D., Herman, K., Pegram, T.: Are corporate climate efforts genuine? An empirical analysis of the climate 'talk-walk' hypothesis. Bus. Strateg. Environ. **31**(7), 3040–3059 (2022)
36. Raghupathi, W., Wu, S., Raghupathi, V.: Understanding corporate sustainability disclosures from the securities exchange commission filings. Sustainability **15**(5), 4134 (2023)
37. Deweerdt, T., Caltabiano, K., Dargusch, P.: Original research: how will the TNFD impact the health sector's nature-risks management? Int. J. Environ. Res. Public Health **19**(20), 13345 (2022)
38. Gahramanova, G.: Corporate climate change disclosures and capital structure strategies: evidence from türkiye. J. Capital Markets Stud. **7**(2), 140–155 (2023)

Ecologically Optimal Quantity Extraction: Conflicting Cases

George Emm. Halkos, George J. Papageorgiou, Emmanuel G. Halkos, and John G. Papageorgiou

Abstract The extraction of environmental resources poses significant challenges for both developed and developing countries. Designing effective policies to mitigate environmental degradation requires accounting for society's long-term responses, for which dynamic modeling provides a valuable tool. This study employs game-theoretic approaches to analyze interactions between resource exploiters and policymakers under Nash and Stackelberg frameworks. In the Nash game, we first examine the shadow price of environmental quality and find that exploiters internalize only part of the future environmental loss. We also show that policymakers tend to act more impatiently than exploiters, a behavioral asymmetry that shapes long-term environmental outcomes. In the hierarchical (Stackelberg) setting, we derive explicit expressions for strategic choices, steady-state environmental quality, and policymakers' payoffs. Finally, we identify the conditions under which conflicts between exploiters and regulators intensify, depending on the costs associated with environmental damage.

Keywords Differential games · Environmental degradation · Nash game · Stackelberg game

1 Introduction

In this paper, we consider a fundamental environmental economics problem, i.e., the problem of extraction of natural resources. We care about how the social planner can regulate this exploitation of natural capital optimally. For that purpose, we choose to study the conflicting case between the representative extracting firm and the social planner of a country. In other words, we consider a dynamic game between the two above players in the Nash (simultaneous move) case and the Stackelberg (hierarchical move) case. In the environmental economics literature, there are many papers

G. Emm. Halkos (✉) · G. J. Papageorgiou · E. G. Halkos · J. G. Papageorgiou
Department of Economics, Laboratory of Operations Research, University of Thessaly, Volos, Greece
e-mail: halkos@uth.gr

G. Halkos et al. (eds.), *Environmental Risk Analysis*, Springer Proceedings in Mathematics & Statistics 540, https://doi.org/10.1007/978-3-032-24155-9_14

concerning both cases of equilibrium. Ulph and Folie [10] in a cartel–fringe type model derive the open-loop equilibrium under constant marginal extraction costs. Groot et al. [11] in the field of exhaustible resources propose a model of cartel and fringe under a feedback Stackelberg assumption. Boucekkine et al. [12], in a Nash model, investigate the extent to which the irreversibility of pollution shapes the free-riding problems inherent in pollution. Benchekroun et al. [13] consider a Nash game between oligopolistic and fringe suppliers of fossil fuel from an exhaustible resource, and producers of a renewable perfect substitute. Antoniadou et al. [14] construct a game of noncooperative common-resource exploitation which delivers analytical solutions for its symmetric Markov-perfect Nash equilibrium. Athanasoglou [15] studies a class of differential games of pollution control with profit functions that are polynomial in the global pollution stock. Wagener [16] concerns about the fundamental question in the problem of externalities: costs that are generated by one group of actors, which have to be borne by another group. Martín-Herrán and Rincón-Zapatero [17] present a method for the characterization of Markov-perfect Nash equilibria being Pareto-efficient in nonlinear differential games.

In our model, the game proceeds as follows: the social planner determines a specified quantity of natural resources to be exploited by the representative firm, ensuring the social optimality of the natural resource stock [1, 4]. The firm is therefore constrained to extract exactly this planner-determined quantity, adjusting only through the intensity of its performance index. Hence, the two strategic instruments are (i) the exploitation quota set by the planner and (ii) the performance intensity chosen by the firm. In the Nash game, our focus lies on periodic equilibria—that is, cyclical solutions—rather than static point equilibria. Such equilibria, often referred to as limit cycle equilibria, are well documented in environmental economics (see Skiba [8], Wirl [9], Dockner et al. [3], and Halkos et al. [5, 6]). The intuition follows from the Poincaré–Bendixson theorem, which applies in planar dynamical systems: if the trajectory of a continuous system remains forever within a bounded region, it must eventually converge to either a steady point or a cycle. If convergence to a point is excluded, the system necessarily evolves into a cycle. This mathematical property translates naturally into economic interpretation: cyclical policies emerge whenever a policy trajectory—such as an environmental subsidy or regulation—is constrained within a bounded two-dimensional space. In such cases, the policy path will inevitably retrace its steps, giving rise to recurrent, cyclical dynamics in resource use and regulation.

Moreover, we extend the analysis by exploring the state separability properties of the model, which allow us to derive more general results. State-separable games offer an important advantage: every open-loop equilibrium coincides with a closed-loop equilibrium, thereby ensuring time-consistent strategies. Within this framework, we identify the conditions under which optimal equilibrium strategies exist for both players.

We then turn to the hierarchical (Stackelberg) setting, which represents the sequential decision-making process typically followed in society. Stackelberg differential games have been widely applied across various fields to capture such leader–follower dynamics. In this context, we derive several propositions that provide explicit analytical expressions for the optimal strategies of both players, together with the parameter conditions under which these strategies hold.

The rest of the paper is organized as follows. Section 2 sets up the Nash game, Sect. 3 concerns the Stackelberg game, while Sect. 4 concludes the paper.

2 Setting up the Nash Game

In this setting, we suppose that the function that depletes the environmental natural resources (thought as a natural capital) is a function of the intensity of the performance index U of the whole extracting firms and of the extracted quantity of the overall resources Q, denoted as the harvesting function $h(U, Q)$. Additionally, we assume that nature grows its resources, therefore the natural resources are augmented without human intervention. We denote this natural growth as the function $G(E(t))$. Putting growth and harvesting together, the instant change of environmental dynamics has the following form:

$$\frac{dE}{dt} = \dot{E} = G(E(t)) - h(U, Q) \tag{1}$$

The nonnegativity constraint is imposed to the environmental index, i.e., $E(t) \geq 0 (\forall t \geq 0)$.

For the harvesting function $h(U, Q)$ we impose the following valid assumptions. Since the higher performance over the environmental optimal leads to degradation, the partial derivative of the function could be positive, i.e., $h_U(U, Q) > 0$. The law of diminishing returns is applied only for the performance index, therefore $h_{UU}(U, Q) < 0$. For the remainder quantity of the natural resources extracted, it is supposed that $h_{QQ}(U, Q) > 0$. Moreover, the two policy instruments U, Q reinforced each other that is $h_{UQ}(U, Q) > 0$.

Player 1

The representative firm that exploits environmental resources derives utility on the one hand from its harvesting activity $h(U, Q)$. However, the intensity of harvesting, denoted by U, is not costless: the firm incurs costs through a cost function $C(U)$. In addition, maintaining the socially optimal level of environmental quality imposes compliance costs (e.g., subsidy withdrawals or environmental fees), which depend on the extraction quantity and environmental quality. These costs are represented by a function $D(E)$ also depending on the environmental quality E [2]. Accordingly, the representative firm seeks to maximize the following objective functional:

$$J_1 = \int_0^\infty e^{-\rho_1 t} \left[h(U, Q) - C(U) - D(E) \right] dt \tag{2}$$

ρ_1 is the firm's discount rate.

The firm's control instrument is the exploitation intensity U. In the absence of regulation by the social planner, greater intensity leads to greater environmental degradation. Put differently, unregulated exploitation depletes environmental quality—contrary to the notion of social optimality emphasized since Hotelling's seminal contribution [7].

Player 2

The social planner regulates exploitation by determining the quantity of resources Q that may be extracted. This policy instrument is set to ensure that extraction supports social optimality. More specifically, the planner, knowing the demand, estimates the quantity that can be sustainably extracted without causing excessive environmental damage, thereby protecting societal welfare. The planner derives instantaneous utility from two sources:

(a) Environmental quality, through $W_1(E)$.
(b) The extracted resource stock, through $W_2(Q)$ provided the extraction occurs in a socially optimal manner.

The social planner's objective is therefore to maximize

$$J_2 = \int_0^\infty e^{-\rho_2 t}\left[W_1(E) + W_2(Q)\right]dt \tag{3}$$

ρ_2 is the social planner's discount rate.

The Game

The strategic interaction between the firm and the social planner thus consists of maximizing (2) and (3) subject to the system dynamics (1). The firm's instrument is the exploitation intensity U while the planner's instrument is the extraction quantity Q.

The Nash Equilibrium of the Game

We proceed by forming the Hamiltonians of the two players:

$$H_1 = h(U, Q) - C(U) - D(E) + \lambda(G(E(t)) - h(U, Q)) \tag{4}$$

$$H_2 = W_1(E) + W_2(Q) + \mu(G(E(t)) - h(U, Q)) \tag{5}$$

and the first order conditions are

$$\frac{\partial H_1}{\partial U} = (1 - \lambda)h_U(U, Q) - C'(U) = 0 \tag{6}$$

$$\frac{\partial H_2}{\partial Q} = W_2'(Q) - \mu h_Q(U, Q) = 0 \tag{7}$$

The evolution of the shadow values λ, μ is defined as follows:

$$\frac{d\lambda}{dt} = \dot{\lambda} = \rho_1 \lambda - \frac{\partial H_1}{\partial E} = \rho_1 \lambda - \lambda G'(E) + D'(E) \tag{8}$$

$$\frac{d\mu}{dt} = \dot{\mu} = \rho_2 \mu - \frac{\partial H_2}{\partial E} = \rho_2 \mu - \mu G'(E) - W_1'(E) \tag{9}$$

According to (6) the Hamiltonian of Player 1, H_1, is concave in the control variable (the performance index) U as far as $\lambda < 1$ and it is guaranteed by the assumptions on derivatives, i.e., $h_{UU} < 0$, $h_{QQ} > 0$ and from the decreasing marginal returns on the society's utility, i.e.,$W_1''(E) < 0$, $W_2''(Q) < 0$. Solving the optimality condition (6) implies that $\lambda = \frac{1}{h_U(U,Q)}\big(h_U(U, Q) - C'(U)\big)$. This result is recorded as the next proposition.

Proposition 1 The shadow value of environmental quality for the firm that extracts, λ, is a positive quantity only if the marginal utility accrued from her exploitation activity exceeds the marginal costs incurred by the same activity. Moreover, the firm's Hamiltonian is maximized only if the same shadow value λ is less than unity.

Proof The result is easily obtained, taking the value of λ and having in mind the assumption $h_U(U, Q)>0$, i.e., the fact that high-intensity performance leads to environmental degradation.

Precisely, the second derivative of the Hamiltonian with respect to the control U is

$$\frac{\partial^2 H_1}{\partial U^2} = (1 - \lambda) h_{UU}(U, Q) - C''(U)$$

For an interior maximizer, we need the Hessian w.r.t. U to be negative, i.e., $(1 - \lambda)h_{UU}(U, Q) - C''(U) < 0$. Under the model assumptions $h_{UU}(U, Q) < 0$ (diminishing returns of the extraction function in the control U) and increasing marginal cost $C''(U) > 0$. Due to diminishing returns of the extraction function, if $(1 - \lambda)$ were negative the term $(1 - \lambda)h_{UU}(U, Q)$ would be positive and could violate the negativity of the Hessian. To ensure the left-hand side is strictly negative in the standard parametrization, it is sufficient (and economically natural) that $(1 - \lambda) > 0$.

Remarks in Economics

(a) **Player 1**: According to (6) the left-hand side of that first order condition (FOC) $(1 - \lambda)h_U(U, Q)$ is the effective marginal revenue of increasing intensity U: h_U is the marginal exploitation activity produced by raising intensity and the factor $(1 - \lambda)$ captures that increased harvesting not only gives immediate benefit h_U, but also reduces the future environment which, through the costate λ, alters the value. This equals marginal cost $C'(U)$.

Player 2: FOC (7) equates marginal social benefit from permitting more extraction $W_2'(Q)$ to the social shadow cost $\mu h_Q(U, Q)$ (the social planner internalizes the damage to the environment through μ).

(b) The necessity for the costate $\lambda > 0$ in (6) implies $C'(U) < h_U(U, Q)$ and because $h_U(U, Q) > 0$ the inequality is well defined. So, the costate λ is greater than zero when the marginal benefit of increasing intensity (the marginal harvest h_U) exceeds the marginal cost $C'(U)$.

Condition $1 > \lambda > 0$ implies $1 > 1 - \lambda > 0$ and therefore $0 < C'(U) < h_U(U, Q)$, i.e., the extracting firm faces positive marginal cost and the marginal extraction of natural resources is larger than that cost. This economically means the firm places positive value on environmental quality only when the immediate marginal gains from raising exploitation intensity exceed marginal costs; and that shadow price must be less than one so that the firm's instantaneous trade-off between private extraction gains and environmental damage yields an internal optimum (concavity).

Periodic Solutions

In this subsection we investigate whether periodic solutions are possible, starting at the steady state and stability analysis. We take the following Jacobian matrix:

$$J = \begin{pmatrix} \frac{\partial \dot{E}}{\partial E} & \frac{\partial \dot{E}}{\partial \lambda} & \frac{\partial \dot{E}}{\partial \mu} \\ \frac{\partial \dot{\lambda}}{\partial \mathrm{E}} & \frac{\partial \dot{\lambda}}{\partial \lambda} & \frac{\partial \dot{\lambda}}{\partial \mu} \\ \frac{\partial \dot{\mu}}{\partial \mathrm{E}} & \frac{\partial \dot{\mu}}{\partial \lambda} & \frac{\partial \dot{\mu}}{\partial \mu} \end{pmatrix} = \begin{pmatrix} G'(E) & \frac{-\partial h(U,Q)}{\partial \lambda} & \frac{-\partial h(U,Q)}{\partial \mu} \\ -\lambda G''(E) + D''(E) & \rho_1 - G'(E) & 0 \\ -\mu G''(E) - W_1''(E) & 0 & \rho_2 - G'(E) \end{pmatrix}$$

This Jacobian gives trace: $tr(J) = \rho_1 + \rho_2 - G'(E)$ and determinant:

$$det(J) = G'(E)\big(\rho_1 - G'(E)\big(\rho_2 - G'(E)\big)\big) - \frac{\partial h(U, Q)}{\partial \lambda}$$

$$\big(\lambda G''(E) + D''(E)\big)\big(\rho_2 - G'(E)\big) - \frac{\partial h(U, Q)}{\partial \lambda}\big(\big(\rho_1 - G'(E)\big(\mu G''(E) + W_1''(E)\big)\big)$$

According to Feichtinger and Novak (1983) (Lemma 4.1), the existence of a pair of purely imaginary eigenvalues requires the following conditions to satisfy:

$$tr(J) > 0, det(J) > 0, w > 0, det(J) = w tr(J)$$

$$\text{where} w = \begin{vmatrix} G'(E) & \frac{-\partial h(U,Q)}{\partial \lambda} \\ -\lambda G''(E) - D''(E) & \rho_1 - G'(E) \end{vmatrix} +$$

$$\begin{vmatrix} \rho_1 - G'(E) & 0 \\ 0 & \rho_2 - G'(E) \end{vmatrix} + \begin{vmatrix} G'(E) & \frac{-\partial h(U,Q)}{\partial \mu} \\ -\mu G''(E) - W_1''(E) & \rho_2 - G'(E) \end{vmatrix} =$$

$= \rho_1\rho_2 - [G'(E)]^2 - \frac{\partial h(U,Q)}{\partial \lambda}[\lambda G''(E) + D''(E)] - \frac{\partial h(U,Q)}{\partial \mu}[\mu G''(E) + W_1''(E)]$
The condition for cyclical strategies (Hopf bifurcations) is

$$w > 0, w = \frac{\det(J)}{tr(J)}$$

which after simple algebraic simplifications reduces into the following:

$$\rho_1\rho_2[\rho_1 + \rho_2 - 2G'(E)] = \frac{\partial h(U,Q)}{\partial \lambda}\left[\lambda G''(E) + D''(E)\right]\rho_1 + \frac{\partial h(U,Q)}{\partial \mu}\left[\mu G''(E) + W_1''(E)\right]\rho_2 \quad (10)$$

Specifications

We specify the functions involved as follows. A diffusion process for the growth function, i.e., $G(E) = RE(1 - E)$, a Cobb–Douglas type specification for the harvesting, i.e., $h(U,Q) = U^\gamma Q$ and for the utility which the society enjoy, arisen from the natural resources remainder stock, in the form $W_2(Q) = \Gamma - \frac{Q^{\xi-1}}{\xi-1}$. The latter function exhibits constant relative risk aversion in the sense of the Arrow–Pratt measure. The utility that the society enjoys from a good environmental quality is assumed in a linear form as well, i.e., $W_1(E) = W_1E$. All the costs are assumed in linear fashion, i.e., the extraction cost $C(U) = CU$, while the environmental degradation cost is specified as $D(E) = DE$.

It is worth noting that all the constant slopes C, D, W_1 and the intrinsic growth rate R are assumed positive real numbers, the coefficients $\gamma \in (0,1)$, $\xi \in (0,1)$, while $\Gamma > 0$.

With the above assumptions we record the following result.

Proposition 2 A necessary condition for cyclical strategies in the game between the exploiters of environmental resources and the social planner who cares about environmental degradation is the first player (the extracting firm) acts less risky than the second (social planner) (i.e., $\rho_2 > \rho_1$).

Proof On request.

Remark on Proposition 2

It is important to understand that $\rho_2 > \rho_1$ opens the possibility that the social planner's term can dominate and the Hopf equality may be satisfied, but other parameter combinations (magnitudes of$D$, W_1, R, γ, and the steady state E) determine whether equality and the additional Hopf technical conditions actually hold.

State Separability of the Game

State separability is an attractive game theoretical property because every open-loop solution of the game is also a closed-loop solution. Therefore, linearity and the induced state separability become the game's crucial assumption which makes economic sense. That is, some functions of the game are slightly modified to be in the following form:

1. The environmental growth function assumed it grows at exponential form, i.e., in the form $G(E) = \omega \mathrm{E}$, ω being the growth rate.
2. The extraction cost is assumed linear, $C(U) = CU$ as in previous subsection.
3. The compliance costs are linear as well $D(E) = DE$.
4. The utilities the society enjoys are left in a linear form as

$$W_1(E) = BE, W_2(Q) = AQ$$

All the involved constants are positive real numbers $\omega, C, D, A, B > 0$.
After all, the system of Eqs. (6)–(9) becomes

$$\frac{\partial H_1}{\partial U} = (1 - \lambda)h_U(U, Q) - C = 0 \tag{11}$$

$$\frac{\partial H_2}{\partial Q} = A - \mu h_Q(U, Q) = 0 \tag{12}$$

$$\frac{d\lambda}{dt} = \dot{\lambda} = \rho_1\lambda - \frac{\partial H_1}{\partial E} = \rho_1\lambda - \lambda G'(E) + D'(E) = \lambda(\rho_1 - \omega) + D \tag{13}$$

$$\frac{d\mu}{dt} = \dot{\mu} = \rho_2\mu - \frac{\partial H_2}{\partial E} = \rho_2\mu - \mu G'(E) - W_1'(E) = \mu(\rho_2 - \omega) - \mathrm{B} \tag{14}$$

and the following limiting transversality conditions have to hold:

$$\lim_{t\to\infty} e^{-\rho_1 t}\mathrm{E}(t)\lambda(t) = 0,\ \lim_{t\to\infty} e^{-\rho_2 t}\mathrm{E}(t)\mu(t) = 0 \tag{15}$$

The analytical expressions of the costate variables (i.e., the shadow prices of the environmental quality for the extracting firm and the society respectively) $\lambda(t)$, $\mu(t)$ are respectively:

$$\lambda(t) = \frac{D}{-\rho_1 + \omega} + e^{(\rho_1-\omega)t}\Omega_1 \tag{16}$$

$$\mu(t) = -\frac{\mathrm{B}}{-\rho_2 + \omega} + e^{(\rho_2-\omega)t}\Omega_2 \tag{17}$$

As the transversality conditions need to satisfy, it is convenient to choose the constant steady-state values, and therefore the adjoint variables collapse to the following constants:

$$\lambda(t) = \frac{-D}{\rho_1 - \omega},\ \mu(t) = \frac{\mathrm{B}}{\rho_2 - \omega} \tag{18}$$

To ensure certain signs for the costate variables (17) we impose another condition on the discount rates, which claim that discount rates are greater than the natural resource's growth, i.e., the condition

$$\rho_i > \omega \tag{19}$$

The above inequality (19) seems restrictive, but, in the opposite case, the maximization of Hamiltonian of the society and the shadow price of natural resources stock for the extracting firm wouldn't occur. Indeed, supposing that $\rho_1 < \omega$ the shadow price of the natural resource stock $\lambda(t)$ for the extracting firm would be positive in the long run, which is impossible. In the case $\rho_2 < \omega$ the shadow price of natural resources stock for the social planner becomes negative which is again impossible, and for the same choice, the value of the objective of the society becomes unbounded when choosing to levy no restrictions to the remainder stock of the natural resources.

We record these findings in the following proposition.

Proposition 3 Under the assumptions:

$G(E) = \omega\mathrm{E}$, $C(U) = CU$, $D(E) = DE$, $W_1(E) = BE$, $W_2(Q) = AQ$ and $h(U, Q)$ satisfying $h_U > 0$, $h_Q > 0$ every open-loop equilibrium is a closed-loop equilibrium. Moreover, the best responses (and thus the equilibrium) feature constant costates $\lambda(t) = \frac{-D}{\rho_1-\omega}$, $\mu(t) = \frac{\mathrm{B}}{\rho_2-\omega}$ and constant controls. These are economically admissible if and only if $\rho_i > \omega$, $i = 1{,}2$.

Proof Sketch

State separability implies open-loop equilibrium is closed loop [18]. The values of the costates are already taken as in (18); admissibility requires $\lambda^* \leq 0$ (firm faces only $D'(E) > 0$ and $\mu^* > 0$) (the social planner values E via $B > 0$) which is equivalent to $\rho_i > \omega$. With constant costates the FOCs give constant controls; the instant change of the environment function $\dot{E} = G(E(t)) - h(U, Q)$ converges and the transversality condition holds.

Economic Interpretation of the Costate Signs

(a) $\lambda^* < 0$ if and only if $\rho_1 > \omega$. This is economically consistent for the extracting firm: a higher environmental quality E raises the firm's flow cost D(E) but gives no direct firm benefit, so the shadow value of E should be nonpositive. If $\rho_1 < \omega$ then the shadow value of environmental quality $\lambda^* > 0$ meaning the firm values E positively—at odds with $\mathrm{D'(E)} = \mathrm{D} > 0$ and no private environmental benefit—producing a perverse comparative statics and, with linear payoffs, scope for unbounded incentives.

(b) $\mu^* > 0$ if and only if $\rho_2 > \omega$. This is economically required for the planner: $W_1'(E) = B > 0$ so an extra unit of environmental resources E should a have positive shadow value. If $\rho_2 < \omega$ the linear structure drives $\mu^* < 0$, contradicting the social planner's valuation of environmental quality, and in the linear objective it opens the door to unbounded improvements by letting E drift.

(c) With $\rho_i > \omega$, $i = 1{,}2$ the stationary costates are constant and the control strategies U^*, Q^* are constant. Then $\dot{E} = \omega(\mathrm{E} - \mathrm{E}^*)$ implies $E(t)$ is bounded and the transversality condition holds, i.e., $\lim_{t\to\infty} e^{-\rho_1 t}\mathrm{E}(t)\lambda^* = 0$.

So, $\rho_i > \omega$ are not merely convenient: they are the admissibility conditions that deliver economically sensible costate signs and well-behaved objectives.

Economic Intuition

If the resource regenerates at rate ω but a player discounts more slowly $\rho_i < \omega$, the present value of leaving the stock in situ grows too quickly relative to discounting, breaking the transversality condition and pushing best responses to corners (delay extraction indefinitely or deplete aggressively). When $\rho_i > \omega$, the discounted shadow value of the stock is bounded, yielding a unique, interior stationary response, and a well-behaved Nash outcome.

3 The Stackelberg Setting

In this section, we analyze the case in which the two players of the game move hierarchically and the intensity of the effort is chosen by the extracting firm before the society via the social planner decides about the volume of the remainder of the stock of the natural resources: i.e., the extracting firm is the leader of the game.

3.1 The Society as a Follower

First, we consider the problem of the follower, i.e., the society, which takes the decision of the leader as given. Therefore, the society, via the social planner, has to maximize the following functional:

$$max_Q \int_0^\infty e^{-\rho_2 t}\big[W_1(E) + W_2(Q)\big]dt \tag{20}$$

Note that the maximization takes place concerning the remainder quantity of the resource stock, whose volume is estimated so that it can be extracted optimally in the future. The environmental quality evolves according to the equation:

$$\frac{dE}{dt} = \dot{E} = G(E(t)) - h(U, Q) \tag{21}$$

already mentioned above, but the specification for its growth is linear; therefore the equation of motion becomes

$$\dot{E} = \omega E - h(U, Q) \tag{22}$$

ω being the growth rate. Moreover, the linearity of both utility functions, which the society enjoy, is assumed, i.e., the utility stemming from the good environmental quality is $W_1(E) = W_1 E$, and the potential utility which will be accrued from the volume of the remainder stock, since it is intended to extracted optimally, is $W_2(Q) = W_2 Q$.

After all, the current value of Hamiltonian of the society will be

$$H_2 = W_1 E + W_2 Q + \mu\big[\omega E - h(U, Q)\big]$$

and the first order condition:

$$\frac{\partial H_2}{\partial Q} = W_2 - \mu h_Q(U, Q) = 0 \tag{23}$$

Assume a Cobb–Douglas specification for the function that depletes the natural resources stock, i.e.,

$$h(U, Q) = U^{\pi} Q^{\varepsilon}, \varepsilon > 1 > \pi > 0$$

Now we adapt the above specification into the maximization condition (23); we get

$$h_Q(U, Q) = \varepsilon U^{\pi} Q^{\varepsilon - 1} \tag{24}$$

Therefore, the optimal strategy that the social planner follows becomes a function of environmental quality. The above discussion is recorded as in the next proposition.

Proposition 4 The equilibrium strategy of the social planner reacting as a follower in the Stackelberg game is.

$$Q^*(U) = \left(\frac{W_2}{\varepsilon\mu}\right)^{\frac{1}{\varepsilon-1}} U^{\frac{\pi}{1-\varepsilon}} \tag{25}$$

Proof Substituting the partial derivative $h_Q(U, Q) = \varepsilon U^{\pi} Q^{\varepsilon-1}$ into the maximization condition (23), we have $W_2 - \mu\varepsilon U^{\pi} Q^{\varepsilon-1} = 0$ which implies the policy (25).

Remarks and Economic Intuition

Note that the social planner's optimal policy, i.e., the policy of the extracted quantity, is dependent only on the model constant parameters and on the extracting firm's intensity of effort.

Since $\varepsilon > 1$ the exponent $(1/\varepsilon - 1) > 0$. Because $\pi > 0$, the exponent $(-\pi/(\varepsilon - 1))$ on U is negative, which means as the leader's extraction effort U increases the follower's optimal remaining-stock policy $Q^*(U)$ decreases. In economic sense, higher current exploitation effort by the leader lowers the socially optimal remainder to hold for future extraction.

Moreover, μ is the society's current-value costate (shadow value of the resource). For an interior solution, we need $\mu > 0$ and the right-hand side to be well defined; we claim these regularity/transversality conditions in the paper.

The policy $Q^*(U)$ decreases with the planner's shadow value μ (higher social valuation of environmental quality implies tighter allowances) and increases with the firm's intensity U (for $\varepsilon \in (0,1)$, higher leader intensity calls for a higher allowance to keep the induced extraction on the planner's optimal trajectory). A larger W_2 (higher marginal social utility from extracted resources) raises the permitted quantity. Overall, the follower's rule trades off marginal social benefits from extraction against the marginal shadow cost of environmental degradation.

If the strategy $Q^*(U)$ implied by the expression (23) is negative or violates feasibility constraints, the social planner will choose a corner solution (e.g., $Q = 0$ or the largest feasible Q); therefore it is clear that $Q \geq 0$ and any upper bounds on Q are respected.

Continuing with solution process, we simplify the differential equation of the social planner's costate variable, i.e., the equation:

$$\dot{\mu} = \rho_2\mu - \frac{\partial \mathrm{H}_2}{\partial E} = \mu(\rho_2 - \omega) - W_1 \tag{26}$$

Substituting the optimal policy of the social planner into the depletion (harvesting) function $h(U, Q)$, we get the analytical form of that function, i.e., the depletion function, when the social planner follows the optimal strategy (25), becomes a function of the model constants and the intensity of performance of the extracting firm. The next proposition is a result of the above discussion.

Proposition 5 When the social planner follows the follower rule $Q^*(U) = \left(\frac{W_2}{\varepsilon\mu}\right)^{\frac{1}{\varepsilon-1}} U^{\frac{\pi}{1-\varepsilon}}$, the harvesting function becomes.

$$h\big(U, Q^*(U)\big) = U^{\pi}\left[Q^*(U)\right]^{\varepsilon} = U^{\frac{\pi}{1-\varepsilon}}\left(\frac{\varepsilon\mu}{W_2}\right)^{\frac{\varepsilon}{1-\varepsilon}} \tag{27}$$

Proof Substituting the optimal policy $Q^*(U)$ (25) into the harvesting function $h(U, Q) = U^{\pi} Q^{\varepsilon}$,

$$h\big(U, Q^*(U)\big) = U^{\pi}\left[\left(\frac{W_2}{\varepsilon\mu}\right)^{\frac{1}{\varepsilon-1}} U^{\frac{\pi}{1-\varepsilon}}\right]^{\varepsilon} = U^{\pi+\frac{\pi\varepsilon}{1-\varepsilon}}\left(\frac{W_2}{\varepsilon\mu}\right)^{\frac{\varepsilon}{\varepsilon-1}}$$

Noting that $\pi + \frac{\pi\varepsilon}{1-\varepsilon} = \frac{\pi}{1-\varepsilon}$ and $\left(\frac{W_2}{\varepsilon\mu}\right)^{\frac{\varepsilon}{\varepsilon-1}} = \left(\frac{\varepsilon\mu}{W_2}\right)^{\frac{\varepsilon}{1-\varepsilon}}$ we get (27).

Short Comment

The extraction function (27) is crucial: it reduces the depletion flow to a single-state-dependent term driven by the firm's intensity U and the social planner's shadow value μ. This substitution is what converts the leader's problem into the two-state optimal control problem solved in Proposition 7.

The economic interpretation of the above research is as in Proposition 6.

Proposition 6 (Cautionary follower response when $\varepsilon > 1$).

Assume $\varepsilon > 1$ and $\pi > 0$. Then the social planner's best-response policy $Q^*(U) = \left(\frac{W_2}{\varepsilon\mu}\right)^{\frac{1}{\varepsilon-1}} U^{\frac{\pi}{1-\varepsilon}}$ satisfies $\frac{\partial Q^*}{\partial U} < 0$, i.e., the planner tightens (reduces) the allowed extraction Q^* when the firm increases its extraction intensity U. Consequently, substituting Q^* into the harvesting flow

$$h\big(U, Q^*(U)\big) = U^{\frac{\pi}{1-\varepsilon}} \left(\frac{W_2}{\varepsilon\mu}\right)^{\frac{\varepsilon}{\varepsilon-1}}$$

yields a harvesting term that is decreasing in U (because $\frac{\pi}{1-\varepsilon} < 0$), so that an increase in the firm's effort can lead—through the planner's tightened allowance—to a lower net depletion flow.

Proof Differentiate $Q^*(U)$ with respect to U. Using the power form,

$Q^*(U, \mu) = \Xi(\mu)U^{\varphi}$, $\Xi(\mu) = \left(\frac{W_2}{\varepsilon\mu}\right)^{\frac{1}{\varepsilon-1}}$, and $\varphi = \frac{\pi}{1-\varepsilon}$.

Hence,

$$\frac{\partial Q^*}{\partial U} = \Xi(\mu)\varphi U^{\varphi-1} = \frac{\varphi}{U} Q^*(U, \mu)$$

Under the assumption $\varepsilon > 1$ we have $1 - \varepsilon < 0$, so $\varphi = \frac{\pi}{1-\varepsilon} < 0$(since $\pi > 0$). Therefore $\frac{\partial Q^*}{\partial U} < 0$, proving the claim. Substituting Q^* into $h(U, Q^*(U))$ gives the stated harvesting function; because the exponent $\frac{\pi}{1-\varepsilon}$ is negative $h(U, Q^*(U))$ is decreasing in U, as asserted, Q.E.D.

Economic Intuition: When the elasticity parameter ε exceeds one ($\varepsilon > 1$), the social planner's optimal rule reacts strongly to changes in U. As the extracting firm intensifies effort, the social planner tightens the quota Q^* sufficiently that overall harvesting may fall. This reflects a strong corrective (precautionary) response by the follower: higher private pressure triggers a stricter public constraint, which can reduce net depletion despite the firm's higher effort.

In order to have better interpretation for the results, we compute the stationary costate $\overline{\mu}$ which satisfies the condition $\dot{\mu} = 0$.

To do this, we apply stationarity to (26), and $0 = \overline{\mu}(\rho_2 - \omega) - W_1$ implying the stationary costate $\overline{\mu} = \frac{W_1}{\rho_2-\omega}$. The existence requires $\rho_2 \neq \omega$ and for a positive shadow value $\overline{\mu} > 0$ it suffices $\rho_2 > \omega$ and of course positive utility W_1. This matches the usual transversality requirement that the society discounts sufficiently faster than the resource grow.

Now we recall the society's best response strategy (23) and derive the stationary best response setting $\overline{\mu} = \mu$. Therefore, the society's best strategy becomes

$$Q^*(U) = \left(\frac{W_2}{\varepsilon}\frac{\rho_2 - \omega}{W_1}\right)^{\frac{1}{\varepsilon-1}} U^{\frac{\pi}{1-\varepsilon}} \tag{28}$$

Policy Implications

(a) The steady-state costate $\overline{\mu}$ equals the marginal environmental utility W_1 discounted by the excess discount rate $\rho_2 - \omega$.
(b) A larger social planner discount rate ρ_2 (relative to growth rate ω) lowers $\overline{\mu}$ which —through the formula—raises the best response Q^*.
(c) Since $\overline{\mu}$ appears in the denominator inside a positive power, larger ρ_2 (implies larger $\rho_2 - \omega$ which in turn implies larger $\overline{\mu}$) actually reduces Q^*. Economically, a planner who discounts more heavily values future stock less (higher ρ_2 decreases the weight on future environmental quality) raising the shadow price $\overline{\mu}$ provided $W_1 > 0$ and $\rho_2 > \omega$.
(d) The society's best response strategy $Q^*(U)$ is fully explicit in model's parameters.

Applying the steady-state costate $\overline{\mu} = \frac{W_1}{\rho_2-\omega}$ into the harvesting function (27) we have

$$h\big(U, Q^*(U)\big) = U^{\frac{\pi}{1-\varepsilon}} \left[\frac{\frac{\varepsilon W_1}{\rho_2-\omega}}{W_2}\right]^{\frac{\varepsilon}{1-\varepsilon}} \tag{29}$$

Since $\varepsilon > 1$ and the exponent $\frac{\pi}{1-\varepsilon}$<0, so the steady-state harvesting $h(U, Q^*(U))$ decreases with U, which means that higher leader effort reduces the follower-chosen harvesting term, evaluated at the follower's optimal remainder.

The Behavior of the Extracting Firm

Following the solution process for hierarchical games (Dockner et al., 2000) we formulate the problem of the leader, i.e., the problem of the firm, which has to take into consideration the dynamics of the optimal decision of the follower, i.e., of the society, which is expressed by the differential equation (26). As a result, equation (26) becomes the second equation of motion for the optimization problem of the firm. In other words, we have to solve an optimal control problem for the leader (the extracting firm) subject to the two differential equations as constraints.

In this setting, we suppose that the extracting firm suffers from a cost which is due to the intensive use of the control policy, i.e., the allowed quantity to extract, and therefore we add the new disutility term as a linear function of the social planner policy Q, i.e., $\Psi(Q) = \Psi Q$, and therefore the firm's functional under maximization becomes

$$J_1 = \int_0^\infty e^{-\rho_1 t}\big[h(U, Q) - C(U) - D(E) - \Psi(Q)\big]dt \tag{30}$$

Substituting the analytical form of the harvesting function (27) and the analytical form of the social planner's optimal policy (23), the problem of the firm that exploits the natural resources is the following (assuming all the costs are in linear form again):

$$max_U \int_0^\infty e^{-\rho_1 t}\left[U^{\frac{\pi}{1-\varepsilon}}\left(\frac{\varepsilon\mu}{W_2}\right)^{\frac{\varepsilon}{1-\varepsilon}} - CU - DE - \Psi\left(\frac{W_2}{\varepsilon\mu}\right)^{\frac{1}{\varepsilon-1}} U^{\frac{\pi}{1-\varepsilon}}\right]dt \tag{31}$$

subject to

$$\dot{E} = \omega E - U^{\frac{\pi}{1-\varepsilon}}\left(\frac{\varepsilon\mu}{W_2}\right)^{\frac{\varepsilon}{1-\varepsilon}} \tag{32}$$

And

$$\dot{\mu} = \mu(\rho_2 - \omega) - W_1 \tag{33}$$

The resulting strategy of the firm that extracts (leader of the game) is recorded as in the next proposition.

Proposition 7 In the Stackelberg setting in which the social planner reacts as a follower of the game, the optimal strategy U^* of the firm that extracts natural resources is the following:

$$U^* = \left[\frac{W_2 C(\varepsilon - 1)}{\pi(\Psi\varepsilon\mu + W_2(\lambda - 1))}\right]^{\frac{1-\varepsilon}{\pi+\varepsilon-1}}\left(\frac{W_2}{\varepsilon\mu}\right)^{\frac{\varepsilon}{\pi+\varepsilon-1}}$$

Proof The Hamiltonian current value of the final maximization integral (30) under the constraints (28) and (29) is.

$$\begin{aligned} H_1 = \quad & U^{\frac{\pi}{1-\varepsilon}}\left[(1-\lambda)\left(\frac{\varepsilon\mu}{W_2}\right)^{\frac{\varepsilon}{1-\varepsilon}} - \Psi\left(\frac{\varepsilon\mu}{W_2}\right)^{\frac{1}{1-\varepsilon}}\right] \\ & +\lambda\omega E - CU - DE - \xi(\mu(\rho_2 - \omega) - W_1) \end{aligned} \tag{34}$$

Note that the new costate variables are λ and ξ corresponding to the (new) states which are the environmental quality E and the costate variable μ of the social planner (the follower of the game). Thus, the problem of the firm that extracts natural resources (leader of the game) becomes an optimal control problem with two state variables.

Continuing with the solution, we take the first order conditions of the Hamiltonian function, i.e.,

$$\frac{\partial H_1}{\partial U} = \frac{\pi}{1-\varepsilon} U^{\frac{\pi+\varepsilon-1}{1-\varepsilon}} \left(\frac{\varepsilon\mu}{W_2}\right)^{\frac{\varepsilon}{1-\varepsilon}} \left(1 - \lambda - \Psi\frac{\varepsilon\mu}{W_2}\right) - C = 0 \tag{35}$$

Solving the first order condition (35), with respect to the optimal extraction policy of the firm U^*, the resulting control can be expressed as a function of the shadow prices of the state E (the environmental quality), the shadow price of the social planner μ, and the constant parameters of the model, Q.E.D.

Second Order Condition

For the solution to be an interior maximizer, we need the second derivative of the Hamiltonian with respect to U to be negative at the FOC. Setting $k = \pi/(1-\varepsilon)$, $k > 0$ and $k - 1 = (\pi + \varepsilon - 1)/(1-\varepsilon)$ may be positive or negative depending on the parameters, a standard sufficient condition is the following (setting $A = \frac{\varepsilon\mu}{W_2}$):

$$(1-\lambda)\mathrm{A}^{\varepsilon/(1-\varepsilon)} - \Psi\mathrm{A}^{1/(1-\varepsilon)} > 0$$

Observing Hamiltonian (34) we split this function in terms depending on control U and term independent of U; therefore the dependent part is the function $\Phi(U) = U^k\big[(1-\lambda)A^{\varepsilon/(1-\varepsilon)} - \Psi A^{1/(1-\varepsilon)}\big] - CU$.

So, the coefficient multiplying U^k in $\Phi(U)$ is positive, because then the left-hand side derivative equation yields $U > 0$. This inequality is equivalently (multiplying by $A^{-\varepsilon/(1-\varepsilon)}$), $(1-\lambda) - \Psi A > 0$ implying

$$\Psi\varepsilon\mu + W_2(\lambda - 1) < 0$$

This is the sign condition implicitly required for the base inside the bracket in the displayed formula to be positive.

Economic Interpretation

The (marginal) net private benefit captured by $(1-\lambda)$ must exceed the marginal disutility the firm suffers from the social planner's extraction allowance ΨA. If that inequality fails, the interior solution for the control policy U does not exist (corner solutions or no maximum).

Proposition 7 shows that the firm's optimal extraction intensity U^* depends not only on technological and cost parameters but also on the shadow values of the environmental state λ and the planner's state variable μ. A higher shadow price of environmental quality λ makes extraction less attractive, reflecting stronger implicit value placed on environmental preservation. Similarly, an increase in μ—which captures the planner's valuation of environmental dynamics—raises the effective cost of extraction through the term $\Psi\varepsilon\mu$, thereby discouraging excessive exploitation. The disutility from the regulatory instrument $\Psi(Q)$ further internalizes the social cost of extraction. Together, these mechanisms imply that stricter regulatory pressure and higher social valuation of the environment shift the firm toward more conservative exploitation strategies, aligning private incentives more closely with social optimality.

The next result ensures the existence of feasible equilibrium policies.

Proposition 8 In the hierarchical (Stackelberg) game with the extracting firm as leader and the social planner as follower, a feasible optimal policy (a feasible interior optimum) $U^* > 0$ for the extracting firm exists only if.

$$\Psi > \frac{W_2(1-\lambda)}{\varepsilon\mu}$$

i.e., the per-unit cost rate Ψ (the disutility the firm suffers from the social planner's allowed quantity Q) must exceed the quantity $\frac{W_2(1-\lambda)}{\varepsilon\mu}$.

Proof Taking the second derivative of the Hamiltonian (34), i.e., the function.

$$\frac{\partial^2 H_1}{\partial U^2} = \frac{\pi(\pi+\varepsilon-1)}{(1-\varepsilon)^2} U^{\frac{\pi+\varepsilon-1}{1-\varepsilon}-1} \left(\frac{\varepsilon\mu}{W_2}\right)^{\frac{\varepsilon}{1-\varepsilon}} \left(1-\lambda-\Psi\frac{\varepsilon\mu}{W_2}\right)$$

the concavity of the Hamiltonian is ensured only if the inequality $\left(1-\lambda-\Psi\frac{\varepsilon\mu}{W_2}\right) < 0$ holds, since the constants π, W_2 and the costate μ are all positive quantities and $\varepsilon > 1$.

Therefore, $\Psi > \frac{W_2(1-\lambda)}{\varepsilon\mu}$.

Economic Interpretation

The inequality is a feasibility/concavity condition: it ensures the Hamiltonian is strictly concave in U and the FOC yields an interior maximum. Economically it says the marginal disutility Ψ faced by the extracting firm from the social planner's extraction allowance must be sufficiently large relative to the firm's net marginal private gain $(1-\lambda)$ (scaled by the endogenous multipliers W_2, ε, μ).

If the inequality fails (i.e., disutility Ψ is too small), the second derivative may be nonnegative and the FOC could correspond to a minimum or no interior maximum—the extracting firm's best response may be a corner solution (e.g., extreme U) rather than the interior U^* in Proposition 7.

4 Conclusions

This paper studies a dynamic game between resource-extracting firms and a social planner who regulate extraction through a quantity cap. The environmental capital is modeled as a stock that grows naturally and is depleted by harvesting. We analyze both noncooperative (open-loop Nash) and hierarchical (Stackelberg) formulations. The main findings are as follows.

Shadow-Price Bound in Nash Equilibrium

In the Nash game the firm's current-value shadow price of environmental quality λ must be strictly less than one for the firm's Hamiltonian to be maximized. Economically, this means the firm internalizes only part of the future environmental loss; if $\lambda \geq 1$ the extracting firm would over-weight future environmental value and the first-order condition cannot deliver an interior maximum.

Positive Shadow Value Condition

The firm's shadow value λ is positive only when the marginal harvest benefit of raising exploitation intensity exceeds the marginal private cost of doing so. Thus a positive valuation of the environment by the firm arises only when immediate private returns dominate operating costs.

Conditions for Cyclical (Periodic) Strategies

Linear stability analysis of the three-dimensional dynamical system (environment, firm costate, planner costate) yields a Hopf condition identifying parameter regions where purely imaginary eigenvalues appear. In the specified nonlinear example (logistic growth, Cobb–Douglas harvesting, and CRRA social utility), a necessary condition for periodic equilibria is that the social planner discount rate exceed the extracting firms' discount rate ($\rho_2 > \rho_1$). Intuitively, when society is relatively more impatient (or the planner's effective responsiveness is higher), the dynamic interaction can produce endogenous cycles in extraction and environmental quality.

State Separability: Open Loop Coincides Closed Loop and Admissibility

Under linear (exponential) environmental growth and linear payoffs, the game becomes state-separable. In this case every open-loop equilibrium is also a closed-loop (feedback) equilibrium. Admissible (economically meaningful) stationary solutions require discount rates above the natural growth rate of the environmental stock: $\rho_i > \omega$, $i = 1, 2$. This ensures the costates have the correct signs and that the transversality conditions hold.

Stackelberg (Leader–Follower) Characterization

Treating the extracting firm as leader and the social planner as follower yields closed-form policy rules. We derive the planner's best response as a function of model parameters and show how the depletion function (the steady-state/flow of environmental stock) depends on the firm's intensity. In the Stackelberg equilibrium, the social planner reduces (or "holds down") its allowed extraction in response to stronger effort by the extracting firm—an adaptive response that mitigates environmental loss.

Firm's Stackelberg Strategy

Solving the leader's control problem subject to the follower's response yields an explicit expression for the firm's optimal intensity U^* as a function of primitives $\pi, \varepsilon, \Psi, \lambda, \mu, W_2, C$. The interior solution exists under a sign condition that links the extracting firm's marginal net benefit and the disutility parameter Ψ.

Feasibility Threshold for Planner Costs (Proposition 8)

An interior feasible policy profile for both players exists iff the planner's per-unit cost rate Ψ is sufficiently large:

$$\Psi > \frac{W_2(1-\lambda)}{\varepsilon\mu}$$

Economically, the planner must impose a sufficiently strong marginal disutility on the firm's permitted extraction (or equivalently adopt sufficiently stringent enforcement/costs) to ensure the firm's Hamiltonian is concave and an interior optimum exists.

Summarizing the policy takeaways:

(a) The social planner can prevent overexploitation only if her policy instruments are appropriately calibrated relative to firms' private incentives (costs and discounting).
(b) Patience and discounting matter: relative discounting ($\rho_2 vs. \rho_1$ influences not only steady states but also the possibility of cycles in exploitation).
(c) Linear state separability simplifies policy design (Open Loop coincides Closed Loop) but imposes admissibility restrictions (discount rates larger than growth rates) that must be respected in calibration.

Disclosure of Interests The authors have no competing interests to declare that are relevant to the content of this article.

References

1. Becker, G., Murphy, K.: A theory of rational addiction. J. Polit. Econ. **96**, 675–700 (1988)
2. Benchekroun, H., Long, N.V.: A class of performance-based subsidy rules. Japanese Econ. Rev. **59**(4), 381–400 (2008)
3. Dockner, E.J. et al.: Differential Games in Economics and Management Science. Cambridge University Press (2000)
4. Gatto, M., et al.: On the optimality of the logistic growth. J. Optim. Theory Appl. **57**(3), 513–517 (1988). https://doi.org/10.1007/BF02346168
5. Halkos, G.E., et al.: Environmental regulation and economic cycles. Econ. Anal. Policy. **64**, 172–177 (2019). https://doi.org/10.1016/j.eap.2019.07.005

6. Halkos, G.E., et al.: Public debt games with corruption and tax evasion. Econ. Anal. Policy. **66**, 250–261 (2020). https://doi.org/10.1016/j.eap.2020.04.007
7. Hotelling, H.: The economics of exhaustible resources. J. Polit. Econ. **39**(2), 137–175 (1931)
8. Skiba, A.K.: Optimal growth with a convex-concave production function. Econometrica **64**(3), 527–539 (1978)
9. Wirl, F.: The cyclical exploitation of renewable resource stocs may be optimal. J. Environ. Econ. Manage. **29**, 252–261 (1995)
10. Ulph, A., Folie, G.M.: Exhaustible resources and cartels: an intertemporal Nash-Cournot model Can. J. Econ. **13**, 645–658 (1980)
11. Groot, F., Withagen, C., de Zeeuw, A.: Strong time–consistency in the cartel–versus–fringe model. J. Econ. Dyn. Control **28**, 287–306 (2003)
12. Boucekkine, R., Ruan, W., Zou, B.: The irreversible pollution game. J. Environ. Econ. Manag. **120** (2023)
13. Benchekroun, H., van der Meijden G., Withagen C.: Economically exhaustible resources in an oligopoly-fringe model with renewables. J. Environ. Econ. Manag. **121** (2023)
14. Antoniadou, E., Koulovatianos, C., Mirman, L.: Strategic Exploitation of a common property resource under uncertainty. J. Environ. Econ. Manag. **65** (2013)
15. Athanassoglou, S.: Dynamic nonpoint-source pollution control policy: ambient transfers and uncertainty. J. Econ. Dyn. Control. **34** (2010)
16. Wagener, F.: Dynamic games in environmental economics and management. Dyn. Games Appl. **12**, 747–750 (2022)
17. Martín-Herrán, G., Rincón-Zapatero, J.P.: Efficient Markov perfect Nash equilibria: theory and application to dynamic fishery games. J. Econ. Dyn. Control. **29** (2005)
18. Olsder, C., Basar, T.: Dynamic Noncooperative Game Theory, 2nd. New York (1995)

Challenges in Achieving Inclusive Development in South Asia

Shuning Chen and Shunsuke Managi

Abstract Despite global economic expansion, benefits remain unevenly distributed, widening income and wealth gaps, and coinciding with socioeconomic instability and environmental degradation that threaten human welfare. This study assesses whether growth in South Asia is inclusive in a welfare sense by examining how different forms of capital contribute to sustainable development. We apply the Inclusive Wealth Index (IWI) to evaluate the level and interplay of natural, produced, and human capital for India, Bangladesh, and Nepal (1990–2020), moving beyond GDP to a welfare-consistent accounting of assets. These three countries are selected because they present representative contrasts in economic structure, demographic pressure, and environmental vulnerability within South Asia, making them informative cases for regional policy. Statistically, we summarise long-run changes with average growth rates and a simple cross-country regression that links 2019 social-capital scores to 1990–2019 component-wise wealth accumulation. Results show a notable lag in per-capita wealth accumulation in all three countries relative to global trends, revealing gaps in local development strategies. We advocate a collaborative approach among policymakers, civil society, and development partners to align investment and governance with long-run wealth, supporting sustainable, inclusive growth suited to South Asia's diverse contexts. The analysis underscores that inclusive-wealth investment and management, especially the restoration of natural capital, and the quality of social capital are central to sustaining well-being. We (i) provide the first comparative IW assessment that statistically links social-capital conditions to component-wise wealth accumulation in South Asia, and (ii) implement a risk-sensitive IW adjustment for CO_2 damages.

Keywords Inclusive wealth index · South Asia · India; Bangladesh · Nepal · Natural capital · Social capital · Welfare accounting

S. Chen
The IDEC Institute, Hiroshima University, Higashi-Hiroshima, Hiroshima, Japan

S. Managi (✉)
Kyushu University, Fukuoka City, Japan
e-mail: managi@doc.kyushu-u.ac.jp

G. Halkos et al. (eds.), *Environmental Risk Analysis*, Springer Proceedings in Mathematics & Statistics 540, https://doi.org/10.1007/978-3-032-24155-9_15

1 Introduction

In academic discourse and international policy, increasing attention is being paid to the concept of inclusive wealth or inclusive growth, also referred to as the growth of comprehensive wealth. This shift in focus—spearheaded by researchers such as Shunsuke Managi, director of the UN Inclusive Wealth Report, and Partha Dasgupta, its advisor—emphasises a more holistic measure of national progress beyond traditional indicators such as gross domestic product (GDP). Their work and the subsequent Inclusive Wealth Reports—a joint initiative of UNEP, UNESCO, UN-Habitat, and UNU—advocate for an accounting framework that integrates produced capital, human capital (e.g. education, health), and, critically, natural capital.

The emphasis on natural capital reflects the growing recognition that depletion of environmental resources poses a major threat to long-term sustainability and human well-being. By incorporating the value of natural assets into national wealth assessments, countries can better understand the true costs and benefits of their development pathways. This approach enables a more accurate evaluation of whether current economic activities are genuinely sustainable and equitable, and it informs national strategies aiming to align growth with the United Nations' Sustainable Development Goals (SDGs). Looking ahead, inclusive wealth, natural capital, and their contribution to well-being are increasingly recognised as indicators for future global agendas, sometimes framed as Sustainable Well-Being Goals or Inclusive Development Goals for 2030.

South Asia—home to the world's densest populations—illustrates the tension between economic potential and sustainability constraints. India, Bangladesh, and Nepal have achieved substantial growth, yet the region continues to face pronounced intra- and inter-country disparities, slower convergence compared with other world regions, and extreme demographic pressures. High population density magnifies opportunities for productivity and learning but also strains infrastructure, public services, and ecosystems. We focus on these three countries because they represent contrasting combinations of economic structure, demographic pressure, and environmental vulnerability within South Asia.

Environmental sustainability and resource management are therefore pivotal. The region faces ecosystem degradation, resource scarcity, and climate-related risks—challenges that GDP alone cannot capture. The Inclusive Wealth Index (IWI), as developed in successive Inclusive Wealth Reports, integrates the three core pillars of produced, human, and natural capital, enabling assessment of whether growth is occurring alongside asset maintenance rather than asset depletion.

This paper extends that agenda by critically examining the pursuit of inclusive and sustainable development in South Asia. We focus on India, Bangladesh, and Nepal as representative cases combining economic diversity, demographic intensity, and environmental vulnerability. We analyse the dynamics of inclusive-wealth accumulation and situate these trajectories within global patterns. Our analysis also considers the roles of natural and social capital, recognising that deficits in either can impede progress.

Natural capital provides essential resources but also sets biophysical boundaries for human activities, highlighting the need for investment portfolios that safeguard ecological integrity. In parallel, social capital—trust, institutions, norms, and cooperative capacity—acts as an enabling condition for converting income into sustained wealth. Building on the conceptual foundations laid out in the Inclusive Wealth Reports, our study investigates how these factors interact in shaping development outcomes in South Asia.

Our central finding is that per-capita inclusive-wealth accumulation in India, Bangladesh, and Nepal lags global trends, largely due to pressures on natural capital. This underscores the urgency of re-aligning development strategies towards restoring natural assets, strengthening human capital, and investing in social capital.

Methodologically, our approach combines inclusive-wealth accounting with statistical analysis. We compute long-run average growth rates of capital stocks and apply generalised linear models to examine how social capital relates to inclusive-wealth accumulation. Environmental risks are explicitly considered through the CO_2-damage adjustment to natural capital, making the framework both statistically rigorous and risk-sensitive.

The remainder of the paper proceeds as follows: Sect. 2 reviews the literature on inclusive growth and the sustainability challenges facing South Asia. Section 3 presents the theoretical basis of inclusive wealth, including the role of enabling social capital. Section 4 describes the data and methodology used. Section 5 reports the empirical results and policy implications. Section 6 concludes.

2 Literature Review

2.1 Inclusive Growth Achieves Sustainability

Economic growth has delivered substantial improvements in human well-being. At the same time, the intensification of globalised economic activity has generated human impacts that may exceed the safe operating space of Earth's natural systems. The concept of planetary limits underscores the need for more sustainable and equitable development models [1–4]. The United Nations' Sustainable Development Goals (SDGs) offer a comprehensive and integrated framework spanning economic, social, and environmental dimensions [3, 5–7].

Assessing progress towards sustainability, however, requires going beyond GDP (OECD, 2021). The "Beyond GDP" movement has gained momentum over recent decades [8], and leading economists have argued for broader welfare metrics for at least half a century [9, 10]. GDP has three well-known shortcomings. First, it does not adequately reflect changes in human welfare, which can misguide policy [11]. Second, as a flow measure, GDP ignores changes in the underlying asset base—including natural capital—that generates income [12, 13]. Third, official GDP statistics struggle to keep pace with structural shifts such as deepening globalisation [8,

14]. These limitations are interrelated and contribute to uncertainty about sustainable development.

Identifying alternatives to GDP is challenging. Many proposals within the current System of National Accounts (SNA) remain closely correlated with GDP and thus add limited new information. Moreover, the stability and convergence of HDI with GDP suggest that GDP per capita still captures a substantial share of cross-country variation [15]. In the SDG Index and Dashboard [16], GDP per capita is the strongest single predictor of SDG performance, yet it explains only about half of the variance across countries [17].

In recent decades, members of the South Asian Association for Regional Cooperation (SAARC) have liberalised and globalised to accelerate convergence with the world economy. India became the world's fifth largest economy by 2022 (GDP $\approx$ US$3.94 trillion). Its reforms expanded trade and FDI and generated spillovers for Bangladesh, Sri Lanka, Nepal, and Bhutan [18]. Nevertheless, SAARC remains less integrated than other trading blocs [19], which has implications for regional growth dynamics [20].

Demography is central to South Asia's development. The region holds the largest share of Asia's population, and population dynamics strongly influence local prosperity. Many economies are in transition, facing continued fertility decline, rising life expectancy, and falling dependency ratios [21].

Environmental pressures are acute. South Asia confronts ecosystem degradation that threatens the well-being of more than 1.7 billion people. Rapid urbanisation, population growth, and industrialisation compound vulnerability and are associated with elevated climate-risk exposure [22]. Although the region currently accounts for a modest share of global energy use ($\approx$ 5.9%), demand is projected to more than double in coming decades; resource constraints are already evident [23]. Thus, despite rapid economic expansion, growth is constrained by both conventional and renewable energy supplies. Although several studies assess sustainability in South Asia, indicators differ widely, and integrating multiple dimensions remains challenging. This motivates the use of a framework that links economic performance to the underlying stocks of wealth.

2.2 *Inclusive Wealth as Sustainability Index*

The Inclusive Wealth Index (IWI) provides a framework for assessing national sustainable development beyond GDP [24–26], and we employ it to measure sustainability in South Asia. IWI's distinctiveness lies in its equity (capital-based) perspective: it tracks changes in national wealth (capital stocks) rather than only income flows. IWI and GDP are therefore complementary—GDP reflects current production, while IWI reflects the productive base that sustains future well-being.

IWI measures the social value of the capital assets that underpin human well-being: human capital, produced (physical) capital, and natural capital. The framework recognises the limited substitutability of finite natural resources and emphasises the importance of efficient and effective capital management for sustainable development. Evidence from wealth accounts links progress towards the SDGs to environmental policy and capital management [27, 28].

Wealth accounting complements the SNA in two fundamental ways. First, it offers a forward-looking perspective by quantifying the productive base required for future goods and services, whereas the SNA records flow that have already occurred. Second, natural-capital accounts provide the flexibility to represent assets—such as climate and biodiversity—that cross national boundaries and are only partially captured in standard accounts. Because these are global public goods, natural-capital accounting extends the geographic and institutional scope relevant for sustainability analysis.

The next section outlines the specific theory of inclusive wealth that underpins our empirical approach.

3 Theoretical Framework

The Inclusive Wealth Index (IWI) is an indicator that measures wealth by combining biophysical data and monetary calculations for multiple capital components [14, 24–26]. Its theoretical foundation is based on the stock and flow approach and has been developed through extensive theoretical and empirical research [12, 13, 29–32]. Demonstrating the equivalence between assets and social welfare is the basis of IWI's assessment of intergenerational sustainability.

In this section, we introduce the theoretical framework of inclusive wealth within the theoretical foundations of the measurement of intergenerational sustainability. It then explains the connotation of the three critical assets and their relationship to social capital.

3.1 Non-Decline Inclusive Wealth as Measures Intergenerational Sustainability

To illustrate the theoretical framework of IWI, consider an arbitrary initial time point s ≥ 0. Let $\underline{C}(s)$ denote a vector of consumption flow at time t, and $U(\underline{C}(s))$ denote economy-wide utility flow. Then let V(t) denote the social welfare for current and the future at s and can be presented as

$$V(t) = \int_{t=s}^{\infty} \left[U\left(\underline{C}(s)\right)e^{\delta(s-t)}\right]ds, \delta > 0 \tag{1}$$

where δ is the discount rate of the utility flow; maximising intergenerational welfare $V(t)$ requires forecasting future utility flows. This task is complex due to market imperfections such as price distortions and externalities. Thus, considering a counterfactual resource reallocation mechanism, we can map from the set of all possible capitals to the set of possible pairs of utility flows for all t > s [30, 33]. This allows us to forecast intergenerational welfare based on whether the initial capital goods stock inherited at s differs from the current time.

Let $\underline{K}(t)$ be the initial capital goods stock; assume that the resource reallocation mechanism is time-autonomous, and then $V(t)$ is a function of t and $\underline{K}(t)$. We have

$$V(t) = V(\underline{K}(t)) = \int_{t=s}^{\infty} [U(\underline{C}(s))e^{\delta(s-t)}]ds, \delta > 0 \tag{2}$$

Then, we discuss sustainable development under the principle of resource allocation mechanism. Let $\Delta V(t)$ denote the perturbation at time t and assume that $\Delta V(t)$ is differentiable. Sustainability is expressed as non-decline welfare through intertemporal changes, so sustainability is maintained if

$$\Delta V(t) = \Delta V(\underline{K}(t)) = V(\underline{K}(t) + \Delta\underline{K}(t)) - V(\underline{K}(t)) > 0 \tag{3}$$

Since $V(t)$ is differentiable, we have

$$\Delta V(K(t)) = \frac{\partial V}{\partial t} + \sum_i \left[\frac{\partial V(K(t))}{\partial K_i}\right]\Delta K_i(t) \tag{4}$$

To link the welfare concept to the price and investment principles, we define the terms on the right-hand side: $p_i(t) \equiv \frac{\partial V}{\partial K_i}$ represents the social value or shadow price of capital assets, and the price over time is expressed as $r(t) = \frac{\partial V}{\partial t}$. We can now use shadow prices as weights to construct an aggregate index of the economy's stock of capital assets. We use W to denote the inclusive wealth index (IWI), where welfare can be expressed as the equivalent of wealth:

$$W(t) = \sum_i p_i K_i(t) + r(t) \tag{5}$$

IWI provides a capital measure of sustainable development. It links the discounted present value of all future consumption possibilities to the weighted sum of the capital-asset (or wealth) profile, which is the productive base of the economic outcome. Capital assets in the IW accounting framework are both intertemporal means of production and direct sources of human well-being, meeting the consumption needs of the current population.

The Inclusive Wealth Index (IWI) is a practical sustainability indicator showing the equivalence between human well-being and wealth. The sustainability of a socioeconomic system is assessed based on the non-decline of wealth over time. By integrating biophysical quantities and the shadow prices of capital assets, the IWI provides a comprehensive understanding of both the means and the ends of achieving sustainability. Thus, the connotation of inclusive-wealth capital assets needs to be clarified to be suitable for sustainability analysis. To discuss the measurability of capital assets, we discuss below the three pillars of capital assets that build inclusive wealth and their relationship to social capital.

3.2 Three-Pillar Capital Assets and Social (Enabling) Capital

Sustainable development aims to meet present and future needs while improving human well-being. The Inclusive Wealth Index (IWI) operationalises this by aggregating produced, human, and natural capital into a welfare-consistent stock measure [25, 34]. In the classical and neoclassical traditions, capital and labour (with land/resources in resource economics) form the productive base; modern growth accounting further highlights human capital as a key driver of output and welfare.

IWI adopts a portfolio view with complementarities and limited substitutability among assets. Using natural capital need not be unsustainable if it raises produced and human capital sufficiently; however, because economic life is embedded in the biosphere, biophysical limits constraint-feasible development paths. Accordingly, IWI treats natural capital as a core asset and evaluates sustainability via the non-decline of inclusive wealth under shadow prices.

Beyond these three stocks, Dasgupta distinguishes enabling assets—public knowledge, institutional and social trust, biodiversity—that do not enter as a separate stock but shift accounting (shadow) prices and production possibilities [27, 29, 31]. Formally, enabling assets affect the shadow-price vector:

$$\mathbf{p}(t) = \frac{\partial V(t)}{\partial \mathbf{K}(t)},$$

And the technology set, thereby altering the value of the three pillars without being directly counted as an additional pillar. They are theoretically central but hard to measure directly at scale.

Operational choice: We therefore do not add "social capital" to the stock vector. Instead, we proxy social enabling conditions using the World Social Capital Monitor and study associations with component-wise per-capita wealth growth [34]. Foundational definitions are provided by Coleman and OECD [35, 36]; known measurement caveats (subjectivity, cross-country comparability) are mitigated by z-standardising items and estimating within income groups (see Methods) [37, 38].

Having established the three essential capital assets and clarified the role of enabling (social) capital, Sect. 4 operationalises this framework: we build the wealth accounts, compute per-capita stocks and growth rates, and implement the statistical models used in the empirical analysis.

4 Methodology

Our empirical design is explicitly statistical. Inclusive-wealth accounting generates per-capita capital stocks and growth rates, while the role of social capital is examined through generalised linear models (identity link, robust errors) estimated across income groups. Environmental risks enter the framework through the CO_2-damage adjustment applied to natural-capital accounts, allowing the statistical analysis to capture the interaction between social enabling conditions and ecological constraints.

4.1 *The Empirical Accounting of Inclusive Wealth*

Measuring wealth levels and changes at the national scale requires detailed accounts of each capital stock and the application of shadow prices. The Inclusive Wealth Index (IWI) aggregates components from the social, economic, and natural domains into a unified index to estimate inclusive wealth and investment aggregates. Figure 1 (to be inserted) summarises the accounting categories under the three-pillar capital-asset framework.

Natural capital: Accounting distinguishes renewable and non-renewable resources. Renewable resources include agricultural land (arable land, pasture), forests (timber and non-market ecosystem values), and marine fish stocks. Non-renewable capital encompasses energy and mineral resources: energy sources

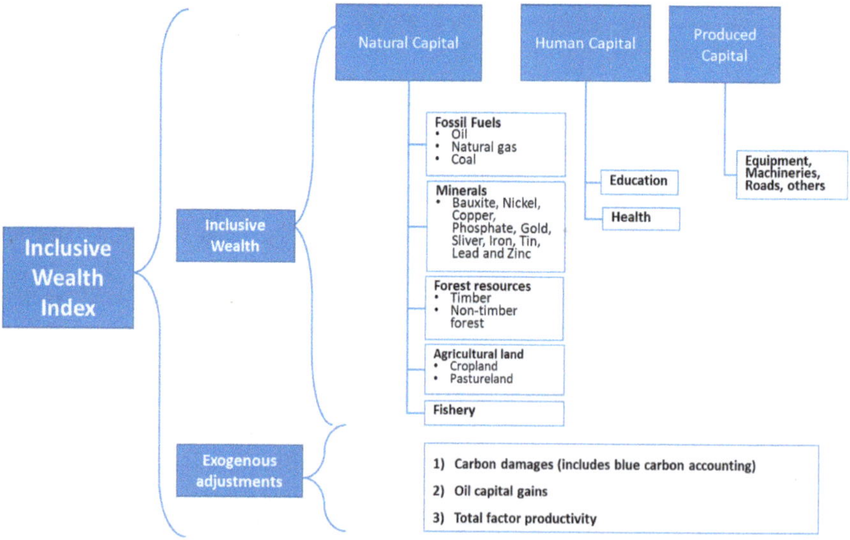

Fig. 1 Schematic representation of the inclusive wealth index and the adjusted inclusive wealth index

comprise oil, natural gas, and coal; minerals include aluminium, nickel, copper, phosphorus, gold, silver, iron, tin, lead, and zinc.

Human capital: We calculate human capital using the lifetime income method and provide gender-disaggregated statistics. Assessment incorporates country-specific health and population dynamics. Because education investments in developing countries often exhibit long lags before returns materialise, we evaluate education using school life expectancy, adjust compensation for education forwardly, and assess human capital by work-life expectancy combined with population dynamics and vital statistics, following the J–F lifetime income approach [39, 40].

Produced capital: We estimate produced capital via the perpetual inventory method (PIM), covering major categories (structures, transportation equipment, communication equipment, and other assets).

Shadow prices: A critical aspect of IWI is assigning shadow prices to capital assets to reflect their social value and contribution to intergenerational well-being. These prices are measured by marginal social substitution rates between assets. Valuation relies primarily on market prices, adjusted to reflect long-term sustainability rather than short-term volatility. An exception is the value of ecosystem services, which is drawn from the Ecosystem Services Valuation Database (ESVD) [41].

Population and time-varying adjustments: Given rapid population growth and uncertainty in future demographics, sustainability must be assessed on a per-capita basis [42, 43]. We therefore apply non-declining IWI per capita at constant shadow prices to measure real sustainable development [13, 31]. In addition to population growth, we incorporate three time-varying adjustments: total factor productivity (TFP), CO_2 damage, and oil-price gains. TFP reflects technological and skill advancements during capital accumulation, while the latter two capture the effects of transnational externalities on national wealth.

Prior evidence and scope expansion: The United Nations Environment Programme (UNEP) has led three Inclusive Wealth Reports, which document a significant global decline in natural capital and the independence of IWI from GDP and other welfare indicators such as HDI [24, 44, 45]. The third update also reports strong correlations between IWI and SDG indicators, especially SDGs 12, 13, and 7 [46]. Building on these reports, we expand natural-capital accounting by updating data, increasing country coverage from 140 (2018) to 163, and extending the time range from 1990 to 2019. We also integrate gender, education, and health dimensions to estimate human capital more directly linked to the SDGs, and we enhance produced-capital estimates by disaggregating buildings, roads, and transport equipment by country and year, applying aggregated depreciation rates.

Despite substantial improvements in asset accounting, data limitations remain—particularly the need for more comprehensive ecosystem-service information. Nevertheless, analysing trends across capital assets provides essential insights for evaluating sustainability performance.

4.2 Social Capital and Inclusive Growth

We use the World Social Capital Monitor (2019), which reports eight country-level items on a 1–10 scale (social climate, interpersonal trust, acceptance of austerity measures, willingness to pay taxes, support for local investment, helpfulness, friendliness, and hospitality). For comparability, each item is z-standardised (mean 0, SD 1). We relate average per-capita capital growth to social capital across 67 countries. Because social-capital data end in 2019, regressions use 1990–2019 growth; descriptive accounting elsewhere covers 1990–2020. We estimate linear models (GLM, identity link ≡ OLS) by income group, and report heteroscedasticity-robust (HC) standard errors. The model is as follows:

$$g_c^X(1990, 2019) = \alpha + \sum_{j=1}^{8} \beta_j z\left(SC_{c,j,2019}\right) + \theta g_c^{GDP}(1990{,}2019) + \varepsilon_C \quad (6)$$

where c indexes countries, $X \in \{\text{IW, PC, HC, NC}\}$ is the capital component; $g_c^X(t_0, t_1) = \frac{\ln X_{c,t_1} - \ln X_{c,t_0}}{t_1 - t_0}$ indicates the average log growth; $z\left(SC_{c,j,2019}\right)$ are the eight standardised social-capital items; and $g_c^{GDP}(1990{,}2019)$ is (optional) GDP-per-capita growth as a control. Models are estimated separately by income group with HC SEs. The descriptive statistical results for the sample countries are presented in Table 1.

4.3 Focus on Sampling Countries in South Asia

Our selection of India, Nepal, and Bangladesh is grounded in geographical, economic, and policy considerations.

Geography: These countries reflect South Asia's diverse landscapes: India spans mountainous regions to extensive coastlines; Nepal is landlocked and predominantly mountainous, encompassing significant portions of the Himalayas; Bangladesh is characterised by deltaic plains with major rivers and low-lying terrain. This diversity enables analysis of how geographical vulnerabilities—the fragility of Nepal's Himalayan ecosystems, Bangladesh's exposure to cyclones and floods, and India's air-pollution and water-scarcity challenges—shape sustainable development and inclusive growth.

Economy: The three economies differ in structure and stage: India's rapid growth has produced a diversified economy combining high-tech industries with substantial agriculture; Bangladesh has advanced in textiles and apparel while maintaining a strong agricultural base; Nepal's smaller economy relies heavily on agriculture, remittances, and tourism. This variation provides a panoramic view of South Asia's heterogeneous growth constraints and opportunities.

Policy: The countries are navigating distinct structural transformations and employ different policy approaches to inclusive growth. Comparing these approaches

Table 1 Descriptive statistics

	N	Mean	Standard deviation	Min	Max
Low income					
Per capita PC growth	16	0.28%	0.64%	−1.18%	1.58%
Per capita HC growth	16	0.38%	0.33%	−0.07%	1.05%
Per capita NC growth	16	−1.76%	1.05%	−4.41%	−0.33%
Per capita IW	16	13,071	13,358	3198	50,563
Per capita GDP	16	612.04	230.84	99.48	984.91
Lower-middle income					
Per capita PC growth	22	0.43%	0.44%	−0.86%	1.27%
Per capita HC growth	22	0.38%	0.28%	0.00%	1.02%
Per capita NC growth	22	−1.46%	1.06%	−4.92%	−0.01%
Per capita IW	22	27,125	19,030	11,515	77,804
Per capita GDP	22	2049	1120	999	5232
Upper-middle income					
Per capita PC growth	14	0.6%	0.42%	−0.04%	1.26%
Per capita HC growth	14	0.70%	0.46%	−0.05%	1.41%
Per capita NC growth	14	−0.73%	0.80%	−2.69%	0.10%
Per capita IW	14	157,596	147,544	59,197	570,328
Per capita GDP	14	7458	2596	4084	12,039
High income					
Per capita PC growth	15	0.77%	0.49%	0.30%	2.01%
Per capita HC growth	15	0.71%	0.28%	0.32%	1.28%
Per capita NC growth	15	−0.44%	1.19%	−4.64%	0.02%
Per capita IW	15	595,446	256,877	225,321	1,320,564
Per capita GDP	15	40,849	19,402	12,803	79,464

offers lessons on the effectiveness of policy interventions in fostering inclusive development.

In conclusion, focusing on India, Nepal, and Bangladesh provides a comprehensive and multifaceted perspective on the development challenges facing South Asia. The geographical, economic, and policy diversity of these countries provides a rich backdrop for examining the determinants of sustainable and inclusive growth in the region. Next, we present results based on inclusive-wealth accounting and the social-capital index, focusing on the roles of natural and social capital in shaping sustainability under the updated IW accounts.

5 Results and Discussion

This section examines levels and changes in wealth for the three selected countries and compares them with global trends over the past 30 years. We focus in particular on the roles of natural capital and social capital in shaping inclusive-wealth outcomes, and conclude with a discussion of the implications for sustainable development in South Asia.

5.1 Global Wealth Trends and South Asia

As shown in Fig. 2, the trajectory of global per-capita inclusive wealth since 1990 has not been sustainable. Between 1990 and 2020, global per-capita wealth fell by 2.7%, driven primarily by the depletion of natural capital, which is now nearly 50% lower than in 1990. Although there has been some recovery in recent years, this has largely reflected a slowing pace of natural-capital decline rather than genuine asset restoration.

Decomposition of global wealth change by capital type and country highlights heterogeneity in accumulation patterns. In South Asia, divergent trends are evident (Fig. 3). Nepal recorded the most severe decline in wealth, attributable to heavy losses in natural capital. By contrast, Bangladesh achieved the strongest per-capita

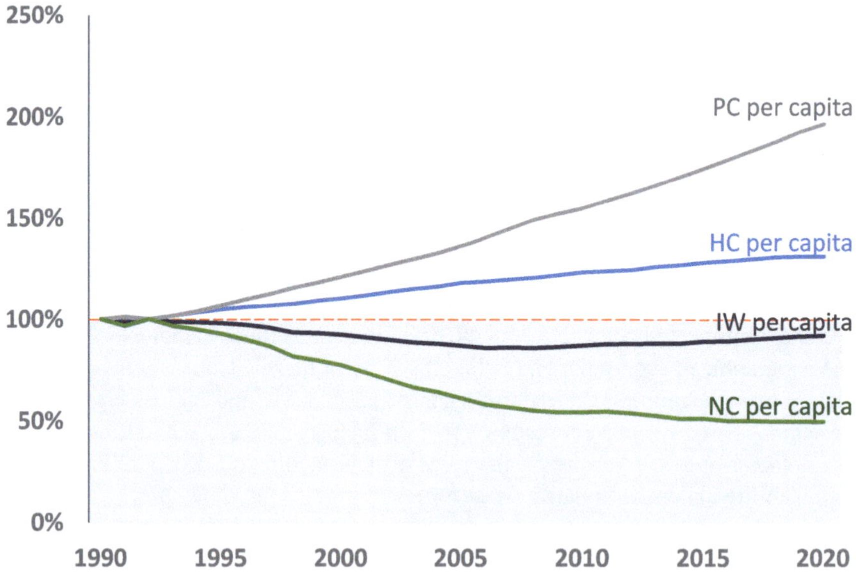

Fig. 2 Overview of the changes in global per-capita wealth referring to 1990

increase, supported by rapid growth in produced and human capital. India also experienced rising wealth, though, as in Bangladesh, these gains were partially offset by ongoing natural-capital depletion. Overall, despite differences in composition, all three economies show evidence of natural-capital exhaustion, which remains the main constraint on sustainable wealth growth.

Figure 4 provides a detailed comparison of India, Bangladesh, and Nepal over the past three decades. All three countries achieved substantial GDP growth, with

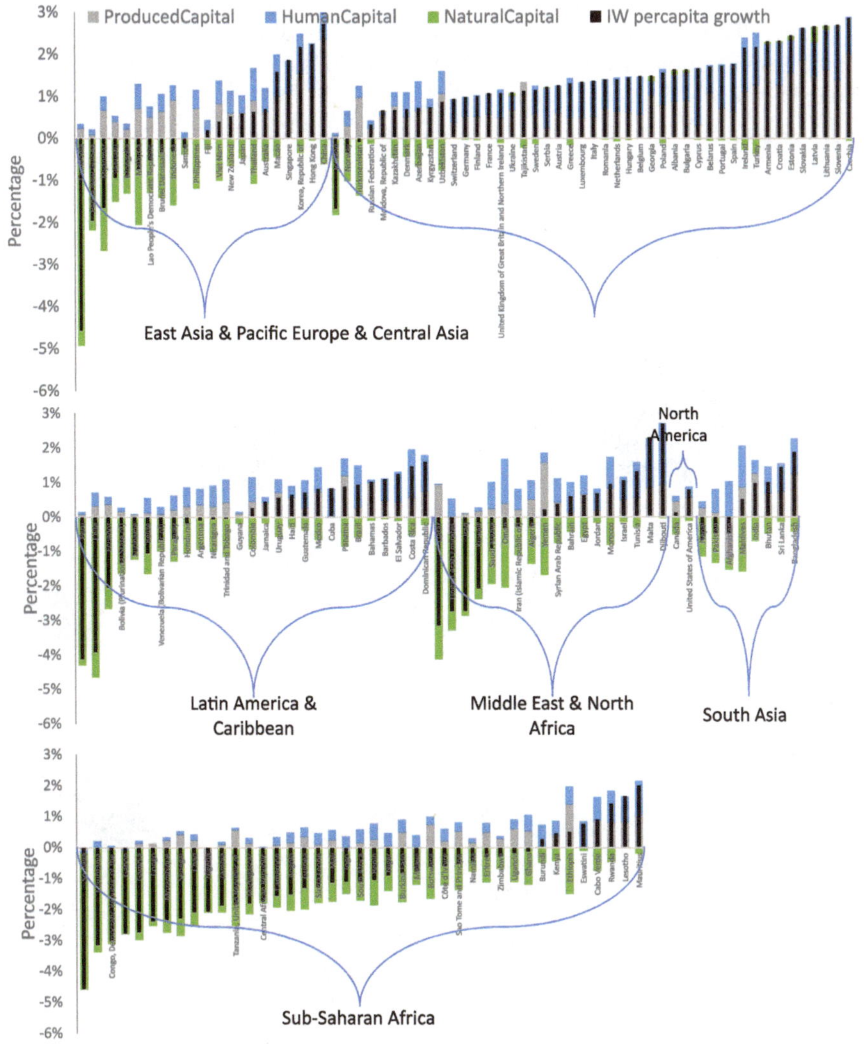

Fig. 3 Global per-capita wealth growth rates disaggregated and by capital form, the annual average for 1990–2020

India leading at an average annual rate of 4.6%, followed by Bangladesh (4.0%) and Nepal (3.0%) (Fig. 4a).

Differences are more pronounced when examining per-capita inclusive-wealth levels. In 2020, Nepal reported the highest per-capita inclusive wealth (US$27,272 in 2015 prices), followed by India and Bangladesh. Notably, Bangladesh's level was less than half of Nepal's (Fig. 4b).

Turning to inclusive-wealth growth rates over 1990–2020, the ranking shifts. Bangladesh recorded the highest average annual growth (3.7%), while India (2.54%) and Nepal (0.76%) trailed behind (Fig. 4c).

Finally, the per-capita growth rates of inclusive wealth reveal a stark contrast. Bangladesh's per-capita wealth rose by 1.8%, India's by 0.97%, whereas Nepal's declined by 0.7% (Fig. 4d).

In sum, Nepal's inclusive wealth actually declined on a per-capita basis, while India and Bangladesh saw only modest gains. These increases were substantially

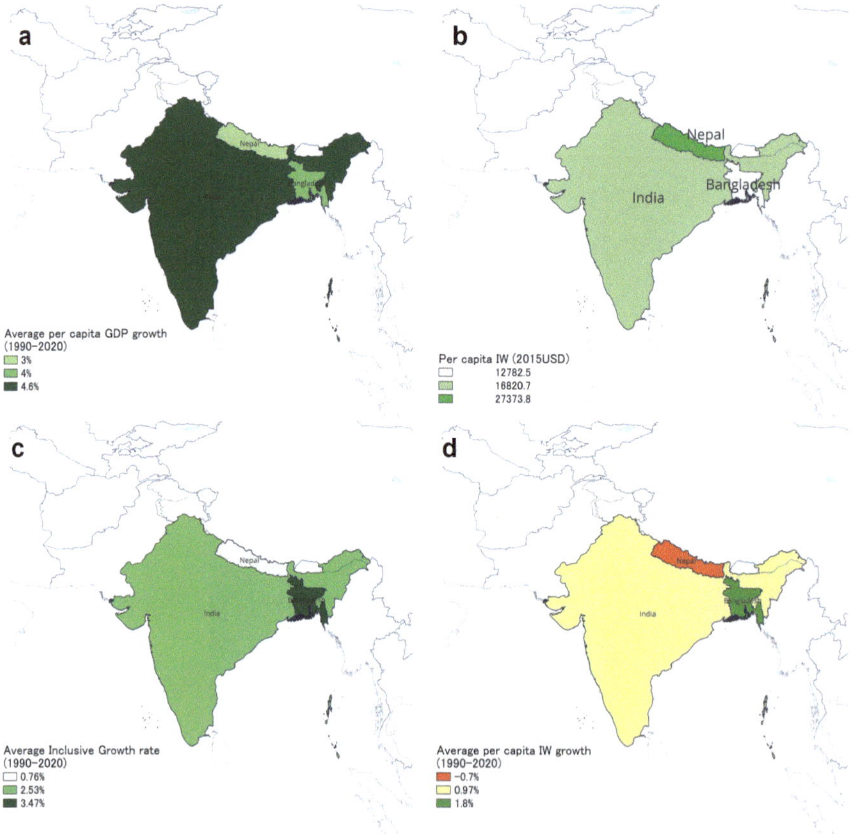

Fig. 4 Inclusive wealth, average GDP growth, and IW and IW per-capita growth in India, Bangladesh, and Nepal (2020)

lower than their corresponding GDP growth rates, underscoring a disconnect between income growth and wealth accumulation. The gap is especially marked in Nepal, where economic growth has not translated into sustained wealth gains, suggesting an unsustainable development trajectory.

5.2 Factors Influencing Sustainability: Capital Composition and Per-Capita Dynamics

We next examine the factors shaping sustainability in the three countries. Figure 5 shows changes in the structure of wealth for India, Bangladesh, and Nepal alongside global shifts. In the base year (1990), all three countries had a lower share of produced capital than the global level. By contrast, India and Bangladesh exhibited higher shares of human capital in wealth—approximately 44% and 63%, respectively. Natural capital accounted for 89% of total wealth in Nepal and 45.8% in India, while Bangladesh's natural-capital share was well below the global average (about 24% versus a global 30%).

By 2020, three decades of accumulation in human and produced capital substantially increased total wealth: India by 1.75 $\times$, Bangladesh by 2.28 $\times$, and Nepal by 1.24 $\times$. Over the same period, both the level and share of natural capital declined markedly. By 2020, natural capital comprised only 22.4% of India's wealth, 7.6% of Bangladesh's, and 74% of Nepal's (down from nearly 90%).

A per-capita comparison is more revealing. As shown in Fig. 6, very low produced-capital per capita depresses productivity and leaves GDP per capita well below the world average. Although increased investment in produced capital lifted GDP per capita in all three countries, wealth accumulation per capita has been much slower. The primary reason is a larger decline in natural-capital per capita than observed globally. In Bangladesh and India, natural-capital per capita continued to fall even after 2010, when the global rate of depletion moderated. Nepal followed a different path: very rapid natural-capital depletion in the first two decades, combined with slow human- and produced-capital accumulation, led to a decline in per-capita wealth. In the last decade, depletion slowed, and other capital stocks grew, producing a modest upturn in per-capita wealth, albeit at a slow pace.

In sum, while South Asian countries recorded notable GDP-per-capita growth driven by rapid produced-capital accumulation, per-capita wealth growth did not keep pace. This gap is largely attributable to substantial natural-capital depletion, most pronounced in Nepal. Moreover, in India and Bangladesh, already low natural-capital levels relative to the global average amplify the constraint. Under these conditions, further resource extraction is an impractical route to matching global GDP growth rates. Emphasising produced capital to raise GDP has been largely ineffective for sustainable development when it coincides with significant natural-capital losses.

Building other forms of capital at the expense of natural capital undermines inclusive wealth and signals an unsustainable trajectory. Over-exploitation pushes these

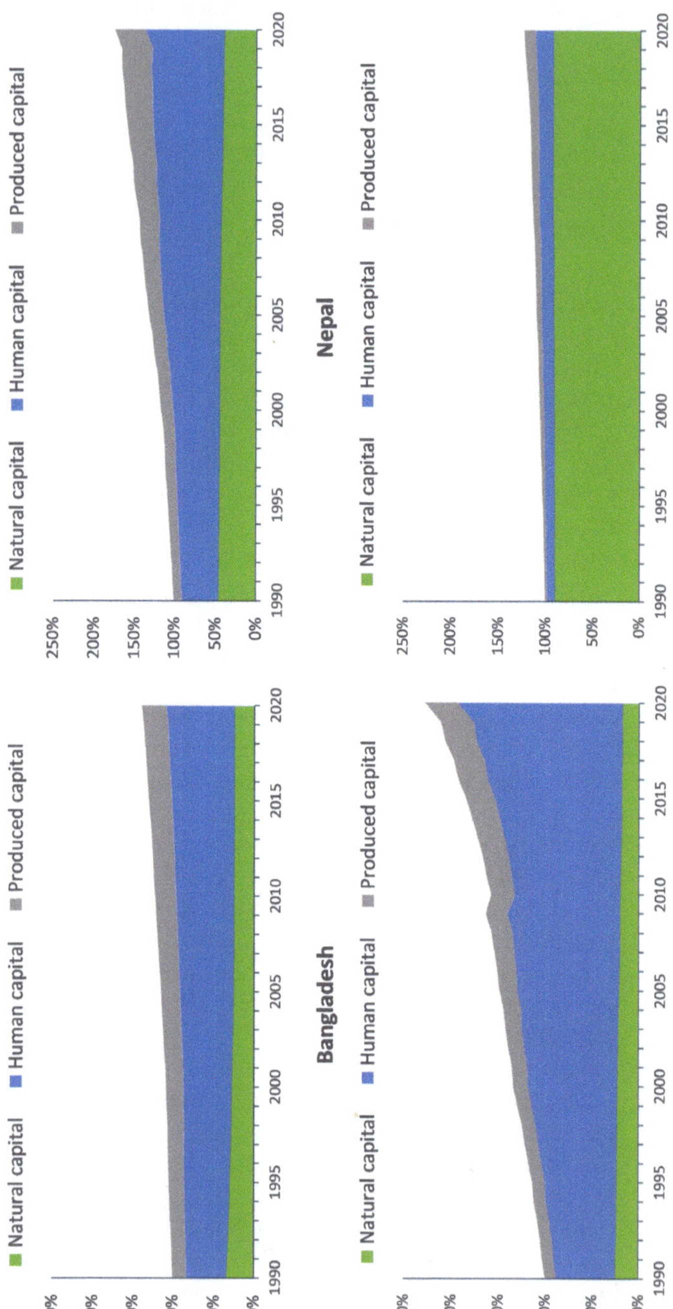

Fig. 5 Changes in world and South Asian countries' share of capital assets in inclusive wealth relative to 1990

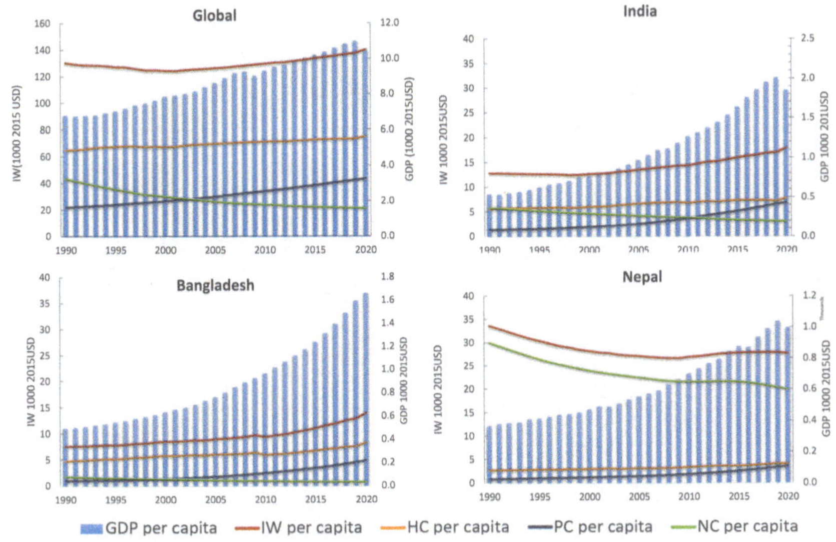

Fig. 6 Per capita IW, capital, and GDP change in three economies during 1990–2020

societies towards sustainability limits. In addition, rapid population growth intensifies pressure on the carrying capacities of all capitals, further shaping the path of per-capita wealth.

Having examined the dynamics of produced, human, and natural capital, we now turn to the role of social capital in shaping inclusive-wealth accumulation, with a focus on how enabling conditions mediate sustainability outcomes in South Asia.

5.3 *Wealth Investment for Inclusive Growth: The Role of Social Capital*

Evidence from global and country-level patterns suggests that growth in produced capital need not always come at the expense of natural capital. In principle, sustainable growth can be achieved through inclusive investment that strengthens complementarities among capital stocks—so that investment in one form supports accumulation in others. Realising these complementarities typically requires changes in social attitudes, institutions, and collective norms—that is, improvements in the enabling conditions captured by social capital.

As a configuration of social relations and institutional capabilities, social capital can influence wealth accumulation at different stages of development. Our analysis compares subjective assessments of social capital across 67 countries and examines their associations with the accumulation of produced, human, and natural capital over three decades (Table 2). Results by income group are.

Table 2 GLM/OLS of average per-capita capital growth (1990–2019) in 2019 social-capital items, estimated within World Bank income groups. Coefficients are per 1-SD increase in each z-standardised item; heteroscedasticity-robust standard errors in parentheses

	Social Climate	Interpersonal Trust	Accepting Austerity Measures	Willingness to pay taxes	Support for local economy	Helpfulness	Friendliness	Hospitality
Natural capital								
Low-income	**0.0129***	0.0016	-0.0016	-0.0086	**0.0088***	-0.0109	**-0.0116****	0.0062
	(0.0067)	(0.0111)	(0.0057)	(0.0057)	(0.005)	(0.007)	(0.0053)	(0.007)
Lower middle-income	-0.0025	0.0015	0.0034	**-0.0155****	**0.0074*****	-0.0055	0.0027	0.0050
	(0.0048)	(0.0046)	(0.0028)	(0.0079)	(0.0027)	(0.0081)	(0.0041)	(0.0036)
Upper middle-income	-0.0005	-0.0111	0.0010	-0.0005	-0.0026	0.0075	-0.0018	-0.0003
	(0.0047)	(0.0082)	(0.0051)	(0.0049)	(0.0036)	(0.006)	(0.0046)	(0.0068)
High-income	0.0180	-0.0089	-0.0073	**0.0217****	**-0.0109***	-0.0126	0.0113	-0.0009
	(0.012)	(0.0092)	(0.0071)	(0.0093)	(0.0058)	(0.0124)	(0.01)	(0.0066)
Produced capital								
Low-income	-0.0051	-0.0033	0.0052	-0.0019	-0.0045	0.0073	0.0015	0.0040
	(0.0074)	(0.0121)	(0.0063)	(0.0062)	(0.0054)	(0.0076)	(0.0058)	(0.0077)
Lower middle-income	0.0005	0.0019	-0.0002	-0.0022	0.0014	-0.0030	-0.0005	0.0018
	(0.0025)	(0.0024)	(0.0015)	(0.0041)	(0.0014)	(0.0042)	(0.0021)	(0.0019)
Upper-middle-income	-0.0012	**0.0072****	-0.0001	-0.0019	0.0010	-0.0025	**-0.0029***	0.0038
	(0.0016)	(0.0028)	(0.0018)	(0.0017)	(0.0012)	(0.0021)	(0.0016)	(0.0024)
High-income	**0.0086***	-0.0007	-0.0009	0.0036	**-0.0063****	**-0.0088***	0.0015	0.0043
	(0.0052)	(0.004)	(0.003)	(0.004)	(0.0025)	(0.0053)	(0.0043)	(0.0028)
Human capital								

(continued)

Table 2 (continued)

	Social Climate	Interpersonal Trust	Accepting Austerity Measures	Willingness to pay taxes	Support for local economy	Helpfulness	Friendliness	Hospitality
Low-income	**0.0036***	0.0019	0.0018	-0.0021	-0.0001	-0.0029	**-0.0055*****	**0.0045****
	(0.0022)	(0.0036)	(0.0019)	(0.0019)	(0.0016)	(0.0023)	(0.0017)	(0.0023)
Lower middle-income	-0.0021	-0.0001	-0.0004	-0.0004	**0.0012***	0.0014	-0.0019	**0.0019***
	(0.0013)	(0.0013)	(0.0008)	(0.0022)	(0.0008)	(0.0023)	(0.0011)	(0.001)
Upper middle-income	-0.0023	0.0061	-0.0010	-0.0033	0.0030	-0.0003	-0.0039	0.0030
	(0.0037)	(0.0064)	(0.004)	(0.0038)	(0.0028)	(0.0047)	(0.0036)	(0.0053)
High-income	0.0022	-0.0020	-0.0008	0.0043	**-0.0038****	0.0005	0.0005	-0.0011
	(0.0036)	(0.0028)	(0.0022)	(0.0028)	(0.0018)	(0.0037)	(0.003)	(0.002)

Notes Outcome $\boldsymbol{g}_{c}^{X} = \left(\mathbf{ln}X_{c,2019} - \mathbf{ln}X_{c,1990}\right)/29$, with $X \in \{Natural, Produced, Human\}$. Predictors are 2019 World Social Capital Monitor items, z-standardised (mean 0, SD 1). Models estimated within income groups; robust (HC) SEs in parentheses. Sample size differs by group due to data availability. * p < 0.10, ** p < 0.05, *** p < 0.01

- Low-income countries: Ratings of social climate, support for the local economy, friendliness, and hospitality correlate significantly with component-wise accumulation. Social climate and support for the local economy are positively associated with natural (and human) capital growth. Hospitality is positively associated with capital accumulation, whereas friendliness shows a negative association.
- Lower middle-income countries (the group including our South Asian cases): Support for the local economy is positively associated with natural and human-capital accumulation. Willingness to pay taxes exhibits a significant negative association with natural-capital growth, while hospitality remains positively associated with human capital. Social climate and friendliness are not statistically related to capital accumulation.
- Upper middle-income countries: No social-capital item relates to natural or human-capital growth; only interpersonal trust shows a positive association with produced-capital investment.
- High-income countries: Perceived support for the local economy is negatively associated with natural and human-capital growth, whereas willingness to pay taxes is positively associated with natural capital. These patterns are consistent with contexts where resource extraction prompts local demands for compensation and redistribution, while tax systems may finance environmental protection.

Taken together, these findings suggest that social capital correlates more strongly with capital accumulation in low- and lower middle-income settings. Institutional support for local investment and a supportive social climate—rather than purely private relational ties—appear especially important for investment in these contexts. In upper-middle- and high-income groups, the relationship is weaker or mixed, likely reflecting more mature institutional frameworks and different policy regimes (e.g., local support for extraction coinciding with natural-capital losses, while greater tax capacity relates to natural-capital gains via funded regulation).

In conclusion, strengthening institutional quality and local-investment support is likely to contribute to wealth accumulation in low- and middle-income countries, including India, Bangladesh, and Nepal. A supportive social environment can also facilitate inclusive investment. These results are associational and should be interpreted as correlations rather than causal effects; nevertheless, they point to social-institutional levers consistent with sustainable, inclusive growth strategies.

6 Discussion

This section examines the divergent development paths of India, Bangladesh, and Nepal, combining inclusive-wealth accounting with the role of social capital.

The results indicate that the three countries are following different asset-accumulation trajectories and face distinct sustainability constraints, yet inclusive development in all cases requires avoiding further depletion of natural capital. In India, sustained GDP growth has been supported by gains in human capital, while natural capital has continued to decline, implying a mounting constraint from natural-asset erosion. Our income-group regressions show that in low- and lower middle-income settings, support for local investment is positively correlated with the accumulation of both natural and human capital, whereas willingness to pay taxes is negatively correlated with natural-capital growth. This suggests that India's most promising social-capital lever is to strengthen local-investment support and the social climate—for example, through transparent community co-financing of watershed restoration or urban green infrastructure—while improving fiscal linkages so tax effort is credibly tied to environmental outcomes (earmarked grants, performance-based budgeting), thereby mitigating the negative association observed in the cross-country data.

Bangladesh faces tighter constraints: high reliance on human capital alongside comparatively low levels of produced and natural capital. The same income-group pattern holds—support for local investment aligns with gains in natural and human capital—pointing to the importance of trust-based local-investment mechanisms (e.g. municipal services, coastal protection, pollution control) that crowd in private and civic co-investment. The negative correlation between tax willingness and natural-capital growth cautions that raising revenue alone is insufficient; governance must ensure that resources are directed to ecosystem protection rather than facilitating additional extraction.

Nepal's trajectory reflects heavy dependence on natural capital with only modest increases in produced and human capital, yielding stagnation or decline in per-capita wealth. In the low-income panel, social climate and support for the local economy are positively correlated with natural (and human) capital growth, hospitality is also positive, while friendliness shows a negative association. Read together, these patterns suggest that strengthening collective, outward-looking cooperation (local-investment support; inclusive social climate) is more conducive to asset building than relying on purely relational ties that can slide into nepotism. For Nepal, institutionalising community-based resource management with explicit benefit-sharing, and leveraging hospitality in ecosystem-compatible tourism, are consistent with social-capital characteristics associated with wealth accumulation.

Overall, these country narratives reinforce the central empirical message: enabling social conditions matter most where institutional capacity is thin. In low- and lower-middle-income contexts, building a supportive, constructive social climate for local investment appears to mediate the trade-off between produced/human-capital growth and natural-capital conservation—a balance that personal ties alone do not achieve. Post-2020 shocks may have directly affected these channels. The COVID-19 pandemic disrupted schooling, health, and labour markets (human-capital formation) and may have altered institutional trust and local cooperation; political and economic instability in Bangladesh and Nepal may also have weakened governance and social cohesion. Although these developments lie outside our dataset (ending 2019/2020), they underscore the need to update inclusive-wealth accounts and reassess social-capital indicators to understand how enabling conditions have shifted and what this implies for sustainable, inclusive growth.

7 Conclusion

This study examined the pursuit of inclusive and sustainable development in South Asia through the cases of India, Bangladesh, and Nepal. These conclusions pertain to India, Bangladesh, and Nepal; regional implications are illustrative rather than universal for South Asia. Using the Inclusive Wealth Index (IWI), we assessed whether per-capita wealth accumulation has kept pace with GDP growth and found that all three countries lag behind global trends. The principal reason is the depletion of natural capital, which constrains long-run welfare even as income rises.

We also analysed the enabling role of social capital, proxied by the World Social Capital Monitor. Our statistical results, based on generalised linear models of long-run capital growth, indicate that social capital is positively associated with wealth accumulation, particularly in lower income contexts where institutional capacity is limited. These findings underscore that both natural and social capital are critical in shaping sustainable development paths.

From an environmental-risk perspective, our analysis incorporates CO_2-damage adjustments into natural-capital accounts, ensuring that sustainability is evaluated within a risk-sensitive statistical framework. This highlights the importance of restoring ecosystems, improving health and education, and strengthening institutional trust as complementary strategies for inclusive-wealth investment.

In conclusion, South Asian countries must realign development strategies to narrow the gap with global wealth trajectories. We recommend coordinated efforts by policymakers, civil society, and development partners to promote inclusive-wealth governance, balancing economic growth with the protection of natural assets and the reinforcement of social capital. By embedding environmental risk and statistical accountability into policy design, the region can advance towards a more balanced and sustainable growth trajectory. Overall, our findings demonstrate that evaluating South Asia's development through inclusive-wealth accounting, supported by statistical analysis and explicit adjustments for environmental risks, provides a rigorous basis for re-aligning policies towards sustainable and equitable growth.

Data Availability The analysis uses inclusive-wealth accounts (produced, human, natural capital; 1990–2020) and 2019 social-capital items from the World Social Capital Monitor. The regression sample comprises 67 countries (see Appendix A). Derived growth rates and the country list are available upon request from the corresponding author.

Ethics/Conflicts Competing interests: The authors declare no competing interests.

Ethical approval Not applicable (public, aggregate country-level data).

Appendix A. Countries in the regression sample (N = 67)

Income group	N	Country
High income	15	Austria; Belgium; Chile; Croatia; Czechia; Finland; France; Germany; Ireland; Italy, Republic of Korea; Netherlands; Norway; United Kingdom; United States of America
Upper middle income	16	Albania; Botswana; Brazil; Belarus; Costa Rica; Gabon; Mauritius; Mexico; Namibia; RussianFederation; Serbia; South Africa; Thailand; Turkey
Lower middle income	22	Algeria; Bangladesh; Cambodia; Cameroon; Congo; Benin; Ghana; India; Islamic Republic of Iran; Kenya; Lao People's Democratic Republic; Lesotho, Nepal; Nigeria; Pakistan; Philippines; Senegal; Viet Nam; Zimbabwe; Egypt; Tanzania; United Republic of; Zambia
Low income	14	Afghanistan; Burundi; Central African Republic; Democratic Republic of theCongo; Ethiopia; Gambia; Guinea; Madagascar; Malawi; Mali; Rwanda; Sierra Leone; Somalia; Uganda; Burkina Faso; Yemen

References

1. Leach, J., Scott, P.: Designing and evaluating science teaching sequences: an approach drawing upon the concept of learning demand and a social constructivist perspective on learning, Routledge, pp. 81–102 (2013)
2. Steffen, W., Rockström, J., Richardson, K., Lenton, T.M., Folke, C., Liverman, D., Summerhayes, C.P., Barnosky,.A.D., Cornell, S.E., Crucifix, M.: Trajectories of the earth system in the anthropocene. In: Proceedings of the National Academy of Sciences, vol. 115, pp. 8252–8259 (2018)
3. Stafford-Smith, M., Griggs, D., Gaffney, O., Ullah, F., Reyers, B., Kanie, N., Stigson, B., Shrivastava, P., Leach, M., O'Connell, D.: Integration: the key to implementing the sustainable development goals. Sustain. Sci. **12**, 911–919 (2017)
4. Kortetmäki, T., Puurtinen, M., Salo, M., Aro, R., Baumeister, S., Duflot, R., Kotiaho, J.S.: Planetary well-being. Jyväskylä, Finland, covering topics from macroecology to conservation biology and restoration ecology, pp 26 (2021)
5. Obersteiner, M., Walsh, B., Frank, S., Havlík, P., Cantele, M., Liu, J., Palazzo, A., Herrero, M., Lu, Y., Mosnier, A.: Assessing the land resource–food price nexus of the sustainable development goals. Sci. Adv. **2**, e1501499 (2016)
6. Brewer, M.K.: Slow fashion in a fast fashion world: promoting sustainability and responsibility. Laws **8**, 24 (2019)
7. Glaser, M., Krause, G., Ratter, B.M.W., Welp, M.: Human-nature interactions in the Anthropocene: potentials of social-ecological systems analysis. Routledge (2012)
8. Stiglitz, J.E., Sen, A., Fitoussi, J.P.: Report by the commission on the measurement of economic performance and social progress. The Commission Paris (2009)
9. Fleurbaey, M., Blanchet, D.: Beyond GDP: Measuring welfare and assessing sustainability, Oxford University Press (2013)
10. Colglazier, W.: Sustainable development agenda: 2030. Science **349**, 1048–1050 (2015)
11. Botelho, V., Silva, A.D.D.: Employment growth and GDP in the euro area. Econ. Bull. Boxes **2** (2019)

12. Arrow, K., Dasgupta, P., Goulder, L., Daily, G., Ehrlich, P., Heal, G., Levin, S., Mäler, K.-G., Schneider, S., Starrett, D.: Are we consuming too much? J. Econ. Perspect. **18**, 147–172 (2004)
13. Arrow, K.J., Dasgupta, P., Goulder, L.H., Mumford, K.J., Oleson, K.: Sustainability and the measurement of wealth. Environ. Dev. Econ. **17**, 317–353 (2012)
14. van den Bergh, J.C.J.M.: A procedure for globally institutionalizing a 'beyond-GDP'metric. Ecol. Econ. **192**, 107257 (2022)
15. Bergh, J.C.J.M.V.D., Botzen, W.J.W.: Global impact of a climate treaty if the Human Development Index replaces GDP as a welfare proxy. Climate Policy **18**, 76–85 (2018)
16. Sachs, J.D., Schmidt-Traub, G., Mazzucato, M., Messner, D., Nakicenovic, N., Rockström, J.: Six transformations to achieve the sustainable development goals. Nature sustainability **2**, 805–814 (2019)
17. Agarwal, M., Saha, R.: Water and sanitation: achievement of large Indian states. Indian J. Human Devel. **15**, 82–99 (2021)
18. Kumar, R.: India & South Asia: Geopolitics, regional trade and economic growth spillovers. J. Int. Trade Econ. Dev. **29**, 69–88 (2020)
19. Ding, D., Masha, M.I.: India's growth spillovers to South Asia, International Monetary Fund (2012)
20. Kumar, R., Dhankar, R.S.: Financial instability, integration and volatility of emerging South Asian stock markets. South Asian J. Business Stud. **6**, 177–190 (2017)
21. Munir, K., Shahid, F.S.U.: Role of demographic factors in economic growth of South Asian countries. J. Economic Stu. **48**, 557–570 (2021)
22. Intriligator, M.D.: Energy security in the Asia-Pacific region. Contemp. Econ. Policy **33**, 221–227 (2015)
23. Mohsin, M., Rasheed, A.K., Sun, H., Zhang, J., Iram, R., Iqbal, N., Abbas, Q.: Developing low carbon economies: an aggregated composite index based on carbon emissions. Sustainable Energy Technol. Assess. **35**, 365–374 (2019)
24. Managi, S., Kumar, P.: Inclusive wealth report 2018, Taylor & Francis (2018)
25. Managi, S., Chen, S., Kumar, P., Dasgupta, P.: Sustainable matrix beyond GDP: investment for inclusive growth. Hum.Ities Soc. Sci. Commun. **11**, 1–10 (2024)
26. Managi, S.: Wealth, inclusive growth and sustainability. Routledge New York, New York (2019)
27. Dasgupta, P.: The economics of biodiversity: the Dasgupta review., Hm Treasury (2021)
28. Dasgupta, P., Managi, S., Kumar, P.: The inclusive wealth index and sustainable development goals. Sustain. Sci. **17**, 899–903 (2022)
29. Dasgupta, P., Mäler, K.-G.: Net national product, wealth, and social well-being. Environ. Dev. Econ. **5**, 69–93 (2000)
30. Arrow, K.J., Dasgupta, P., Mäler, K.-G.: Evaluating projects and assessing sustainable development in imperfect economies. Environ. Resource Econ. **26**, 647–685 (2003)
31. Dasgupta, P.: Measuring the wealth of nations. Annu. Rev. Resour. Econ. **6**, 17–31 (2014)
32. Bateman, I.J., Mace, G.M.: The natural capital framework for sustainably efficient and equitable decision making. Nature Sustainability **3**, 776–783 (2020)
33. Dasgupta, P.: The Dasgupta review: supplementary notes on investment in conservation and restoration, family planning, and reproductive health. Natl. Inst. Econ. Rev. **256**, 162–177 (2021)
34. Chen, S., Managi.: Tracing the evolution of natural capital in global sustainability metrics: the advance of inclusive wealth. Sustainability Sci 1–14 (2024)
35. World Social Capital Monitor (2019)
36. Coleman, J.S.: Social capital in the creation of human capital. Am. J. Sociol. **94**, S95–S120 (1988)
37. OECD.: Social Capital (2020)
38. Engbers, T.A., Thompson, M.F., Slaper, T.F.: Theory and measurement in social capital research. Soc. Indic. Res. **132**, 537–558 (2017)
39. Markowska-Przybyla, U., Ramsey, D.M.: The association between social capital and membership of organisations amongst Polish students. Economics and Sociology **9**, 305 (2016)
40. Jorgenson, D.W., Fraumeni, B.M.: Investment in education. Educ. Res. **18**, 35–44 (1989)

41. Jorgenson, D.W., Fraumeni, B.M.: The output of the education sector, University of Chicago Press, pp. 303–341 (1992)
42. Brander, L.M., de Groot, R., Guisado Goni, V., Schaegner, P., Solomonides, S., van't Hoff, V., McVittie, A., Sposato, M., Do, L., Ghermandi A.: Ecosystem services valuation database ESVD. Foundation for Sustainable Development and Brander Environmental Economics (2021)
43. Barbier, E.B.: How to make the next green new deal work. Nature **565**, 6–7 (2019)
44. UN, Department of Economic and Social Affairs, Population Division.: World Population Prospects 2019–Fertility Indicators (2019)
45. UNU-IHDP., Inclusive wealth report 2012.: measuring progress toward sustainability, Cambridge University Press (2012)
46. UNU-IHDP and UNEP, Inclusive Wealth Report 2014.: Cambridge University Press (2015)
47. Sugiawan, Y., Kurniawan, R., Managi, S.: Assessing the United Nations sustainable development goals from the inclusive wealth perspective. Sci. Rep. **13**, 1601 (2023)

Information and Communication Technology Adoption and Climate Risk Perception: Evidence from Smallholder Farmers in Rural Vietnam

Matthew Donaldson, Tri-Anh Duc Nguyen, and Trung Thanh Nguyen

Abstract As climate change intensifies extreme weather events, understanding household climate risk perception (RP) becomes critical for building societal resilience, particularly among smallholder farm households in developing regions who have limited adaptive capacity against covariate climate shocks. The study examines the relationship between information and communication technology (ICT) access and smallholder farmers' RP of climate-related threats. Using three waves of longitudinal data in Vietnam from the Thailand-Vietnam Socio Economic Panel spanning 2013–2017, we employ a double-hurdle modeling framework to analyze how ICT access influences household RP across multiple dimensions. Our empirical strategy distinguishes between the probability of reporting positive RP (first hurdle) and the intensity of RP among those reporting positive values (second hurdle). Our findings reveal that ICT access demonstrates no statistically significant relationship with the incidence of positive RP but consistently exhibits positive and significant effects on RP intensity. Among households reporting positive RP, ICT access is associated with 18.06% higher income RP, 15.18% higher asset RP, 10.71% higher perceived likelihood, and 8–11% higher perceived severity measures. This indicates that ICT access enhances intensity rather than mere presence of risk awareness. Female-headed households consistently demonstrate 20–64% lower RP. Religious household heads and those in agricultural occupations exhibit higher RP. These results suggest that ICT access enhances information processing capabilities among households already engaged in risk assessment rather than fundamentally altering baseline risk awareness. Policy implications indicate that expanding ICT access should be complemented with risk awareness education to maximize benefits, while addressing identified gender disparities requires targeted interventions.

M. Donaldson · T.-A. D. Nguyen · T. T. Nguyen (✉)
Institute for Environmental Economics and World Trade, Leibniz University Hannover, Hannover, Germany
e-mail: thanh.nguyen@iuw.uni-hannover.de

M. Donaldson
e-mail: matthew.donaldson@iuw.uni-hannover.de

T.-A. D. Nguyen
e-mail: anh.nguyen@iuw.uni-hannover.de

G. Halkos et al. (eds.), *Environmental Risk Analysis*, Springer Proceedings in Mathematics & Statistics 540, https://doi.org/10.1007/978-3-032-24155-9_16

Keyword Climate change · Double-hurdle · Conditional mixed process

1 Introduction

As climate change intensifies the frequency and severity of extreme weather events, understanding how households perceive and respond to risks becomes critical for building societal resilience. Climate change is more prevalent for smallholder farm households in economically developing regions due to their limited capacity to prepare for, or otherwise mitigate against, the negative consequences of climate shocks [31]. Risks of climate shocks are exogenous events, meaning forecasting the likelihood, timing, and severity of their occurrence is largely unreliable and prohibitively complex. In comparison to other types of risk such as risks of illness, crime, and bankruptcy, a potential victim has no control over its prevention and very limited foresight of its occurrence. Furthermore, the previous examples of risk are idiosyncratic, uniquely affecting each household. However, climate shocks are often covariate, meaning the effects are experienced by many households, albeit to differing degrees of severity. In addition, smallholder farmers are limited in accessing information concerning climate shocks, especially in remote areas. One of the opportunities in this regard is the development of information and communication technologies (ICT) in general and the provision of access to the Internet in particular. For example, the Internet is becoming available and affordable to an increasing share of the global population, especially in the developing world [32]. However, the relationship between access to ICT and climate risk perception (RP) remains understudied.

Existing studies document the importance of access to information in behavior change, wherein RP is a central tenant. Information's effect on RP must be contextualized beyond the domain-specific nature of climate change and further analyzed based on the source. The literature further highlights RP within the context of adaptive behavior against climate shocks. Sources of information may distort the directional relationship between access to information and RP of climate change. [26] study Vietnamese farmers living near the Mekong Delta and find that the RP decreases more when climate change information is received from public media and local authorities than from friends, relatives, and neighbors. When these farmers receive climate information from "other" sources (including from the Internet, monks, and priests), this tends to increase RP. Furthermore, [12] examine the RP of smallholder farmers in Mali and find its positive correlation with the number of information sources of climate change accessed by mobile phone. These sources were categorized separately as Internet sources, information received through phone calls, messaging, and other mobile-enabled services. Yang et al. [54] find that internet-based information acquisition significantly promotes climate-adaptive behavior among Chinese farmers, with risk perception playing a mediating role—yet the determinants of risk perception itself, particularly the role of ICT access, remain understudied.

This study examines the effects of access to ICT on smallholder farm households' RP with regard to (i) aggregated RP index for all climate shocks, and (ii) separated RP indices regarding income and assets of rural households. Two risk perception indices (RPI) of covariate shocks with respect to household income and household assets are our dependent variables, and ICT access is our main treatment variable. We also examine the heterogeneity in the effects based on religion, gender, and main occupation of household heads. We utilize three waves of the longitudinal TVSEP in Vietnam between 2013 and 2017 and employ a double-hurdle modeling framework for our analysis. Our empirical analysis reveals that while ICT access demonstrates no statistically significant relationship with the probability of reporting positive RP in first-hurdle estimations, it consistently exhibits positive and significant effects in second-hurdle analyses. These results suggest that ICT access enhances information processing capabilities among households already engaged in risk assessment rather than fundamentally altering baseline risk awareness.

The article proceeds as follows. Section 2 reviews the literature and provides the theoretical background along with the conceptual framework illustrating the criteria and rationale behind our model specification. Section 3 describes the data and our estimation strategy. Section 4 reports the results and discusses our findings. Section 5 concludes.

2 Literature Review and Theoretical Background

2.1 Literature Review

The definition of risk is highly contextual and is seldomly recognized as a universally defined term. It is generally expressed as expected loss or disutility, within the context of uncertainty or a probability distribution [3]. Herein, we consider the definition of risk in [21] as a "(set of) scenario(s), s_i, which has a probability of occurring, p_i, and a consequence, c_i." By adopting this definition of risk, we differentiate several definitions of risk within the literature which focus merely on the consequence of an event under the presupposed condition that the event will occur or has already occurred (such that p_i is not a consideration). This is an important distinction, as our analysis incorporates dimensions of both probability and consequence. In other words, we consider RP as the subjective evaluation of the probability and consequence of different shocks.

Risk perception is important because it directly influences adaptation behavior. In Protection Motivation Theory (PMT), agents' action or inaction in response to a threat is determined by the cognitive mediating processes initiated by receipt of such information. In response to external stimuli indicating a threat, an agent's appraisal of the threat and coping abilities directly affect their protection motivation which ultimately result in either action or inaction [28].

Building on PMT, the Extended Parallel Process Model (EPPM) posits that threat and efficacy appraisals determine an agent's course of action and further differentiates the determinants of whether an agent may take adaptive changes versus maladaptive changes versus no response whatsoever [53]. One of the significant departures from PMT inherent in the EPPM is the claim that an individual will not take any action if there is no perceived threat (i.e., either perceived susceptibility, perceived severity, or both are low). Therefore, before appraisals of efficacy may be initiated, the agent must appraise the threat as response-worthy to begin with.

Further building on the EPPM, the Risk Perception Attitude (RPA) Framework differentiates "appraised threat" from "perceived risk" and illustrates the causal relationship between RP and proactive behaviors [34]. Though borrowing heavily from the EPPM, the RPA Framework posits that RP is mediated by Self-Efficacy (SE) such that

1. High RP and High SE result in a responsive attitude (*Responsive Group*).
2. High RP and Low SE result in an avoidance attitude (*Avoidance Group*).
3. Low RP and High SE result in a proactive attitude (*Proactive Group*).
4. Low RP and Low SE result in an indifference attitude (*Indifference Group*).

Rimal and Real [34] examine how self-efficacy mediates RP and how the mediating relationship ultimately affects behavior. Therefore, the hypothesis presupposes such mediation can only exist where RP is present. The partial support for the hypothesis suggests that efficacy-induced effects are largely absent and that differing RP only seems to have an effect across differing efficacy groups. Consequently, the subjects with combined low self-efficacy and high-risk perception can exhibit rather *more* motivated, *more* intentioned efforts to both seek out and implement mitigating measures compared to high RP, high SE subjects.

While the RPA framework may show that self-efficacy moderates the impact of RP on behaviors, there is limited empirical evidence supporting such hypothesis. Furthermore, models examining channels by which RP determines behavior, particularly climate change mitigation behavior, are complex, exhibit mixed empirical results when applied, and often co-opt similar dimensions for analysis.

Similarly, the relationship between RP and adoption of climate shock mitigation behavior is mixed and highly contextualized.[1] The so-called "Risk Perception Paradox," or RPP [50], asserts the linkage between RP and protective action may not

[1] ONE plausible explanation for the large heterogeneity of RP across the world is the cultural theory of risk by Douglas and Wildavsky [35]. The theory posits that RP varies systematically across cultural worldviews: Hierarchical cultures emphasizing institutional authority and traditional risk management approaches, Individualistic cultures focusing on market-based solutions and personal responsibility for risk management; Egalitarian cultures emphasizing collective action and systemic approaches to risk; Fatalistic cultures viewing risks as largely uncontrollable. Along the same line, [17] directly connect to the cultural theory and find that environmental performance significantly varies with RP across Hofstede's cultural dimensions of their global sample. Hofstede's dimensions used are as follows: **Uncertainty avoidance** correlates with how societies institutionally respond to environmental risks; **Individualism versus Collectivism** affects whether environmental risks are addressed through individual or collective action; **Long-term orientation** influences temporal risk perception and intergenerational environmental concerns.

be clear in its directional relationship due to three primary reasons. The first reason relates to prior experience of a given shock and motivation. It states that individuals who have had experiences with a shock and maintain a higher perceived risk likelihood of a given shock may not take preventive action. For these individuals, the perceived benefits of taking no action outweigh the potential negative impacts of such a shock. Vinh Hung et al. [49] examine excessive floodplain development in Hanoi, Vietnam, and find that community RP of catastrophic flooding remained low, despite many members of the community having experienced catastrophic flooding in the past and even report expected flooding to repeat again. This is partially due to the perceived benefit of the new development and differentiating expectations of future flood-level rise. Those who had not experienced catastrophic flood levels of 1971 did not expect flooding would reach similar levels. Those who had such experience believed current mitigation measures were sufficient against water levels they had previously witnessed. In short, village RP remained low despite high expected likelihood of future flooding because of the counteracting effect of high self-efficacy and high benefit of new development.

The second reason for the RPP is related to trust's role in alleviating agency. Agents may maintain a high RP but attribute the responsibility to some other agent or institution they deem better suited to take measures against a shock. Terpstra [42] conducted experiments in the Netherlands to test the risk-as-feelings hypothesis and examine the affect heuristic: increased RP induced by negative emotions leading to adaptive behavior. More precisely, the author examines whether the affect heuristic provides a causal pathway for predicting citizens' flood preparedness intentions based on three variables: prior flood experience, trust in public flood protection, and flood RP. Terpstra [42] discovers that, among those surveyed who had experienced a flood in the past, perceived consequences did not hold a significant relationship to preparedness intentions whereas perceived likelihood increased such intentions. Furthermore, when presented with negative messaging from previous floods, participants generally anticipated higher likelihood of floods. However, trust in flood protection mitigates perceived likelihood such that the more trust one has in public flood protection, the less they perceive future flooding to occur. This effect was three-fold in comparison to the hypothesized affect heuristic which had been tested by invoking feelings associated with previous floods.

The last reason for the RPP is related to self-efficacy. It states those with a given RP may not engage in adaptive behaviors because they simply do not have the resources to do so. In this regard, the authors do not focus on resource capacity restriction in a monetary sense (such as limited access to credit or low levels of liquid savings) but rather mental capacity. Siegrist et al. [37] and Wachinger et al. [50] claim that the disparity between the RP of Swiss laypersons and Swiss experts regarding the required action to mitigate against flooding in high-risk areas is due knowledge disparity. In this regard, self-efficacy is not represented by higher asset and livestock levels to smooth consumption following a shock nor by controlling for higher income, levels of savings, access to insurance, and diversified income streams—but rather also information.

Building on this theoretical understanding, the empirical literature demonstrates that these theoretical predictions manifest in measurable ways across different contexts and populations. The literature on RP has demonstrated each of these facets are integral to post-shock consumption smoothing strategies following a shock, thus improving self-efficacy. In lesser developed countries (LDCs) where institutional access to credit and insurance markets are limited, smallholder farms have been shown to rely on asset sales to recover from shocks. Livestock sales have been shown to improve consumption smoothing in response to covariate drought shocks in Burkina Faso [25], though consumption following a shock had been shown to be less than pre-shock levels, despite ample household stocks allowing for total smoothing. The reason for this may be deliberate: smallholder farmers who rely on livestock as primary income may limit sales and post-shock consumption to hedge against excessive production loss in the future. Evidence suggests that while livestock sales may increase after an income shock, the degree to which sales increase is further limited when the shock simultaneously impacts assets directly [30]. Therefore, evidence of a poverty trap wherein surpassing a given threshold of livestock sales following a shock leading to drastically reduced household income in the future [11] may serve as an argument to rather shift toward inclusion of formal insurance markets. Nevertheless, higher levels of livestock serve as a strong vehicle for overall poverty reduction for households [14] and may vary together with household income and other measures regarding self-efficacy.

The literature of RP demonstrates limited evidence on how levels of income, assets, savings, insurance market participation, and diversified income streams affect shock mitigation behavior through its relationship with self-efficacy. Additionally, self-efficacy has varying relationships with both adaptive behavior and RP [29, 39, 50]. That said, RP has been shown to be significantly correlated with these controls throughout the literature in addition to gender and education [12], age and religion [27], and insurance market participation [33], with the latter being disputed in the direction of its relationship.

Where EPPM and RPA focus on deriving RP as an emotional process through an affect response, they largely fail to analyze how information and logic-driven heuristics interact with them. The affect heuristic, a process of decision-making based on a positive or negative emotional affect response, does not incorporate domain-specific knowledge or expertise of risk. A large body of work on affective decision-making has been done in response to the overt focus on cognitive functioning within the risk literature [46]. As a result, the "dual-process" theories of decision-making outlining two separate modes of decision-making simultaneously mediating each other have been largely accepted as a progression from the risk-as-feelings versus rational choice theory behavioral model debate [20].

Knowledge, expertise, and access to information have all been found within the literature to have significant effects on RP concerning climate change and related adaptive behaviors. It is important to underscore that knowledge and expertise both differ from education. Knowledge and expertise are domain-specific and not necessarily correlated with school attendance. Sundblad et al. [40] study how climate change knowledge can predict judgment of probability and consequences of climate change. The authors find that subjects with more domain-specific education related to environmental topics (irrespective of total years of education) experienced both higher cognitive and affective risk judgments toward climate change. Incidentally, the less educated subjects generally had marginally higher mean knowledge of climate change than those with higher overall education (i.e., with "Above secondary"). Finally, Sundblad et al. [40] demonstrate that specific knowledge of causes and specific knowledge of consequences of climate change predicted both higher cognitive and affective risk judgments of climate change, irrespective of number of years of total schooling.

Studying the impact of access to information on adaptive behaviors, Yang et al. [54] on apple growers in the Shaanxi Province of China show that Internet information acquisition promoted anti-frost adaptive measures to be taken. The IV Probit model used in the study had shown that Internet information yielded statistically significant, positive predicting power of adopting anti-frost measures where "traditional media" and "face-to-face communication" yielded no such significance. This significance is further examined by income thresholds and is shown to yield similarly significant and positive predicting power for low- and middle-income households, however, this is not replicated for high-income households. Finally, Yang et al. [54] purport that RP has a mediating effect on the decision to adopt anti-frost measures when introduced in the model as a separate control. Their results present Internet information having a lower, yet significant, positive impact on the decision to adopt anti-frost measures.

Quantifying risk perception has been implemented in various ways throughout the RP literature. For example, Sullivan-Wiley et al. [39] collect 5-point Likert scale data through survey constructs measuring perceived risk likelihood, severity, and holistic concern of varying climate hazard events. They compose their index as half the sum from adding one-third the product of likelihood of experiencing the risk and severity of outcome to the general holistic concern about the issue. Sullivan-Wiley et al. [39] consider a mixed product and additive approach in composing their index.

Moreover, RPI has been imputed using dimension reduction through principal component analysis (PCA) to identify common modes of expression for RP as worry toward weather events and conflict [12]. Weber et al. [52] conducted PCA across five separate risk domains of (1) financial, (2) health/safety, (3) recreational, (4) ethical, and (5) social decisions with responses related to risk aversion and expected benefits across the domains to quantify and analyze RPI's relationship to risk behaviors.

Some studies have used prospect theory as the framework for estimating RPI as a function of risk aversion, loss aversion, and probability weighting [48]. Prospect theory considers the difference between an objective expected value (EV) and an agent's expected utility (EU) of an event to be attributable to agents' distortion assignment of value (distorting EU) and corresponding probabilities of outcome (distorting EV). Agents accomplish this distortion through a subjective value function and probability weighting function, respectively [44]. The inherent problem in utilizing this strategy of RPI imputation is that it presupposes objective expected values of probability and outcomes exist. This is generally not the case within a climate shock context and is best utilized under a lab setting.

2.2 Theoretical Background in Behavioral Economics and Statistics

While the RPP illustrates the complex relationship between RP and adaptive behavior, it also highlights the critical role of information in mediating these relationships. The advent of ICT has fundamentally altered how agents acquire, process, and act upon risk-related information, necessitating a more nuanced theoretical understanding of these mechanisms within the context of climate RP. ICT fundamentally transform how economic agents perceive climate risks by altering information acquisition, processing, and dissemination mechanisms. This transformation operates through several established theoretical frameworks in behavioral economics and statistics, providing crucial insights into the mechanisms underlying the Risk Perception Paradox discussed above.

Pioneering the literature on the impacts of information aggregation through Bayesian learning framework, Kyle [24] establishes the mathematical foundation for how private information affects market liquidity and price formation in markets with asymmetric information. His model shows how informed traders' signals get incorporated into prices through strategic trading, directly informing the $\pi(s|\theta)$ component—how signals relate to true states. Foster and Viswanathan [16] extend the baseline framework by Kyle's [24] for understanding information revelation to multiple informed traders who must forecast others' information. Thus, ICT adoption can revolutionize the Bayesian updating process through which agents revise climate risk beliefs [2, 15].

In the standard framework, posterior beliefs follow:

$$\pi(\theta|s) = \frac{\pi(s|\theta)\pi(\theta)}{\pi(s)} \tag{1}$$

where θ represents true climate risk parameters, s represents received signals, and $\pi(\theta)$ is the prior belief distribution. ICT can affect this mechanism through two main channels. First, it increases prior precision by providing access to comprehensive climate news and information. Second, continuous information streams enable more frequent belief updating, accelerating convergence toward accurate risk assessments.

However, ICT also introduces network effects that complicate pure Bayesian learning. Social media platforms create information cascades where early signals disproportionately influence collective beliefs [1, 23]. The multiplicative nature of information sources requires sophisticated aggregation methods, as agents must assess source credibility and correlation structures across signals [4, 19]. This theoretical framework directly relates to the second component of the RPP, where trust in institutions affects how agents weight different information sources and allocate responsibility for risk mitigation.

Besides its impacts through Bayesian learning framework, ICT profoundly affects the availability heuristic, defined as the cognitive tendency to overweight easily recalled information when assessing probabilities [43]. High-resolution imagery and video content of climate disasters increase the emotional salience of extreme events, making them disproportionately influential in RP [6, 7]. This operates through what Bordalo et al. [5] formalize as salience-weighted probabilities:

$$\tilde{\pi}(x) = \frac{\pi(x).\sigma(x, C)}{\sum \pi(x').\sigma(x', C)} \tag{2}$$

where σ(x, C) represents event salience in context C.

Paradoxically, while ICT provides access to comprehensive historical climate data, the continuous news cycle and emphasis on recent events may strengthen recency bias [4, 22, 41, 51]. Spatial compression—whereby distant climate events feel proximate through digital media—alters the geographical weighting of climate risks in individual decision-making [9, 38]. These mechanisms provide theoretical grounding for understanding the first component of the RPP, where prior experience with shocks may not translate to higher RP due to competing cognitive processes.

ICT adoption creates complex, non-monotonic effects on climate RP that intersect with each component of the Risk Perception Paradox. While improving information access and processing capabilities, ICT simultaneously amplifies cognitive biases and creates new channels for misperception. The net effect depends critically on how ICT infrastructure develops—whether toward centralized platforms that amplify biases or decentralized systems that facilitate robust information aggregation. This theoretical framework suggests that the relationship between information access and self-efficacy (the third component of the RPP) is mediated not just by resource constraints, but by the cognitive architecture through which ICT-delivered information is processed.

Understanding these mechanisms is essential for designing climate communication strategies and policy interventions in an increasingly digital information environment and provides the theoretical foundation for examining how ICT access affects RP and subsequent adaptive behaviors in empirical contexts.

3 Data Description and Estimation Strategy

3.1 Data

The panel data used in this research were taken from Thailand-Vietnam Socio Economic Panel as part of the research project, Poverty Dynamics and Sustainable Development: A long-term panel project in Thailand and Vietnam financed by the German Research Foundation (DFG). The Thailand-Vietnam Socio Economic Panel (TVSEP) represents a comprehensive longitudinal research initiative designed for analyzing socioeconomic development and poverty vulnerability dynamics across rural regions of two Southeast Asian emerging economies. Following United Nations Department of Economic and Social Affairs sampling guidelines [45], the TVSEP employs rigorous selection criteria focusing on provinces characterized by low per capita income, high agricultural dependency, significant risk exposure, and inadequate infrastructure development [18]. Even though Vietnam and Thailand have achieved remarkable progress in reducing poverty over the last decades, households in rural areas, specifically in the TVSEP provinces, are poor in multiple dimensions and mainly rely on agriculture to sustain their livelihood [13]. Additionally, ethnic minorities and households in the remote areas are even more likely to be poor [10]. In Vietnam, the TVSEP selects three provinces (Ha Tinh, Thua Thien Hue, and Dak Lak) as study sites. The TVSEP utilizes a sophisticated three-phase stratified random sampling approach, systematically selecting sub-districts/communes based on population proportions, followed by village selection using probability-proportional-to-size sampling, culminating in random household selection within each village. This procedure yielded 220 villages and about 2,000 households in Vietnam. The panel maintains exceptional data quality with minimal attrition rates of approximately 3% per wave. Thanks to the partnership with local partners, the TVSEP has collected survey data on more than 10,000 individuals from more than 2000 households throughout 220 villages across Ha Tinh, Thua Thien Hue, and Dak Lak (Fig. 1). Vietnamese household data were collected and published from 2007 through 2022 across eight waves with full samples across six. More specifically, the full survey has been undertaken in 2007, 2008, 2010, 2013, 2016, and 2017. The survey data from Waves 5, 6, and 7 capture the reference periods 05/2012–04/2013, 05/2015–04/2016, and 05/2016–04/2017 respectively. These waves were selected due to data availability and sample consistency.

ICT access was imputed for Wave 5 across several constructs in varying sections. Households having ICT access meet one of the following conditions outlined. In Sect. 2.5 in the TVSEP household survey (TVSEP Sect. 2.5 for short) related to social networks, respondents are presented Question 5: "What are the three main sources of information about what is going on in your country? Please specify in order of priority." If respondents included "Internet" (coded 7) as any of the three main sources of information, their respective household was deemed to have access to information and communication technology (ICT). In TVSEP Sect. 6.2, respondents were asked to "Please list all investments [1.5 million Vietnamese Dong] your [household] undertook in the three years (05/10–04/13), start with the newest one." Households which reported they had invested in a computer (coded 74) were assigned ICT access. In TVSEP Sect. 6, Question 5, respondents were required to list all non-farm self-employment of members within the household. If any household member responded they owned or worked at an "Internet-shop" (coded 11), the household was assigned ICT access. In TVSEP Sect. 9.1, respondents are presented with a list of assets and are instructed to indicate how many of each list-item the household owns. All households indicating non-zero, positive values for "Personal Computer" (coded 42) were assigned ICT access. ICT access was imputed for Waves 6 and 7 with a survey measurement not available in earlier waves. In TVSEP Sect. 9.2, Question 19, respondents were asked "When your household uses the Internet what is the major device used?". All households with responses inconsistent with "does not apply" (coded 5) were assigned ICT access. Proportions of all households with

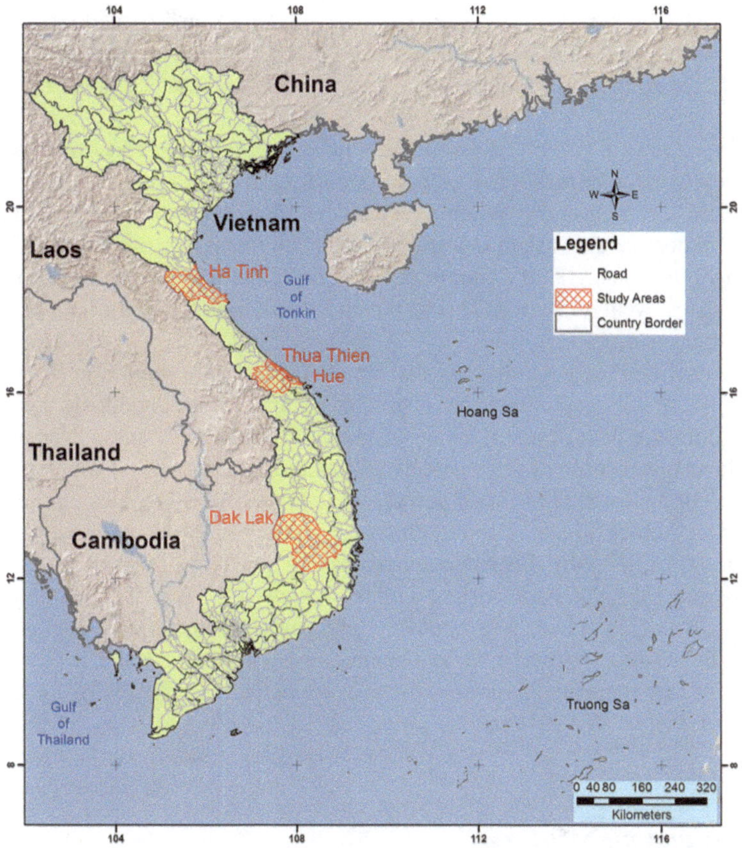

Fig. 1 TVSEP study sites in Vietnam (*Source* Schulte et al. [36])

ICT access in an average village are 17.2% in 2013, 37.9% in 2016, and 44.7% in 2017.

Previous shock experience within the reference period of the survey was imputed directly from TVSEP Sect. 3.1 across all waves in the same manner. Respondents are asked "Was your household affected by any of the following events between [reference period]." Enumerators then read from a list of different shocks, of which 17 are used in our analysis. A household is coded as having prior experience with a given shock if at least one member reports being affected during the reference week.

Section 3.2 is where perceived severity is elicited across two separate survey measurements for each shock type: perceived severity of shock on household income (Question 3) and perceived severity of shock on household assets (Question 3a). Question 3 is posed to the respondent as "If [event] occurred within the next 12 months, what would be the impact on your household: income?" and Question 3a is stated as "If [event] occurred within the next 12 months, what would be the impact on your household: assets." Answers to both questions are coded as ordinal responses: “High” (coded 1), “Moderate” (coded 2), “Low” (coded 3), and “No Impact” (coded 4). The scale is reversed in our dataset for ease of results interpretation such that High = 4, Moderate = 3, Low = 2, No Impact = 1, and No Answer = 0.

Both measures of perceived severity are, however, subject to censoring based on the survey structure. Question 1 of Sect. 3.2 asks “Do you think that [event] will occur in the next 5 years?”. When a respondent answers “No,” they are not asked Question 3 nor are they asked Question 3a. The enumerator is instructed to bypass all further questions related to the respective shock and move on to Question 1 for the next shock event on the list. This results in an observation of 0 for both perceived severity measures.

Perceived likelihood of a shock is also imputed in the same manner across all three waves. In Sect. 3.2, respondents are instructed to consider all previously mentioned shock within the next 5 years. When posed with Question 1, an answer of “No” results in an observation of 0 for perceived likelihood. However, in instances where a respondent answers “Yes,” then perceived likelihood of a shock is elicited from Question 4. Across all shock types, Question 4 is stated to the respondents “How often, do you think, will [event] occur in next 5 years?”. Answers are coded as integers from 1 through 6, consistent with their count (e.g., 1 = “1 time in 5 years”) up to 5 with 6 being coded as “more than 5 in 5 years.” Descriptive statistics of ICT access, climate shock experience, and controls are reported in Table 1. Descriptive statistics of main dependent variables by year are in Table 2 (for 2023), Table 3 (for 2016), and Table 4 (for 2017).

3.2 Estimation Strategy

We construct RPI using categorical Likert responses of respondents’ perception of likelihood and impact [47]. We incorporate dimensions of likelihood and impact following a four-fold pattern of RP in which the relationship between perceived likelihood and impact falls along a continuum within a four-quadrant matrix, thus delineating between arbitrary high and low levels across both dimensions. The matrix has been used to illustrate how the translation and switch effect of a risk may explain increasing risk aversion despite increasing expected gains or decreasing expected losses [8]. To identify potentially counteracting forces, we examine the effects of predictors on the latent components of RPI, represented as multiplicative functions of perceived likelihood and impact (or severity):

$$RPI = I(x) \times L(x), \tag{3}$$

where $I(x)$ is a function representing perceived impact of a shock event and $L(x)$ is a function representing perceived likelihood of a shock.

We specify RPI as separate indices in relation to RPI with respect to income and assets, utilizing separate survey responses related to severity measure of a given

Table 1 Descriptive statistics of key variables

Variable	Description	Mean	SD	Min	Max
ICT access	Dummy = 1 if HH has access to ICT (computer, smartphone device, tablet, or any device connected to Internet)	0.33	0.47	0.00	1.00
Covariate shock experience	Dummy = 1 if HH has experienced any covariate shock	0.51	0.50	0.00	1.00
Flood	Dummy	0.12	0.33	0.00	1.00
Drought	Dummy	0.20	0.40	0.00	1.00
Livestock Disease	Dummy	0.25	0.43	0.00	1.00
Landslide	Dummy	0.01	0.10	0.00	1.00
Storm	Dummy	0.04	0.20	0.00	1.00
Output prices	Dummy	0.09	0.28	0.00	1.00
Input prices	Dummy	0.02	0.13	0.00	1.00
HH size	Number of HH members (person)	3.72	2.60	0.00	18.00
HH head age	Age of HH head (years)	55	13	18.00	105.00
Female head	Dummy = 1 if HH head as female	0.20	0.40	0.00	1.00
Education	Number of years of education of HH head	6.55	4.24	0.00	113.00
Religious	Dummy = 1 if HH head reports he/she is not atheist	0.22	0.42	0.00	1.00
Married	Dummy = 1 if HH head reports she/he is married	0.81	0.39	0.00	1.00
Agricultural occupation	Dummy = 1 if HH head reports main occupation in agriculture	0.65	0.48	0.00	1.00
Annual income (mil VND)	Annual total income of the household and its members	73	121	-249	3,719
Livestock value (mil VND)	Total current value of all livestock held by the household	13	28	0	546

shock on a household's income versus assets, such that

$$RPI_{income} = S(x)_n \times L(x) \tag{4}$$

$$RPI_{assets} = S(x)_a \times L(x), \tag{5}$$

where *RPI* is the risk perception index, $S(x)$ is a function representing perceived shock severity, $L(x)$ is a function representing perceived shock likelihood, n is a subscript indicating a severity measure of income, and a is a subscript indicating a severity measure of assets.

Table 2 Summary statistics for shock level measures of perceived likelihood, severity of income, and severity of assets for 2013

	Summary statistics		Ordinal count						
	Mean	SD	0	1	2	3	4	5	6
2013: Perceived likelihood									
Flood	1.096	1.897	1,393	83	146	80	16	228	64
Drought	1.577	1.926	1,025	121	281	207	49	292	35
Livestock disease	1.587	2.106	1,129	90	217	142	32	289	111
Landslide	0.148	0.836	1,934	18	5	6	2	27	18
Storm	1.233	1.952	1,307	88	174	109	20	237	75
Output prices	0.472	1.328	1,732	29	76	62	15	54	42
Input prices	0.563	1.511	1,725	15	61	42	20	93	54
2013: Perceived severity of income									
Flood	0.960	1.540	1,404	41	67	237	261	–	–
Drought	1.581	1.716	1,030	38	115	399	428	–	–
Livestock disease	1.365	1.643	1,129	43	139	374	325	–	–
Landslide	0.093	0.546	1,946	10	8	23	23	–	–
Storm	1.020	1.508	1,308	81	120	274	227	–	–
Output prices	0.445	1.162	1,740	7	26	113	124	–	–
Input prices	0.423	1.098	1,762	5	40	74	17	–	–
2013: Perceived severity of assets									
Flood	0.808	1.376	1,403	145	79	210	173	–	–
Drought	1.074	1.372	1,031	405	154	234	186	–	–
Livestock disease	0.993	1.355	1,129	339	121	270	151	–	–
Landslide	0.087	0.523	1,946	15	9	18	22	–	–
Storm	1.019	1.511	1,307	92	112	264	235	–	–
Output prices	0.267	0.791	1,740	136	36	64	34	–	–
Input prices	0.246	0.724	1,732	164	31	63	20	–	–
Observations	2,010								

The model estimates the effect of ICT Access on Risk Perception indices. The treatment variable, access to ICT, takes the value of 1 for a household with access to ICT within the reference period of the survey (subscript i refers to household). A dummy variable is included as a control, taking the value of 1 if the household had experienced a covariate shock within the survey reference period. Socio-demographic characteristics of the household heads are also included. Further controls of economic characteristics such as household annual income and livestock value are included as well. The model applied toward the respective RPI indices is

$$RPI_{fi} = \gamma_f A_i + \theta P_i + \alpha_f V_i + \beta_f W_i + u_{fJi}, \tag{6}$$

Table 3 Summary statistics for shock level measures of perceived likelihood, severity of income, and severity of assets, for 2016

	Summary statistics		Ordinal count						
	Mean	SD	0	1	2	3	4	5	6
2016: Perceived likelihood									
Flood	0.688	1.471	1,450	101	131	59	6	138	8
Drought	2.044	2.042	751	137	300	177	46	292	35
Livestock disease	1.419	1.913	1,054	142	233	111	28	317	8
Landslide	0.201	0.853	1,757	40	37	17	3	31	8
Storm	1.188	1.749	1,099	210	228	96	15	235	10
Output prices	0.502	1.227	1,535	80	139	54	7	65	13
Input prices	0.283	1.002	1,707	38	72	18	1	46	11
2016: Perceived severity of income									
Flood	0.698	1.354	1,460	38	57	183	155	–	–
Drought	1.813	1.632	759	72	207	474	381	–	–
Livestock disease	1.267	1.544	1,063	49	211	352	218	–	–
Landslide	0.178	0.715	1,765	22	36	37	33	–	–
Storm	1.111	1.477	1,117	106	210	263	197	–	–
Output prices	0.578	1.268	1,548	10	48	160	127	–	–
Input prices	0.267	0.867	1,718	16	47	128	89	–	–
2016: Perceived severity of assets									
Flood	0.373	0.837	1,461	284	61	47	40	–	–
Drought	0.954	1.073	759	757	162	135	80	–	–
Livestock disease	0.757	1.103	1,064	510	121	111	87	–	–
Landslide	0.136	0.584	1,765	60	25	24	19	–	–
Storm	1.065	1.463	1,120	167	178	219	209	–	–
Output prices	0.256	0.646	1,548	265	35	30	15	–	–
Input prices	0.127	0.470	1,719	136	17	13	8	–	–
Observations	1,893								

where

- RPI_{fi} is the risk perception index related to f severity measure (whether n, income or a, assets).
- A_i is access to ICT.
- P_i is our dummy variable representing previous experience of any indexed covariate shock within the survey reference period.
- V_i is a (Y × 1) vector of household demographic controls.
- W_i is a (Z × 1) vector of household economic controls.
- u_{fi} is the white noise term related to f severity measure.

Table 4 Summary statistics for shock level measures of perceived likelihood, severity of income, and severity of assets, for 2017

	Summary statistics		Ordinal count						
	Mean	SD	0	1	2	3	4	5	6
2017: Perceived likelihood									
Flood	1.144	1.946	1,316	62	121	87	11	243	58
Drought	1.185	1.827	1,188	130	197	91	19	261	12
Livestock disease	1.416	1.973	1,096	123	193	144	22	261	59
Landslide	0.132	0.712	1,820	13	18	18	0	27	2
Storm	1.505	2.078	1,100	101	182	125	24	275	91
Output prices	0.688	1.388	1,426	79	174	109	13	82	15
Input prices	0.218	0.888	1,762	20	46	27	5	28	10
2017: Perceived severity of income									
Flood	0.911	1.436	1,317	15	132	286	148	–	–
Drought	1.063	1.464	1,188	37	193	325	155	–	–
Livestock Disease	1.251	1.539	1,098	9	193	412	186	–	–
Landslide	0.124	0.624	1,820	5	11	39	23	–	–
Storm	1.140	1.444	1,100	66	230	371	131	–	–
Output prices	0.803	1.439	1,426	7	52	239	174	–	–
Input prices	0.198	0.736	1,762	5	40	74	17	–	–
2017: Perceived severity of assets									
Flood	0.525	0.951	1,317	340	96	115	30	–	–
Drought	0.649	1.016	1,188	395	144	136	35	–	–
Livestock disease	0.832	1.196	1,099	382	135	201	81	–	–
Landslide	0.106	0.565	1,820	20	13	24	21	–	–
Storm	1.070	1.411	1,100	134	252	255	157	–	–
Output prices	0.452	0.928	1,426	249	93	97	33	–	–
Input prices	0.119	0.494	1,762	84	21	25	6	–	–
Observations	1,898								

This direct analysis allows us to examine the direct influence of ICT access on RPI; however, the results provide no insights into the direction nor significance of its influence on perceived severity and likelihood.

Perceived severity is specified as a composite average with respect to its measure, f, across all covariate shock types such that

$$S_{fi} = \frac{1}{c} \times \sum_{c=1}^{C} s_{cfi} \tag{7}$$

Therefore, the dimensional analysis of perceived severity requires our model be applied to every individual shock type and repeated for each severity measure. The TVSEP observes seven covariate shock types across the selected panel waves including flooding of agricultural land, drought, livestock disease, landslide, storm, output price shocks, and input prices shocks.

Perceived severity of covariate shocks on income for household i, noted as S_{ni},, is therefore a composite average of observed perceived severity across all individual covariate shocks (lowercase c), with $c \in [1,7]$. Therein we apply our model in matrix format such that

$$S_{ni} = \frac{1}{7} \times \sum_{c=1}^{7} s_{cni} = \beta_n^T A_i + \beta_{cn}^P P_{ci} + \beta_n^V V_i + \beta_n^W W_i + u_{ni} \tag{8}$$

$$s_{cni} = \begin{bmatrix} s_{1ni} \\ s_{2ni} \\ \cdot \\ \cdot \\ \cdot \\ s_{7ni} \end{bmatrix}, \tag{9}$$

where

- $\beta_{cn}^A \in \mathbb{R}^{7\times 1}$ is the coefficient matrix for treatment variable, ICT access.
- A_i is our treatment variable, ICT access.
- β_{cn}^P is the coefficient for previous experience of respective shock, c.
- P_{ci} is our previous shock experience dummy variable, with respect to shock c.
- $\beta_n^V \in \mathbb{R}^{7\times Y}$ is the coefficient matrix for demographic controls.
- V_i is a Y × 1 vector of household demographic controls, $V_i \in \mathbb{R}^{Y\times 1}$.
- $\beta_n^W \in \mathbb{R}^{7\times Z}$ is the coefficient matrix for economic controls.
- W_i is a Z × 1 vector of household characteristic controls, $W_i \in \mathbb{R}^{Z\times 1}$.
- $u_{ni} \in \mathbb{R}^{7\times 1}$ is the vector of error terms.
- Subscript n indicates the perceived severity on income measure.

It is noted that the previous shock experience dummy is not vectorized. Rather, it is re-specified with respect to the associated shock severity measure in each individual estimation of s_{cni}. For example, s_{1ni} representing severity measure of flooding (c=1) with respect to income (subscript n) for household i may be estimated with dummy variable, P_{1i}, taking on value 1 if the respective household experienced flooding of agricultural land within the survey reference period. Accordingly, perceived severity of covariate shocks on assets for household i is specified such that

$$S_{ai} = \frac{1}{7} \times \sum_{1}^{7} s_{cai} = \beta_a^A A_i + \beta_{ca}^P P_{ci} + \beta_a^V V_i + \beta_a^W W_i + u_{ai} \tag{10}$$

Unlike perceived severity, the perceived likelihood dimension has only one measure per shock type and thus is a component of RPI of income and RPI of assets. Its specification mirrors that of perceived severity insofar that it is also a composite average of observations across covariate shocks, such that

$$L_i = \frac{1}{C} \times \sum_{c=1}^{C} l_{ci} \quad (11)$$

Therefore, it is logical to specify perceived likelihood of covariate shocks on household i as

$$L_i = \frac{1}{7} \times \sum_{1}^{7} l_{ci} = \beta_T A_i + \beta_c^P P_{ci} + \beta_V V_i + \beta_W W_i + u_i \quad (12)$$

Our model specifications in scalar format of Eqs. 7–12 are presented in Tables 5, 6, and 7.

Concerning our empirical strategy, our first stage analysis begins with estimating the direct effect ICT access on overall RPI of income and RPI of assets using a double-hurdle estimation strategy. The first hurdle implements a logit estimation to determine the probability of observing a non-zero value for RPI of income and assets. The second hurdle is estimated using the truncated dataset containing non-zero observations for RPI of income and RPI of assets. The estimation strategy implemented on positive values of RP indexes is a Poisson Pseudo Maximum Likelihood (PPML) model with household and year fixed effects, reported with clustered error terms at the household level. After examining the effects of ICT on overall measures of RPI_n and RPI_a and their respective dimensions, we examine the effects of ICT on

Table 5 Equations 7 and 8 in scalar form

Structural equation for perceived severity	$S_{fi} = \frac{1}{C} \times \sum_{c=1}^{C} s_{cfi}$	
Covariate shocks $c \in [1, 7]$	Perceived severity of covariate shocks on income for household i	Scalar form
1 = Flooding 2 = Drought 3 = Livestock disease 4 = Landslides 5 = Storm 6 = Output prices 7 = Input prices	$S_{ni} = \frac{1}{7} \times \sum_{1}^{7} s_{cni}$	$\begin{cases} s_{1ni} = \gamma_{1n} A_i + \theta_{1n} P_{1i} + \alpha_{1n} V_i + \beta_{1n} W_i + u_{1ni} \\ s_{2ni} = \gamma_{2n} A_i + \theta_{2n} P_{2i} + \alpha_{2n} V_i + \beta_{2n} W_i + u_{2ni} \\ s_{3ni} = \gamma_{3n} A_i + \theta_{3n} P_{3i} + \alpha_{3n} V_i + \beta_{3n} W_i + u_{3ni} \\ s_{4ni} = \gamma_{4n} A_i + \theta_{4n} P_{4i} + \alpha_{4n} V_i + \beta_{4n} W_i + u_{4ni} \\ s_{5ni} = \gamma_{5n} A_i + \theta_{5n} P_{5i} + \alpha_{5n} V_i + \beta_{5n} W_i + u_{5ni} \\ s_{6ni} = \gamma_{6n} A_i + \theta_{6n} P_{6i} + \alpha_{6n} V_i + \beta_{6n} W_i + u_{6ni} \\ s_{7ni} = \gamma_{7n} A_i + \theta_{7n} P_{7i} + \alpha_{7n} V_i + \beta_{7n} W_i + u_{7ni} \end{cases}$

Table 6 Equation 10 in scalar form

Covariate shocks $c \in [1, 7]$	Perceived severity of covariate shocks on assets for household i	Scalar form
1 = Flooding 2 = Drought 3 = Livestock disease 4 = Landslides 5 = Storm 6 = Output prices 7 = Input prices	$S_{ai} = \frac{1}{7} \times \sum_1^7 s_{cai}$	$\begin{cases} s_{1ai} = \gamma_{1a}A_i + \theta_{1a}P_{1i} + \alpha_{1a}V_i + \beta_{1a}W_i + u_{1ai} \\ s_{2ai} = \gamma_{2a}A_i + \theta_{2a}P_{2i} + \alpha_{2a}V_i + \beta_{2a}W_i + u_{2ai} \\ s_{3ai} = \gamma_{3a}A_i + \theta_{3a}P_{3i} + \alpha_{3a}V_i + \beta_{3a}W_i + u_{3ai} \\ s_{4ai} = \gamma_{4a}A_i + \theta_{4a}P_{4i} + \alpha_{4a}V_i + \beta_{4a}W_i + u_{4ai} \\ s_{5ai} = \gamma_{5a}A_i + \theta_{5a}P_{5i} + \alpha_{5a}V_i + \beta_{5a}W_i + u_{5ai} \\ s_{6ai} = \gamma_{6a}A_i + \theta_{6a}P_{6i} + \alpha_{6a}V_i + \beta_{6a}W_i + u_{6ai} \\ s_{7ai} = \gamma_{7a}A_i + \theta_{7a}P_{7i} + \alpha_{7a}V_i + \beta_{7a}W_i + u_{7ai} \end{cases}$

Table 7 Equations 11 and 12 in scalar form

Structural equation for perceived likelihood	$L_i = \frac{1}{C} \times \sum_{c=1}^{C} l_{ci}$	
Covariate shocks $c \in [1, 7] \cap \mathbb{Z}$	Perceived likelihood of covariate shocks for household i	Scalar form
1 = Flooding 2 = Drought 3 = Livestock disease 4 = Landslides 5 = Storm 6 = Output prices 7 = Input prices	$L_{Ci} = \frac{1}{7} \times \sum_1^7 l_{ci}$	$\begin{cases} l_{1i} = \gamma_1 A_i + \theta_1 P_{1i} + \alpha_1 V_i + \beta_1 W_i + u_{1i} \\ l_{2i} = \gamma_2 A_i + \theta_2 P_{2i} + \alpha_2 V_i + \beta_2 W_i + u_{2i} \\ l_{3i} = \gamma_3 A_i + \theta_3 P_{3i} + \alpha_3 V_i + \beta_3 W_i + u_{3i} \\ l_{4i} = \gamma_4 A_i + \theta_4 P_{4i} + \alpha_4 V_i + \beta_4 W_i + u_{4i} \\ l_{5i} = \gamma_5 A_i + \theta_5 P_{5i} + \alpha_5 V_i + \beta_5 W_i + u_{5i} \\ l_{6i} = \gamma_6 A_i + \theta_6 P_{6i} + \alpha_6 V_i + \beta_6 W_i + u_{6i} \\ l_{7i} = \gamma_7 A_i + \theta_7 P_{7i} + \alpha_7 V_i + \beta_7 W_i + u_{7i} \end{cases}$

shock-level measures of RPI (rpi_n and rpi_a). The double-hurdle estimation strategy is further applied to the aggregated dimensions of perceived likelihood, perceived severity with respect to income, and perceived severity with respect to assets.

The second stage of analysis is constricted to the second hurdle's non-zero dataset of RPI_n and RPI_a. Utilizing a zero-inflated ordered probit modeling strategy with year fixed effects and clustered errors at the household level, we report latent variable effects of ICT access on shock-level risk perception indices, rpi_n and rpi_a. At this level of analysis, we reveal how ICT access affects shock-level RP indices for those households reporting some level of overall RP.

The third stage reports findings on the effect of ICT access on shock-level latent variables: perceived likelihood, l_{ci}, perceived severity with respect to income, s_{cni}, and with respect to assets, s_{cai}. A two-hurdle joint estimation of the shock-level dimensions via conditional mixed process (CMP) first reports the effect of ICT access on binary selection of a probit model. In such method, we determine how

ICT access affects observing non-zero values of all three components of l_{ci}, s_{cni}, and s_{cai}. The second hurdle then jointly estimates an ordered probit model via CMP estimator to report average marginal effects of ICT access on each component's ordinal category. This allows us to analyze the directional impact that access to ICT has on each shock-level component of RP.

4 Results and Discussion

4.1 Effects of ICT on Aggregated RPI of Income and Assets

Table 8 presents the combined results of both hurdle estimates for the effects of ICT access on aggregated RPI of income and assets. In the first hurdle for RPI of income, ICT access does not hold a statistically significant relationship of predicting a positive observation of RPI. Odds ratio of previous covariate shock experience implies an associated 235% higher likelihood of observing a positive RPI of income than a zero observation. Odds ratio of 0.449 for female head of household implies a 55.1% decrease in the likelihood of observing a positive RPI of income than a zero observation. For those households with religious heads, an odds ratio of 1.368 implies a 36.8% higher likelihood of observing a positive RPI of income than those households without a religious head. Furthermore, an odds ratio of 1.301 for households where their heads mainly engage in agriculture is associated with a 30.1% higher likelihood of observing a non-zero RPI of income. Regarding the second hurdle using the Poisson pseudo maximum likelihood model (PPML) on aggregated RPI of income, ICT access has a statistically significant positive effect. The incidence rate ratio of 1.1806 implies that access to ICT is associated with an 18.06% higher expected RPI of income across those observations where a positive RPI of income was observed. Across those same observations, households experiencing some previous covariate shock within the survey reference period are correlated with a 33.81% higher RPI of income (IRR = 1.3381). For those female-headed households, the IRR of 0.7897 implies a 20.03% lower RPI of income as compared to male-headed households.

Examining the RPI with respect to assets, the first hurdle results show that ICT access does not have a statistically significant effect on observing a positive value. For those households with previous covariate shock experience within the survey reference period, odds ratio of 3.361 implies a 236% higher likelihood of observing a positive RPI of assets than those without any experience of covariate shocks. Odds ratio of 0.456 for those female-headed households implies a 54.4% lower likelihood of observing a positive RPI of assets than households with male heads. Odds ratio of 1.381 for households with their religious head implies a 38.1% higher likelihood of observing a positive RPI of assets than those households with atheist heads. Households where their heads mainly engage in agriculture is shown to have an odds ratio of 1.343, indicating a 34.3% higher likelihood of observing a positive RPI of assets than those without agricultural employment.

Table 8 2-Hurdle estimation of RPI, logit estimation (1st hurdle) and PPML fixed-effects model (2nd hurdle)

	Income			Assets		
	1st hurdle	2nd hurdle		1st hurdle	2nd hurdle	
	Odds ratios of Y > 0	Log coefficients	IRRs	Odds ratios of Y > 0	Log coefficients	IRRs
ICT access	1.065	**0.1660*****	**1.1806*****	1.064	**0.1413****	**1.1518****
	(0.137)	(2.692)		(0.137)	(2.144)	
Covariate shock	**3.351*****	**0.2912*****	**1.3381*****	**3.361*****	**0.2693*****	**1.3090*****
	(0.357)	(6.187)		(0.358)	(5.314)	
HH size	0.967	0.0188	1.0190	0.964	**0.0388****	**1.0396****
	(0.033)	(1.151)		(0.033)	(2.203)	
HH head age	0.995	−0.0037	0.9963	0.997	−0.0066	0.9934
	(0.011)	(−0.667)		(0.0106)	(−1.247)	
Female head	**0.449****	**−0.2361***	**0.7897***	**0.456****	−0.0706	0.9318
	(0.166)	(−1.667)		(0.169)	(−0.458)	
Education	1.004	−0.0096	0.9905	1.003	−0.0120	0.9880
	(0.031)	(−0.703)		(0.0313)	(−0.781)	
Religious	**1.368****	−0.0290	0.9714	**1.381****	−0.0310	0.9695
	(0.217)	(−0.341)		(0.220)	(−0.362)	
Married	0.818	−0.2157	0.8059	0.849	−0.0776	0.9254
	(0.240)	(−1.563)		(0.249)	(−0.511)	
Agricultural occupation	**1.301****	0.0835	1.0870	**1.343****	0.1053	1.1110
	(0.170)	(1.217)		(0.176)	(1.418)	
HH income	1.000	−0.000	1.0000	1.000	−0.0000	1.0000
	(0.000)	(−0.491)		(0.000)	(−0.645)	
Livestock value	1.000	−0.000	1.0000	1.000	−0.0000	1.0000
	(0.000)	(−0.321)		(0.000)	(−0.010)	
Constant		**1.0085*****			**0.6619***	
		(2.751)			(1.759)	
Obs	2,461	4,171		2,460	4,171	

Notes se in parenthesis (Columns 1 and 4), *z* statistics in parentheses (Columns 2, 3, 5, and 6). $^{*}p < 0.10$, $^{**}p < 0.05$, $^{***}p < 0.01$

Second hurdle of RPI of assets estimated with PPML model is constricted to those households reporting positive RPI values. ICT access is associated with a 15.18% higher RPI than those without ICT access, all else equal ($p < 0.05$ level). An IRR of 1.3090 for previous covariate experience indicates an association with a 30.9% higher RPI of assets than those households without any experience of covariate shocks ($p < 0.01$). Households are associated with a 3.96% higher RPI of assets with each additional member, as implied by an IRR of 1.0396 ($p < 0.05$).

4.2 Effects of ICT on Aggregated Perceived Likelihood and Severity with Respect to Income and Assets

Table 9 provides combined double-hurdle results for our model on the effect of ICT access on aggregated perceived likelihood. In the first-hurdle Logit estimation, ICT access holds no statistically significant relationship to observing a positive value for perceived likelihood (PL). Odds ratio of 3.216 for previous covariate shock experience indicates an association of a 221.6% higher likelihood of observing a positive value for perceived likelihood than those households without previous covariate shock experience. With regards to female-headed households, odds ratio of 0.452 implies an association of a 54.8% lower likelihood of observing a non-zero, positive value for perceived likelihood. Households with religious heads are associated with a 36.6% higher likelihood of observing a positive value for perceived likelihood than households with atheist heads, as indicated by an odds ratio of 1.366. For those households where the heads mainly engage in agriculture, an odds ratio of 1.259 implies an association of 25.9% higher likelihood of observing a positive value for perceived likelihood ($p < 0.1$).

Second hurdle results are interpreted across all households where a non-zero, positive value was observed for perceived likelihood. Within this data framework, an incidence rate ratio of 1.1071 for ICT access indicates an association with a 10.71% higher perceived likelihood than those without ICT access, all else held constant ($p < 0.05$). Previous experience of covariate shocks is associated with a 13.73% higher perceived likelihood as indicated by an IRR of 1.1373 ($p < 0.01$). Next, we estimate our double-hurdle model for the effect of ICT on aggregated perceived severity (PS).

Table 10 provides combined findings of double-hurdle estimation of both perceived severity on household income and household assets. The first hurdle logit estimation of observing a non-zero value of perceived severity of income is given in Column 1. Statistically significant coefficients related to odds ratios of observing positive perceived severity of income are head of household age ($p < 0.10$), female head of household ($p < 0.10$), religious head of household ($p < 0.05$), and agricultural occupation household head ($p < 0.01$). It should be noted, however, that the unusual coefficient associated with previous covariate shock is due to near-perfect collinearity. Odds ratio of 0.970 for age of household head implies an association of 3% decrease in the likelihood of observing a positive value with each additional year of age. Odds ratio of 0.356 for female head of household implies an association of a 64.4% lower likelihood of observing a positive perceived severity of income as compared to households with male heads. Households with religious heads are associated with a 53.5% higher likelihood of observing a positive value for perceived severity of income than those with the atheist heads. Households where the heads mainly engage in agriculture are associated with a 70.9% higher likelihood of observing a positive perceived severity of income than those whose heads do not work in agriculture.

The second hurdle estimation with respect to perceived severity of income is constricted to those households where a non-zero, positive value is observed. We

Table 9 2-Hurdle estimation of PL: fixed-effects logit estimation (1st hurdle), PPML fixed-effects model (2nd hurdle)

	1st hurdle	2nd hurdle	
	Odds ratios of Y > 0	Log coefficients	IRRs
ICT access	1.042	**0.1018*****	**1.1071*****
	(0.134)	(2.710)	
Covariate shock	**3.216*****	**0.1287*****	**1.1373*****
	(0.343)	(4.424)	
HH size	0.963	0.0036	1.0036
	(0.033)	(0.353)	
HH head age	0.996	−0.0003	0.9997
	(0.011)	(−0.082)	
Female head	**0.452****	−0.0528	0.9486
	(0.168)	(−0.546)	
Education	1.001	−0.0051	0.9950
	(0.0312)	(−0.579)	
Religious	**1.366****	−0.0218	0.9785
	(0.217)	(−0.419)	
Married	0.783	−0.0838	0.9196
	(0.230)	(−0.897)	
Agricultural occupation	**1.259***	0.0255	1.0258
	(0.163)	(0.588)	
HH income	1.000	0.0000	1.0000
	(0.000)	(0.481)	
Livestock value	1.000	−0.0000	1.0000
	(0.000)	(−0.078)	
Constant		0.2829	
		(1.208)	
Obs	2,444	4,203	

Notes se in parenthesis (results Column 1), *z* statistics in parentheses (results Column 2). $^{*}p < 0.10$, $^{**}p < 0.05$, $^{***}p < 0.01$

can see an incidence rate ratio of 1.1080 for ICT access with statistical significance at the p < 0.01 level. This implies households with ICT access are associated with a 10.8% higher perceived severity of income than those households without access, all else held constant. IRR of 1.2057 for previous experience with a covariate shock indicates an association of 20.57% higher perceived severity of income than those without previous covariate shock experience (p < 0.01). IRR coefficient of 0.9953 for age of head of household indicates an association with a 0.47% decrease in perceived severity of income with each year increase in age (p < 0.10). Households where the

Table 10 2-Hurdle estimation of PS: fixed-effects logit estimation (1st hurdle) and PPML fixed-effects model (2nd hurdle)

	1st hurdle on income	2nd hurdle on income		1st hurdle on assets	2nd hurdle on assets	
	Odds ratios of Y > 0	Log coefficients	IRRs	Odds ratios of Y > 0	Log coefficients	IRRs
ICT access	1.209	**0.1025*** **	**1.1080*** **	1.183	**0.0777** **	**1.0808** **
	(0.235)	(3.635)		(0.214)	(2.487)	
Covariate shock	910,185,705.5	**0.1871*** **	**1.2057*** **	**40.29*** **	**0.1766*** **	**1.1932*** **
	(1.084e + 12)	(8.186)		(11.23)	(6.702)	
HH size	1.012	0.0111	1.0112	1.003	**0.0235*** **	**1.0238*** **
	(0.051)	(1.473)		(0.047)	(2.686)	
HH head age	**0.970* **	**−0.0047* **	**0.9953* **	0.982	**−0.0066** **	**0.9934** **
	(0.017)	(−1.807)		(0.0148)	(−2.305)	
Female head	**0.356* **	−0.0746	0.9281	0.434	0.0195	1.0197
	(0.198)	(−1.012)		(0.222)	(0.221)	
Education	0.947	−0.0012	0.9988	0.976	−0.0016	0.9984
	(0.041)	(−0.202)		(0.031)	(−0.240)	
Religious	**1.535** **	−0.0166	0.9835	**1.584** **	−0.0451	0.9559
	(0.326)	(−0.427)		(0.322)	(−1.077)	
Married	0.707	**−0.1298* **	**0.8782* **	0.811	−0.0497	0.9515
	(0.317)	(−1.922)		(0.332)	(−0.646)	
Agricultural occupation	**1.709*** **	0.0363	1.0370	**1.657*** **	0.0469	1.0480
	(0.337)	(1.147)		(0.302)	(1.252)	
HH income	1.000	−0.0000	1.0000	1.000	−0.0000	1.0000
	(0.000)	(−0.943)		(0.000)	(−1.208)	
Livestock value	1.000	0.0000	1.0000	1.000	0.0000	1.0000
	(0.000)	(0.979)		(0.000)	(1.172)	
Constant		**0.3398** **			0.0037	
		(2.003)			(0.019)	
Obs	1926	4523		1984	4495	

Notes se in parenthesis (results Columns 1 and 4), *z* statistics in parentheses (results Columns 2, 3, 5, and 6). $*p < 0.10$, $**p < 0.05$, $***p < 0.01$

head is married has an IRR of 0.8782, indicating an association of 12.28% lower perceived severity of income than other households.

First hurdle logit estimation of perceived severity of assets examines the likelihood of observing a non-zero, positive value. We observe an odds ratio of 40.29 for covariate shock experienced in the reference period (p < 0.01), implying over 3,000%

higher odds of observing positive perceived severity of assets. The extreme magnitude of this estimate is consistent with quasi-complete separation, where households that experienced a covariate shock report positive perceived severity of assets with near certainty. This estimate should be interpreted with caution. Odds ratio of 1.584 for households with religious heads implies an association of 58.4% higher likelihood of observing positive perceived severity of assets than those with atheist heads ($p < 0.05$). For households with household heads in an agricultural occupation, an odds ratio of 1.657 implies an association of 65.7% higher likelihood of observing a positive perceived severity of assets than those with non-agricultural occupation.

In the second hurdle estimation, the Poisson pseudo maximum likelihood model is applied to those data where perceived severity of assets is strictly non-zero, positive values. We observe a positive, statistically significant relationship for ICT access at the $p < 0.05$ level. The IRR of 1.0808 implies households with ICT access are associated with 8.08% higher perceived severity than those without. Households which experienced covariate shocks within the survey reference period are associated with a 19.32% higher perceived severity of assets than those without previous experience, as indicated by IRR of 1.1932 ($p < 0.01$). Statistically significant ($p < 0.01$) IRR coefficient of 1.0238 for household size implies that households are associated with a 2.38% higher perceived severity of assets with each additional household member. Statistically significant relationship ($p < 0.05$) of household head age of 0.9934 indicates that households are associated with a 0.66% lower perceived severity of assets with each additional year.

5 Conclusion

As climate change intensifies the frequency and severity of extreme weather events, understanding how households perceive and respond to risks becomes critical for building societal resilience. Climate change is more prevalent for smallholder farm households in economically developing regions due to their limited capacity to prepare for, or otherwise mitigate against, the negative consequences of climate shocks. Increasingly recognized as essential infrastructure including for smallholder farmers, ICT access may serve as a crucial tool for enhancing risk communication and preparedness, particularly in vulnerable communities where accurate risk assessment can mean the difference between adaptation and catastrophic loss.

This study employs a double-hurdle modeling framework to examine the relationship between ICT access and household RP across multiple dimensions using three waves of the longitudinal TVSEP in Vietnam between 2013 and 2017. Our empirical analysis reveals that while ICT access demonstrates no statistically significant relationship with the probability of reporting positive RP in first-hurdle estimations, it consistently exhibits positive and significant effects in second-hurdle analyses. Among households reporting positive RP values, ICT access is associated with 18.06% higher RP for income, 15.18% higher for assets, 10.71% higher perceived likelihood, 10.8% higher perceived severity for income, and 8.08% higher perceived severity for assets. This pattern indicates that ICT access primarily influences the intensity rather than the incidence of RP.

Our results reveal consistent demographic patterns. Previous covariate shock experience emerges as the strongest predictor across all models, with substantially elevated effects ranging from 135% to over 3,900% increases. Female-headed households consistently demonstrate 20–64% lower RP across all dimensions, while religious household heads and those in agricultural occupations exhibit higher RP, suggesting cultural and occupational factors significantly influence risk assessment.

These findings align with theoretical frameworks indicating that ICT access enhances information processing capabilities among households already engaged in risk assessment rather than fundamentally altering risk awareness. The consistent gender disparities warrant policy attention, potentially reflecting differential information access or structural constraints limiting women's risk articulation capabilities.

Our empirical strategy demonstrates the value of double-hurdle models in RP studies, providing nuanced insights unavailable in single-equation frameworks. However, despite using longitudinal data, our binary ICT treatment at household level could be refined to consider technology use intensity and types. Additionally, potential endogeneity between ICT access and unobserved household characteristics affecting RP may bias our estimates, suggesting instrumental variable approaches could strengthen future analyses. The subjective nature of RP measures may introduce measurement error that may attenuate estimated effects, while our findings' external validity requires testing across diverse geographical and institutional contexts.

Policy implications suggest that expanding ICT access could enhance household risk assessment capabilities primarily among those already evaluating risks. Complementary interventions promoting both technology access and risk awareness education are needed to maximize digital inclusion benefits. The strong experiential learning effects highlight opportunities for disaster risk reduction programs to leverage ICT platforms for cross-community knowledge sharing. Future research should examine dynamic ICT-RP relationships and explore specific mechanisms through which technology influences household risk assessment capabilities.

Disclosure of Interests The authors have no competing interests to declare that are relevant to the content of this article.

Acknowledgements We would like to thank the Editor and anonymous referees for their constructive comments and suggestions. We thank the farmers in the surveyed provinces in Vietnam for their support and cooperation. This study relies on data from the long-term project No. 20220831434900116103, funded by the Deutsche Forschungsgemeinschaft (DFG). For detailed information, please see www.tvsep.de.

References

1. Acemoglu, D., Dahleh, M.A., Lobel, I., Ozdaglar, A.: Bayesian learning in social networks. Rev. Econ. Stud. **78**(4), 1201–1236 (2011). https://doi.org/10.1093/restud/rdr004
2. Aliakbari, E., McKitrick, R.: Information aggregation in a prediction market for climate outcomes. Energy Econ. **74**, 97–106 (2018). https://doi.org/10.1016/j.eneco.2018.06.002
3. Aven, T., Renn, O.: On risk defined as an event where the outcome is uncertain. J. Risk Res. **12**(1), 1–11 (2009). https://doi.org/10.1080/13669870802488883
4. Ballantyne, A.G., Wibeck, V., Neset, T.S.: Images of climate change—a pilot study of young people's perceptions of ICT-based climate visualization. Clim. Change **134**(1), 73–85 (2016). https://doi.org/10.1007/s10584-015-1533-9
5. Bordalo, P., Gennaioli, N., Shleifer, A.: Salience theory of choice under risk. Quart. J. Econ. **127**(3), 1243–1285 (2012). https://doi.org/10.1093/qje/qjs018
6. Bordalo, P., Gennaioli, N., Shleifer, A.: Salience and consumer choice. J. Polit. Econ. **121**(5), 803–843 (2013). https://doi.org/10.1086/673885
7. Bordalo, P., Gennaioli, N., Shleifer, A.: Salience. Ann. Rev. Econ. **14**(1), 521–544 (2022). https://doi.org/10.1146/annurev-economics-051520-011616
8. Bosch-Domènech, A., Silvestre, J.: Reflections on gains and losses: A $2 \times 2 \times 7$ experiment. J. Risk Uncertain. **33**(3), 217–235 (2006). https://doi.org/10.1007/s11166-006-0333-z
9. Brügger, A., Dessai, S., Devine-Wright, P., Morton, T.A., Pidgeon, N.F.: Psychological responses to the proximity of climate change. Nat. Clim. Chang. **5**(12), 1031–1037 (2015). https://doi.org/10.1038/nclimate2760
10. Burau, O., Nguyen, T.T.: Persistent poverty in rural Vietnam: differential asset dynamics and the role of ethnic minorities. J. Dev. Stud. 1–21 (2025). https://doi.org/10.1080/00220388.2025.2489563
11. Carter, M.R., Lybbert, T.J.: Consumption versus asset smoothing: testing the implications of poverty trap theory in Burkina Faso. J. Dev. Econ. **99**(2), 255–264 (2012). https://doi.org/10.1016/j.jdeveco.2012.02.003
12. Cullen, A.C., Anderson, C.L., Biscaye, P., Reynolds, T.W.: Variability in cross-domain risk perception among smallholder farmers in Mali by gender and other demographic and attitudinal characteristics. Risk Anal. **38**(7), 1361–1377 (2018). https://doi.org/10.1111/risa.12976
13. Do, M.H., Nguyen, T.T., Grote, U.: Insights on household's resilience to shocks and poverty: evidence from panel data for two emerging economies in Southeast Asia. Climate Dev. 1–15 (2025). https://doi.org/10.1080/17565529.2024.2446358
14. Do, T.L., Nguyen, T.T., Grote, U.: Livestock production, rural poverty, and perceived shocks: evidence from panel data for Vietnam. J. Dev. Stud. **55**(1), 99–119 (2019). https://doi.org/10.1080/00220388.2017.1408795
15. Eisele, M., Troost, C., Berger, T.: How Bayesian are farmers when making climate adaptation decisions? A computer laboratory experiment for parameterising models of expectation formation. J. Agric. Econ. **72**(3), 805–828 (2021). https://doi.org/10.1111/1477-9552.12425
16. Foster, F.D., Viswanathan, S.: Strategic trading when agents forecast the forecasts of others. J. Financ. **51**(4), 1437–1478 (1996). https://doi.org/10.1111/j.1540-6261.1996.tb04075.x
17. Halkos, G., Zisiadou, A.: Relating environmental performance with socioeconomic and cultural factors. Environ. Econ. Policy Stud. **20**(1), 69–88 (2018). https://doi.org/10.1007/s41885-018-0037-2
18. Hardeweg, B., Klasen, S., Waibel, H.: Establishing a database for vulnerability assessment. In: Klasen, S., Waibel, H. (eds.) Vulnerability to Poverty: Theory, Measurement and Determinants, with Case Studies from Thailand and Vietnam, pp. 50–79. Palgrave Macmillan, London (2013)
19. Heal, G., Millner, A.: Uncertainty and decision making in climate change economics. Rev. Environ. Econ. Policy **8**(1), 120–137 (2014). https://doi.org/10.1093/reep/ret023
20. Kahneman, D.: A perspective on judgment and choice: mapping bounded rationality. Am. Psychol. **58**(9), 697–720 (2003). https://doi.org/10.1037/0003-066X.58.9.697
21. Kaplan, S., Garrick, B.J.: On the quantitative definition of risk. Risk Anal. **1**(1), 11–27 (1981). https://doi.org/10.1111/j.1539-6924.1981.tb01350.x

22. Klibanoff, P., Marinacci, M., Mukerji, S.: A smooth model of decision making under ambiguity. Econometrica **73**(6), 1849–1892 (2005). https://doi.org/10.1111/j.1468-0262.2005.00640.x
23. Kuran, T., Sunstein, C.R.: Availability cascades and risk regulation. Stanf. Law Rev. **51**(4), 683–768 (1998)
24. Kyle, A.S.: Continuous auctions and insider trading. Econometrica **53**(6), 1315–1335 (1985). https://doi.org/10.2307/1913210
25. Lange, S., Reimers, M.: Livestock as a buffer stock in poorly integrated markets. Econ. Dev. Cult. Change **69**(2), 727–764 (2021). https://doi.org/10.1086/703081
26. Le Dang, H., Li, E., Nuberg, I., Bruwer, J.: Farmers' perceived risks of climate change and influencing factors: a study in the Mekong Delta, Vietnam. Environ. Manag. **54**(2), 331–345 (2014). https://doi.org/10.1007/s00267-014-0299-6
27. Liebenehm, S., Waibel, H.: Simultaneous estimation of risk and time preferences among small-scale cattle farmers in West Africa. Am. J. Agr. Econ. **96**(5), 1420–1438 (2014). https://doi.org/10.1093/ajae/aau056
28. Maddux, J.E., Rogers, R.W.: Protection motivation and self-efficacy: a revised theory of fear appeals and attitude change. J. Exp. Soc. Psychol. **19**(5), 469–479 (1983). https://doi.org/10.1016/0022-1031(83)90023-9
29. Martin, I.M., Bender, H., Raish, C.: What motivates individuals to protect themselves from risks: the case of wildland fires. Risk Anal. **27**(4), 887–900 (2007). https://doi.org/10.1111/j.1539-6924.2007.00930.x
30. McPeak, J.: Contrasting income shocks with asset shocks: livestock sales in Northern Kenya. Oxf. Econ. Pap. **56**(2), 263–284 (2004). https://doi.org/10.1093/oep/gpf040
31. Nguyen, T.T., Grote, U., Neubacher, F., Do, M.H., Paudel, G.P.: Security risks from climate change and environmental degradation: implications for sustainable land use transformation in the Global South. Curr. Opin. Environ. Sustain. **63**, 101322 (2023). https://doi.org/10.1016/j.cosust.2023.101322
32. Nguyen, T.T., Nguyen, T.T., Grote, U.: Internet use, natural resource extraction and poverty reduction in rural Thailand. Ecol. Econ. **196**, 107417 (2022). https://doi.org/10.1016/j.ecolecon.2022.107417
33. Peng, R., Zhao, Y., Elahi, E., Peng, B.: Does disaster shocks affect farmers' willingness for insurance? Mediating effect of risk perception and survey data from risk-prone areas in East China. Nat. Hazards **106**(3), 2883–2899 (2021). https://doi.org/10.1007/s11069-021-04569-0
34. Rimal, R.N., Real, K.: Perceived risk and efficacy beliefs as motivators of change. Hum. Commun. Res. **29**(3), 370–399 (2003). https://doi.org/10.1111/j.1468-2958.2003.tb00844.x
35. Rippl, S.: Cultural theory and risk perception: a proposal for a better measurement. J. Risk Res. **5**(2), 147–165 (2002). https://doi.org/10.1080/13669870110042598
36. Schulte, O., Mumber, J., Nguyen, T.T.: Agricultural commercialisation, asset growth and poverty in rural Vietnam. Aust. J. Agric. Resour. Econ. **67**(3), 388–416 (2023). https://doi.org/10.1111/1467-8489.12517
37. Siegrist, M., Gutscher, H., Earle, T.C.: Perception of risk: the influence of general trust, and general confidence. J. Risk Res. **8**(2), 145–156 (2005). https://doi.org/10.1080/1366987032000105315
38. Spence, A., Poortinga, W., Pidgeon, N.: The psychological distance of climate change. Risk Anal. Int. J. **32**(6), 957–972 (2012). https://doi.org/10.1111/j.1539-6924.2011.01695.x
39. Sullivan-Wiley, K.A., Gianotti, A.G.S.: Risk perception in a multi-hazard environment. World Dev. **97**, 138–152 (2017). https://doi.org/10.1016/j.worlddev.2017.04.002
40. Sundblad, E.L., Biel, A., Gärling, T.: Cognitive and affective risk judgements related to climate change. J. Environ. Psychol. **27**(2), 97–106 (2007). https://doi.org/10.1016/j.jenvp.2007.01.003
41. Sunstein, C.R.: The availability heuristic, intuitive cost-benefit analysis, and climate change. Clim. Change **77**(1), 195–210 (2006). https://doi.org/10.1007/s10584-006-9073-y
42. Terpstra, T.: Emotions, trust, and perceived risk: affective and cognitive routes to flood preparedness behavior. Risk Anal. **31**(10), 1658–1675 (2011). https://doi.org/10.1111/j.1539-6924.2011.01616.x

43. Tversky, A., Kahneman, D.: Availability: a heuristic for judging frequency and probability. Cogn. Psychol. **5**(2), 207–232 (1973). https://doi.org/10.1016/0010-0285(73)90033-9
44. Tversky, A., Kahneman, D.: Advances in prospect theory: cumulative representation of uncertainty. J. Risk Uncertain. **5**(4), 297–323 (1992). https://doi.org/10.1007/BF00122574
45. UN: Designing household survey samples: practical guidelines. The department of Economic and Social Affairs of the United Nations (UN). New York: UN (2005). https://unstats.un.org/unsd/demographic/sources/surveys/Handbook23June05.pdf. Accessed 02 Jan 2025
46. van der Linden S (2017) Determinants and measurement of climate change risk perception, worry, and concern. SSRN J. https://doi.org/10.2139/ssrn.2953631
47. van Winsen, F., de Mey, Y., Lauwers, L., van Passel, S., Vancauteren, M., Wauters, E.: Determinants of risk behaviour: effects of perceived risks and risk attitude on farmer's adoption of risk management strategies. J. Risk Res. **19**(1), 56–78 (2016). https://doi.org/10.1080/13669877.2014.940597
48. Villacis, A.H., Alwang, J.R., Barrera, V.: Linking risk preferences and risk perceptions of climate change: a prospect theory approach. Agric. Econ. **52**(5), 863–877 (2021). https://doi.org/10.1111/agec.12659
49. Vinh Hung, H., Shaw, R., Kobayashi, M.: Flood risk management for the RUA of Hanoi. Disaster Prev. Manag. Int. J. **16**(2), 245–258 (2007). https://doi.org/10.1108/09653560710739568
50. Wachinger, G., Renn, O., Begg, C., Kuhlicke, C.: The risk perception paradox-implications for governance and communication of natural hazards. Risk Anal. **33**(6), 1049–1065 (2013). https://doi.org/10.1111/j.1539-6924.2012.01942.x
51. Weber, E.U.: What shapes perceptions of climate change? Wiley Interdiscip. Rev. Clim. Change **1**(3), 332–342 (2010). https://doi.org/10.1002/wcc.41
52. Weber, E.U., Blais, A.R., Betz, N.E.: A domain-specific risk-attitude scale: measuring risk perceptions and risk behaviors. Behav. Decis. Mak. **15**(4), 263–290 (2002). https://doi.org/10.1002/bdm.414
53. Witte, K.: Fear control and danger control: a test of the extended parallel process model (EPPM). Commun. Monogr. **61**(2), 113–134 (1994). https://doi.org/10.1080/03637759409376328
54. Yang, H., Cai, W., Liu, J., Huo, X.: Impact of internet information on apple growers' adaptive behaviors to frost disasters: theory and empirical research from the perspective of psychological perception. Agriculture **11**(10), 905 (2021). https://doi.org/10.3390/agriculture11100905

FSC
www.fsc.org
MIX
Papier | Fördert gute Waldnutzung
FSC® C083411

Kolibri 360 GmbH
Ferdinand-Jühlke-Straße 7
99095 Erfurt, Deutschland
produktsicherheit@kolibri360.de